CISM COURSES AND LECTURES

Series Editors:

The Rectors of CISM
Sandor Kaliszky - Budapest
Mahir Sayir - Zurich
Wilhelm Schneider - Wien

The Secretary General of CISM
Giovanni Bianchi - Milan

Executive Editor
Carlo Tasso - Udine

The series presents lecture notes, monographs, edited works and proceedings in the field of Mechanics, Engineering, Computer Science and Applied Mathematics.
Purpose of the series is to make known in the international scientific and technical community results obtained in some of the activities organized by CISM, the International Centre for Mechanical Sciences.

INTERNATIONAL CENTRE FOR MECHANICAL SCIENCES

COURSES AND LECTURES - No. 374

TOPOLOGY OPTIMIZATION IN STRUCTURAL MECHANICS

EDITED BY

G.I.N. ROZVANY
ESSEN UNIVERSITY

Springer Wien New York

Le spese di stampa di questo volume sono in parte coperte da
contributi del Consiglio Nazionale delle Ricerche.

This volume contains 131 illustrations

In order to make this volume available as economically and as
rapidly as possible the authors' typescripts have been
reproduced in their original forms. This method unfortunately
has its typographical limitations but it is hoped that they in no
way distract the reader.

ISBN 3-211-82907-5 Springer-Verlag Wien New York

PREFACE

Topology optimization is a relatively new and rapidly expanding field of structural mechanics. It deals with some of the most difficult problems of mechanical sciences but it is also of considerable practical interest, because it can achieve much greater savings than mere cross-section or shape optimization.

Topology optimization has been discussed extensively at recent international meetings (e.g. Udine, Italy 1990; Berchtesgaden, Germany 1991; Sesimbra, Portugal 1992; Lyngby, Denmark 1992; Rio de Janeiro, Brazil 1993; Panama City, FL, USA 1994; Goslar, Germany 1994; Kyoto, Japan 1996; and Bellevue, WA, USA 1996).

As preliminary reading for this course, we can recommend the lecture notes for the previous meeting in Udine on a similar topic (Rozvany, 1992), a long review article on layout optimization (Rozvany, Bendsøe and Kirsch, 1995), a book by Bendsøe (1995) and the proceedings of a NATO workshop on topology design (Bendsøe and Mota Soares, 1993).

Whilst the primary emphasis of this course is on basic theories and fundamental research, numerical methods and practical applications are also covered.

The chapter contributed by the Editor covers aims, scope, basic concepts and methods of topology optimization, showing both recent extensions of exact topology optimization to multipurpose structures and applications of discretized topology optimization to real-world problems in accordance with national and international design standards. These methods are based on generalizations of the optimal layout theory using optimality criteria, which are discussed in greater detail in a book by the Editor (Rozvany, 1989). Some special computational difficulties are also reviewed.

Topology optimization of discrete structures with an emphasis on non-smooth aspects is discussed in a valuable contribution by Achtziger, who illustrates his methods with truss examples.

The problem of generalized or variable topology shape optimization by the homogenisation method is treated in a highly rigorous fashion by Allaire who also presents interesting numerical applications.

Generalized shape optimization by means of the "bubble method" for optimal hole positioning, together with highly practical industrial applications, is presented in a valuable contribution by Eschenauer and Schumacher.

A very general treatment of reduction and expansion processes in topology optimization is given by Kirsch, who discusses a number of design aspects of this field and illustrates lucidly these concepts with simple examples.

Finally, topology and reinforcement layout optimization of disk, plate and shell structures is discussed in an outstanding contribution by Krog and Olhoff, which also contains numerical examples involving stiffness and eigenfrequency criteria.

The Editor wishes to express his gratitude to the International Centre for Mechanical Science, and in particular to its rector, Professor Kaliszky for making this meeting possible; to the lecturers for their devoted efforts; and to the participants for their attention and useful discussions during the course.

References

Bendsøe, M.P. 1995: Methods for the Optimization of Structural Topology, Shape and Material. Springer-Verlag, Berlin.

Bendsøe, M.P.; Mota Soares, C.A. (Eds.) 1993: Topology Design of Structures. Kluwer, Dordrecht.

Rozvany, G.I.N. 1989: Structural Design via Optimality Criteria. Kluwer, Dordrecht.

Rozvany, G.I.N. (Ed.) 1992: Shape and Layout Optimisation of Structural Systems and Optimality Criteria Methods, CISM Courses and Lectures No. 325. Springer-Verlag, Vienna.

Rozvany, G.I.N.; Bendsøe, M.P.; Kirsch, U. 1995: Layout Optimisation of Structures. Review article in Appl. Mech. Rev. (ASME), Vol. 48, No. 2, pp. 41-118.

G.I.N. Rozvany

CONTENTS

**AIMS, SCOPE, BASIC CONCEPTS AND METHODS
OF TOPOLOGY OPTIMIZATION**

G.I.N. Rozvany

Essen University, Essen, Germany

1. THE TWO FUNDAMENTAL PROBLEMS OF TOPOLOGY OPTIMIZATION

Topology means the pattern of connectivity or spatial sequence of members or elements in a structure. Optimization of the topology is involved in two fundamental classes of problems, namely

- layout optimization and
- generalized (variable topology) shape optimization.

1.1 Layout Optimization of Grids and Honeycombs

A *one-dimensional (1D) structure* has the fundamental geometrical property that the dimensions of its cross-sections are small in comparison to its length. By assuming that in the deformed structure the cross-sections remain plane, all strains and stresses at all points of a 1D structure are uniquely determined by the deformed shape of its centroidal axis. This represents a considerable simplification in comparison to three-dimensional solids.

A *grid* is the union of intersecting 1D structures. The intersection of two or more 1D structures is called a *joint* and the segment of a 1D structure in between two joints is termed a *member*. The effect of member intersections on strength, stiffness or cost of a grid is usually neglected in the optimization procedure.

Examples of grids are trusses, grillages, shellgrids and cable-nets. A *truss* is a grid consisting of straight members under purely axial compression or tension and is loaded only by concentrated forces at the intersections of the centroidal axes of members (Fig. 1a). If all centroidal axes and loads of a truss are contained in a plane, then it is termed a *plane truss*, otherwise it is called a *space truss*. The members of trusses are called *bars* or *truss elements*.

A *grillage* (Fig. 1b) consists of a system of intersecting one-dimensional flexural elements termed *beams*, whose centroidal axes are contained in a plane. All loads on a grillage are acting normal to this plane. The simple grillage in Fig. 1b has a clamped end (A), three simple supports (B, C and E) and a free end (D).

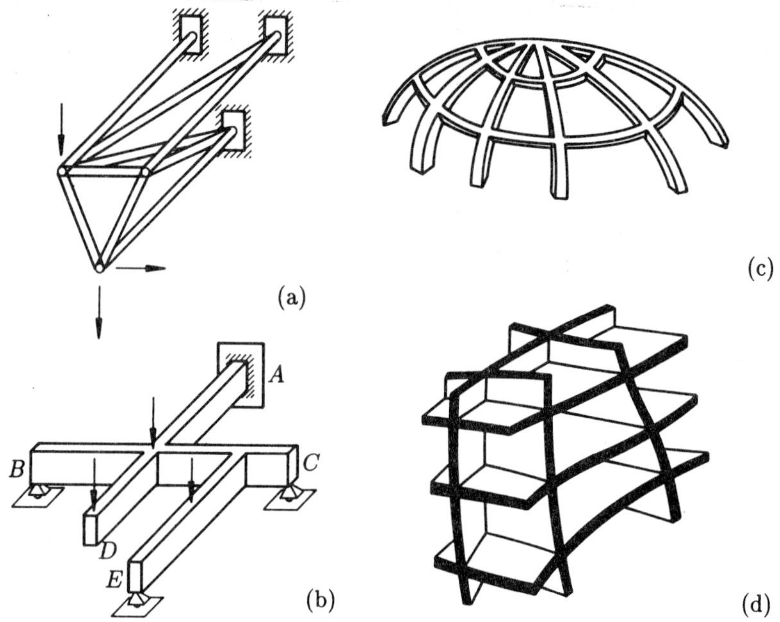

Fig. 1. Examples of grids: (a) truss, (b) grillage, (c) shellgrid and (d) honeycomb.

Shellgrids (Fig. 1c) and *cable-nets* have the defining feature that the centroidal axes of their members are contained in a curved surface. Cable-nets have the distinguishing property that the flexural stiffness of its members is negligibly small and all their members are subject to axial tension.

A *two-dimensional (2D) structure* has the geometrical property that one of its dimensions (termed *thickness*) is small in comparison to other dimensions. By assuming that all line segments normal to the midsurface of the deformed 2D structure remain straight, all stresses and strains can be determined uniquely from the deformed shape of the midsurface. If its midsurface is a plane segment, then a 2D structure is called a *plate*, otherwise it is termed a *shell*. A plate under plane stress is also called a *disk*. The union of intersecting two-dimensional structures is termed a *honeycomb* (Fig. 1d).

Layout optimization of a grid or honeycomb means the simultaneous selection of the optimal

- *topology* (spatial sequence or connectivity of members or elements),
- *geometry* (location of intersections of member axes or midsurfaces), and
- *cross-sectional dimensions* (sizing).

The union of all *potential* members or elements in a topology optimization problem is termed the *ground structure* (Dorn *et al.* 1964). During the optimization procedure, nonoptimal members (termed subsequently "vanishing" members) are eliminated.

The basic concepts of layout optimization are explained conceptually in the context of plane trusses in Fig. 2. The ground structure, *i.e.* the union of all potential members, is shown in Fig. 2a. From this initial topology, an optimal topology is selected in Fig. 2b. This topology is then maintained through Figs. 2c and d, which have the optimal geometry as well (for a given finite number of members). Finally, the sizes or cross-sectional areas (indicated by the line thickness) change from nonoptimal to optimal in Fig. 2d. Naturally, the above three properties must be optimized simultaneously, because a separately optimized topology may no longer be optimal if we change the geometry.

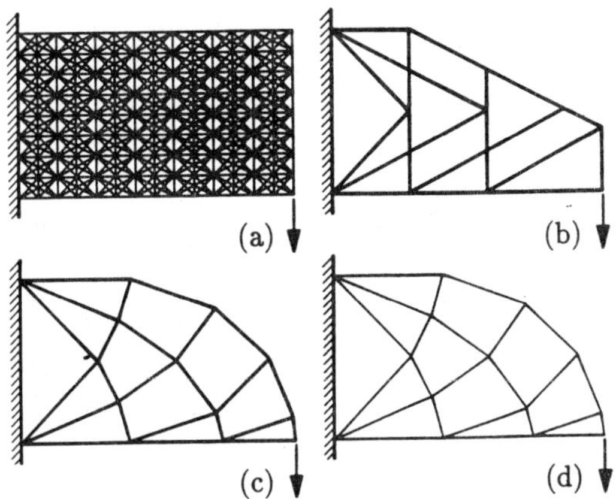

Fig. 2. Layout optimization: (a) ground structure, identical topology in (b)-(d), identical geometry in (c)-(d), and different sizes in (c) and (d).

1.2 Generalized (or Variable Topology) Shape Optimization

A *porous structure* consists, in the context of this book, partially of a given material and partially of empty space (or "cavity" or "void"*). In the case of 2D structures, we may term these voids "perforations", and the corresponding porous structures *perforated disks, plates* or *shells.*

The *volume fraction* (F_V) of a porous structure is the volume occupied by material (V_M), divided by the total or available volume (V_A)

$$F_V = \frac{V_M}{V_A}. \tag{1}$$

In the case of grids and honeycombs, the volume fraction is assumed to tend to zero

$$F_V \to 0. \tag{2}$$

---

*In the theory of porous media, the voids may be filled by fluids. This case is not considered in the present study.

When the available space is occupied by two or more materials, the corresponding structure is termed a *composite*.

Generalized shape optimization or *variable topology shape optimization* involves the simultaneous optimal selection of

- the *shape* of the *external boundaries* of the structure; and
- the *topology and shape* of the *internal boundaries* between material and void in porous structures and between two different materials in composites.

Generalized shape optimization of composites is shown conceptually in Fig. 3, in which shaded areas denote a stronger, stiffer and more expensive material, while dotted regions a weaker, less stiff and cheaper one. In the limiting case (porous structures), the latter denotes voids (with zero strength, stiffness and cost). The initial design is shown in Fig. 3a and the conceptually optimized general shape in Fig. 3b. In Fig. 3, the external boundary is fixed. If the latter is variable and also to be optimized, then the weaker material could be removed from the areas ABC and DEF in Fig. 3b.

Generalized shape optimization has also been called "advanced layout optimization" in the earlier literature (e.g. *Rozvany and Ong* 1987), because it can be regarded as generalization of the layout problem from zero to a nonzero volume fraction.

A brief review of generalized shape optimization is given in Section 12.

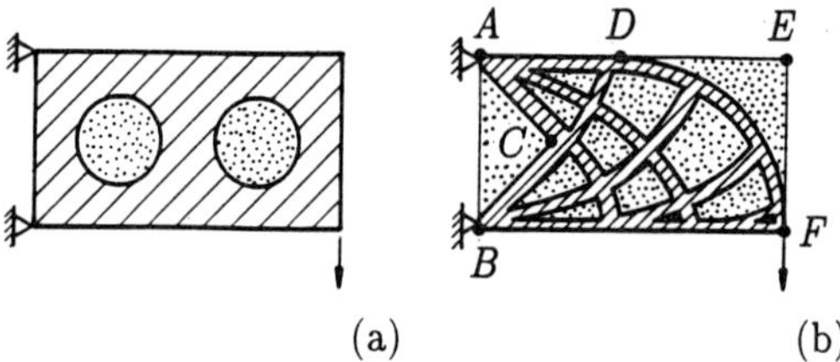

Fig. 3. Generalized (or variable topology) shape optimization: (a) initial design, (b) optimal design. Shaded areas: stronger, stiffer and more expensive material. Dotted areas: weaker, less stiff and cheaper material or void.

2. TWO BASIC FORMULATIONS IN TOPOLOGY OPTIMIZATION

Depending on the type of ground structure for a layout problem, the solution may appear in an *exact-analytical* form or an *approximate-discretized* form. For the former, the ground structure contains an infinite number of members and is also termed *structural universe* (e.g. Rozvany **1989**). For the latter, the ground structure consists of a finite number of members.

In the case of an exact optimal truss layout, for example, the structural universe consists of an infinite number of potential truss elements in all possible directions at all points of the so-called *structural domain D* which is some subset of the two- or three-dimensional Euclidean space ($D \subset I\!\!R^2$ or $I\!\!R^3$).

2.1 Exact-Analytical Formulations

Exact solutions of layout optimization often consist of a dense network containing an infinite number of intersecting members with an infinitesimal spacing. This type of solution will be termed a *gridlike or honeycomblike continuum*, which represents a generalization of Prager's (**1974**) terminology of "trusslike continuum" or "grillagelike continuum". Due to this form of solutions, exact layout optimization is often criticized as being unpractical. In actual fact, *exact layout optimization* has the following *important functions*:

- The exact (explicit) solution often consists of a finite number of members, in which case it is directly applicable in practice (Fig. 4).

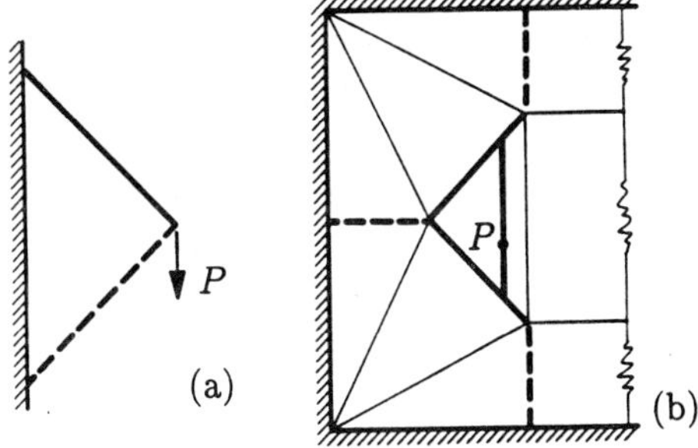

Fig. 4. Examples of exact optimal layouts consisting of a finite number of members: (a) truss, (b) grillage. Continuous thick lines: bars in tension or beams in positive ["sagging"] bending. Broken thick lines: bars in compression or beams in negative ["hogging"] bending. Thin lines in (b) represent optimal region boundaries.

- Highly economical discretized solutions can be obtained by selecting a finite number of joints in an exact optimal gridlike continuum and connecting them by straight members (Fig. 5).

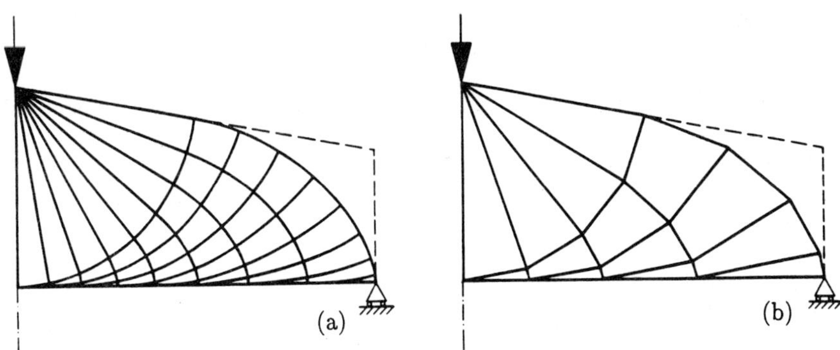

Fig. 5. (a) Exact optimal layout consisting of a gridlike continuum, with a finite number of members shown, (b) discrete truss layout obtained by connecting some nodes of the exact layout with straight members.

- Exact optimal layouts can be used as basis for assessing the relative economy of any discretized solution for the same problem.
- Explicit solutions reveal fundamental features of optimal layouts, which has often proved very useful.
- In many layout problems, multiple optima of equal weight (or equal cost) exist. While exact analytical methods usually outline all these (see Section 9), numerical methods often select randomly one of these optima.

The obvious *advantages of exact analytical layouts* are that

- the results are clear and not obscured by discretization and/or convergence errors;
- regions with indeterminate optimal member directions are identified; and
- sensitivities, i.e. the influence of the magnitude of given design parameters (e.g. level of permissible stresses and displacements) can be determined explicitly.

The *disadvantages* of exact optimal layouts are as follows.

- The basic limitation of this approach is that it is not general but is restricted to somewhat idealized problems. This, however, is changing to some extent with the introduction of exact layouts for multipurpose trusses and grillages.
- Exact analytical solutions cannot, in general, be derived by "black box" methods, since for each class of problems an individual formulation and a deep insight into optimal layouts is required. However, systematic and even computer-aided methods for finding the exact layout for certain classes of problems are being developed (see e.g. Rozvany and Hill 1978; Rozvany, Gerdes and Gollub 1996).
- The exact optimal layout may be unpractical, owing to a complicated topology, the large (often infinite) number of members or instability of the "optimal" structure. However, simplified solutions based on exact layouts are usually highly economical (see Fig. 5).

There are also rigorous analytical methods for deriving optimal layouts with a finite number of members (see papers by Rozvany and Prager 1976; Prager 1978a and b).

2.2 Approximate-Discretized Formulations

Approximate-discretized solutions are usually found by one of the following classes of numerical methods in a finite dimensional design space:

- optimality criteria methods,
- mathematical programming methods, or
- random search methods.

Optimality criteria (OC) methods use the Kuhn-Tucker necessary conditions for cost minimality which are also employed in exact-analytical formulations. The discretized version of these conditions are used for suitable iterative design formulae, using an approximation between iterative cycles (e.g. temporarily assuming that the internal forces are independent of the cross-sectional areas).

Mathematical programming aims at reducing the cost in a locally optimal manner in each iteration until the gain in cost becomes smaller than a specified limit.

Random search methods (e.g. genetic algorithms) generate a larger number of designs and use a systematic search for improved solutions (e.g. by recombinations of better solutions).

Methods of topology optimization can be assessed on the basis of their *efficiency* and *robustness*. In general, *optimality criteria methods* have been found the *most efficient* and *least robust*, while *random search methods* the *most robust* but *least efficient*. However, the latter may represent the only suitable method if the problem is highly nonconvex and hence the number of local minima extremely large.

In view of the ground structure approach, we could also define *layout optimization* as a *special class of cross-section (sizing) optimization* in which the cross-sectional areas may take on a zero value.

An alternative to the ground structure approach is an *incremental synthesis approach*, in which we start with a few members and then progressively add new members (e.g. Kirsch 1995). The difficulties with this method are that

- no simple methods are available at present for finding explicitly the optimal position of new members, and
- a subset of the optimal layout may not represent a stable structure and hence it cannot be used as a starting point for finding the latter.

2.3 A Comparison of Exact-Analytical and Approximate-Discretized Formulations

The fundamental features of exact-analytical and approximate-discretized solutions are compared in Table 1.

3. BASIC FEATURES OF OPTIMAL LAYOUT THEORY

A very efficient approach to both exact-analytical and approximate-discretized formulations of layout optimization is the *theory of optimal layouts*, developed in the late seventies by Prager and Rozvany (e.g. *1977*) as a generalization of Michell's (1904) criteria, and extended considerably by the author in the eighties and nineties. Optimal layout theory is based on four fundamental concepts, namely

- the structural universe (or ground structure),
- continuum type optimality criteria (COC),
- the adjoint structure, and
- the layout criterion function.

3.1 Structural Universe (Ground Structure)

Early applications (e.g. Prager and Rozvany *1977*) of the optimal layout theory dealt with exact-analytical layouts, for which an infinite number of potential members constitute the so-called *structural universe*. Discretized approximate layouts were investigated via optimal layout theory more recently. To be consistent with the literature on discretized layouts, the corresponding set of potential members will be termed *ground structure*, see also Section 1 and Fig. 2a.

Table 1 A comparison of exact-analytical and approximate-discretized formulations.

Formulation	Exact	Approximate
Computational Method	Analytical	Numerical
Structural Model	Continuum	Discretized (Finite Elements)
Solution Procedure	Simultaneous Solution of All Equations	Iterative Solution
Initial Structure	Structural Universe (Infinite Number of Members)	Ground Structure (Finite but Large Number of Members)
Prescribed Minimum Cross-Sectional Area	Zero	Small (or Zero)
Usual Means of Computation	By Hand*	By Computer

*Some computer algorithms for generating analytically the optimal layout are also available.

3.2 Continuum-Type Optimality Criteria (COC)

These criteria are based on the Kuhn-Tucker conditions of cost optimality. In a relaxed formulation of the optimization problem, only equilibrium (and no compatibility) conditions are included. Later it is shown that the optimality criteria also imply kinematic admissibility, provided that some displacement constraints are active. From the above optimality criteria, explicit *design formulae* for the calculation of optimal cross-sectional dimensions can be obtained. For illustration purposes, we consider elastic trusses with one load condition and several displacement constraints. The optimal design formula for these problems can be stated as

$$x^e = \sqrt{\frac{f^e \overline{f}^e}{E^e \varrho^e}}, \tag{3}$$

where x^e is the cross-sectional area, E^e Young's modulus, ϱ^e the specific weight, f^e the member force and $\overline{f}^e$ the *adjoint* member force in the truss element e. The concept of adjoint structures plays a vital role in the theory of optimal layouts and will be explained in the next subsection.

A detailed description of continuum-type optimality criteria can be found in the author's book (Rozvany **1989**) on this topic.

3.3 The Adjoint Structure

Certain quantities in the optimality criteria and design formulae can be interpreted as state variables of a fictitious structure termed *adjoint structure*. The corresponding loads, internal forces, strains and displacements are called *adjoint loads, adjoint forces, adjoint strains and adjoint displacements.*

Considering again an elastic truss with one load condition and several displacement constraints, the *adjoint truss* will have the same topology, geometry, cross-sections and support conditions as the "real" truss, but the loading on the truss will consist of

$$\overline{P}^j = \sum_{d=1}^{D} \nu_d \hat{P}_d^j \qquad (j = 1, \ldots, J), \tag{4}$$

where $\overline{P}^j$ is the "adjoint" load at the degree of freedom j, ν_d is the Lagrange multiplier for the displacement constraint d $(d = 1, \ldots, D)$, and $\hat{P}_d^j$ is the *virtual load* at the degree of freedom j for the constraint d, which can be determined as follows.

If the constraint d puts a limit on the displacement at a single degree of freedom (say at $j = \alpha$), then the constraint d takes the form

$$u^\alpha \leq \Delta_d, \tag{5}$$

where Δ_d is the prescribed upper limit on the displacement u^α (Fig. 6a). The corresponding virtual load $\hat{P}_d$ for the displacement constraint d consists of a unit ("dummy") point load at the degree of freedom α (Fig. 6b).

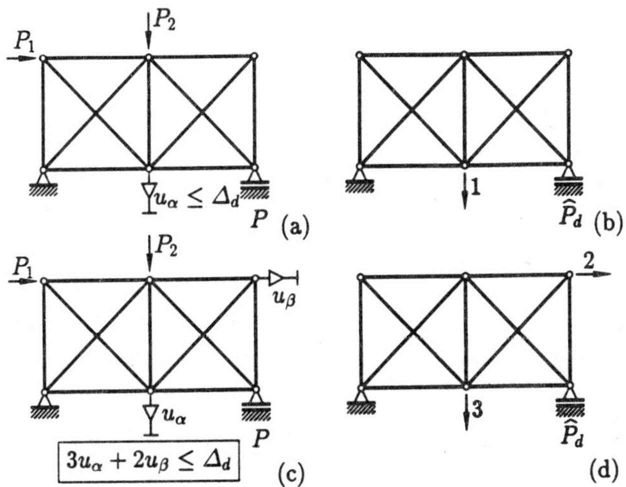

Fig. 6. (a) (c) Real loads and displacement constraints for trusses; (b) (d) the corresponding virtual loads.

The displacement constraint d may also limit the weighted combination of displacements at several degrees of freedom (Fig. 6c), in which case the virtual load consists of point loads corresponding to the weighting factors in the displacement constraint (Fig. 6d).

Moreover, some displacement constraints may also restrict the value of a negative displacement, in the form

$$-u^\alpha \leq \Delta_d. \tag{6}$$

The multipliers ν_d are constant for all degrees of freedom and their values are determined from the active displacement constraints. This procedure is illustrated in Section A.1 in the Appendix. For inactive displacement constraints, the Lagrange multipliers take on a zero value.

Similarly to (4), the *adjoint forces* and *adjoint strains* are defined as

$$\overline{f}^e = \sum_{d=1}^{D} \nu_d \hat{f}_d^e, \quad \overline{\varepsilon}^e = \sum_{d=1}^{D} \nu_d \hat{\varepsilon}_d^e, \tag{7}$$

where $\hat{f}_d^e$ are the *virtual forces* equilibrating the virtual loads $\hat{P}_d^j$ and $\hat{\varepsilon}_d^e$ are the corresponding virtual strains

$$\hat{\varepsilon}_d^e = \frac{\hat{f}_d^e}{E^e x^e}. \tag{8}$$

The adjoint strains $\overline{\varepsilon}^e$ can also be expressed in terms of the adjoint forces $\overline{f}^e$ or virtual forces $\hat{f}_d^e$ as

$$\overline{\varepsilon}^e = \frac{\overline{f}^e}{E^e x^e} = \frac{\sum_{d=1}^{D} \nu_d \hat{f}_d^e}{E^e x^e}. \tag{9}$$

Whilst we have used the simple problem of a truss with one load condition and several displacement constraints for illustration purposes, the adjoint truss can readily be defined for other types of design constraints and their combinations.

The *advantages* of using the mechanical analogy *of adjoint structures* in optimality criteria methods are as follows:

- Certain abstract quantities in the optimality criteria gain a physical meaning, which makes them easier to visualize for those familiar with structural behaviour. This also facilitates error detection.
- Established methods of structural analysis can be used for calculating the adjoint state variables in both analytical and numerical (e.g. finite element) solutions.

In the latter, the same decomposed stiffness matrix can be used for both the real and adjoint structures.

It is to be mentioned that for some simple problems the state variables of the adjoint structure reduce to the so-called "adjoint variables" used in sensitivity analysis. This is by no means so in the case of more complex design constraints.

3.4 The Layout Function ϕ

Another central concept in the theory of the optimal layout is the *layout function* ϕ. It will be shown in detailed proofs that the layout function must take on a *unit value along optimal, nonvanishing members* of the ground structure or structural universe and *must be smaller than or equal to unity along all nonoptimal, vanishing members*:

$$\boxed{\begin{aligned} \phi^e &= 1 \quad (\text{for } x^e > 0), \\ \phi^e &\leq 1 \quad (\text{for } x^e = 0). \end{aligned}}$$

$$\tag{10}$$
$$\tag{11}$$

The relations (10) and (11) will be called the *fundamental conditions of layout optimality*. The layout function usually depends on the real and adjoint state variables, such as strains. Its calculation and applications will be illustrated with two simple examples in Sections 4 and 5.

4. ILLUSTRATION OF THE BASIC CONCEPTS OF LAYOUT THEORY THROUGH A SIMPLE DISCRETIZED LAYOUT PROBLEM

It should be mentioned at the outset that very efficient numerical computer methods are available for automatically deriving the optimal solution for discretized layout problems of this kind (see Chapter 8). However, this simple problem will be solved by hand calculations for didactic reasons, with a view to maintaining clarity of the concepts involved.

4.1 The Underlying Theory of Optimal Truss Layouts for Displacement Constraints

For a truss with a single load condition and several displacement constraints, the layout function becomes (Rozvany 1992; see a proof in Section A.2 in the Appendix)

$$\phi^e = (E^e/\varrho^e)\varepsilon^e\overline{\varepsilon}^e , \tag{12}$$

where

$$\varepsilon^e = \frac{f^e}{x^e E^e} , \tag{13}$$

and $\overline{\varepsilon}^e$ was defined in (9). Then the layout optimality conditions are given by (10) and (11) with (12).

4.2 Problem Statement: Three-Bar Truss with Displacement Constraints

We consider the simplest possible ground structure for a layout problem which is a three-bar truss. For additional simplicity, we consider a single load condition consisting of a horizontal and a vertical point load (Fig. 7a). Both the horizontal and vertical displacements are constrained, but the limits on these displacements are *not* proportional to the corresponding loads.

4.3 Simple Demonstration of the Role of the Layout Function ϕ

In general, a layout will be identified by listing the nonvanishing members in brackets. The optimal layout (a, b) for the considered problem, together with the values of the layout function for the members, is shown in Fig. 7b and for a nonoptimal layout (b, c) in Fig. 7c. Within each layout, the members have been optimized. The weight of these two solutions is, respectively, $W_{a,b} = W_{\mathrm{opt}} = 23.333$ and $W_{b,c} = 69.641$. It will be shown later that the lowest weight for the third layout (a, c) is even higher, $W_{a,c} = 74.641$ (see Fig. 8i).

   Looking now at the values of the layout function ϕ, we can see immediately from Fig. 7b that the layout (a, b) is optimal, because

• for nonvanishing members the layout function ϕ takes on a unit value; and, more importantly,

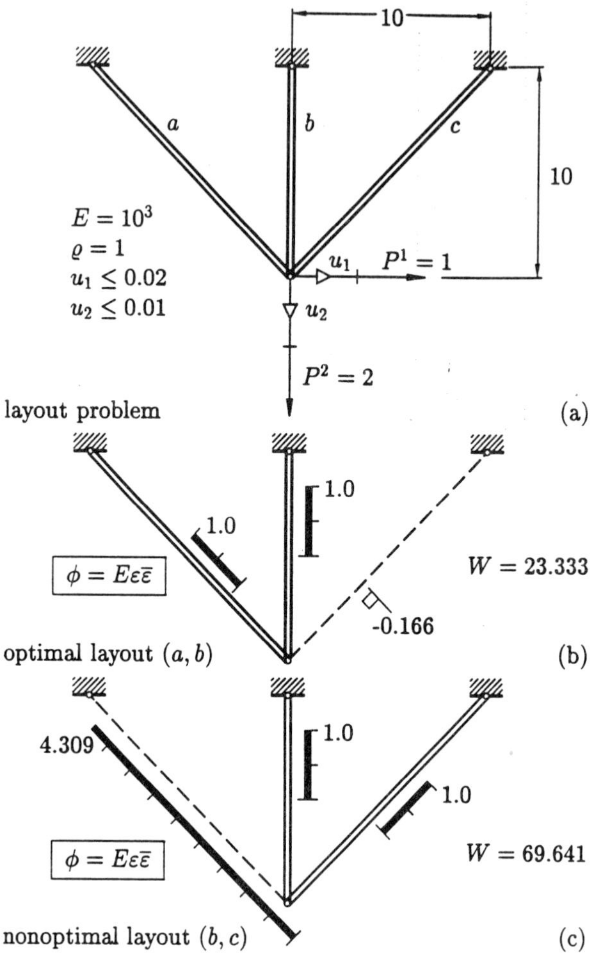

Fig. 7. A simple illustrative example. (a) Problem statement; (b) layout function ϕ values for the optimal solution and (c) for a nonoptimal solution.

- ϕ is *smaller than unity* – in fact negative, *for the vanishing member.* Since *the layout function represents the relative efficiency of a particular member in a layout problem,* the bar "a" is clearly highly inefficient, and as such it is the one that should be eliminated.

On the other hand, one can see immediately from Fig. 7c that the layout (b, c) is highly nonoptimal, because the layout function for member "a" violates quite considerably the inequality condition (10). The fact that $\phi^a = 4.3$ (instead of $\phi^a \leq 1$) indicates that this vanishing bar would be highly efficient and hence it should not have been taken out.

4.4 Further Details of the Solution Process

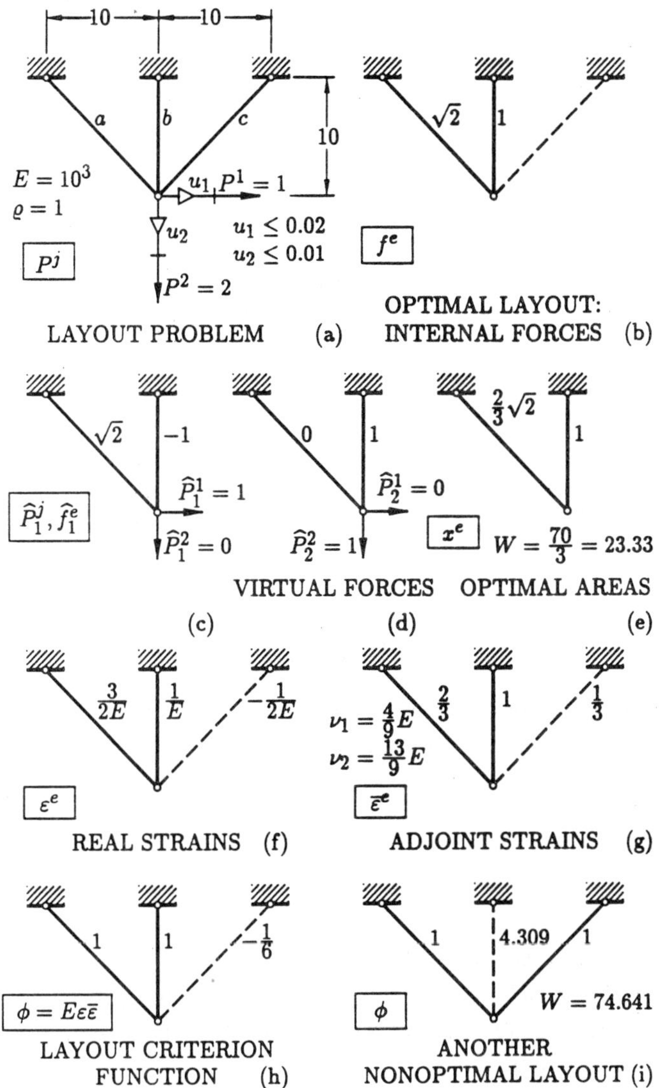

Fig. 8. (a)–(h) Details of the derivation of the optimal solution in Fig. 7b and (i) another nonoptimal solution.

Figure 8 repeats the problem statement (Fig. 8a) and shows the optimal layout with the "real" member forces f^e (Fig. 8b), virtual loads and the corresponding member forces $\widehat{f}_1^e$, $\widehat{f}_2^e$ (Fig. 8c and d), the optimal cross-sectional areas (Fig. 8e), the real strains (Fig. 8f), adjoint straints (Fig. 8g) and the layout function (Fig. 8h) for the members.

The values of the layout function for the members of another highly nonoptimal layout (a, c) and its weight are shown in Fig. 8i.

The derivation of these results involves only elementary statics and kinematics and is given in Section A.1 in the Appendix.

A useful check on optimal solutions for a given topology is the calculation of the total weight W from *both* the primal and dual formulae

$$W = \sum_e x^e L^e, \quad \overline{W} = \sum_d \Delta^d \nu^d, \tag{14}$$

where ν^d are Lagrange multipliers. This check is given for all three layouts in Section A.1.4.

4.5 Comments on Certain Complexities of Layout Problems

It can also be observed from the above example that, even for the simplest possible ground structure (a three-bar truss), the analytical derivation of the exact optimal solution is not simple. Owing to nonconvexity of the problem, three possible two-bar topologies (i.e. ab, bc and ac) had to be investigated. In the case of the two nonoptimal ones (bc and ac), it was found that only one displacement constraint is active for the least-weight solutions within the given topology. Moreover, it is still necessary to prove that the three-bar topology (a, b, c) is nonoptimal. This was done by an iterative procedure (DCOC), which has also shown that the topology (a, b) is indeed optimal.

5. ILLUSTRATION OF THE BASIC CONCEPTS OF LAYOUT THEORY THROUGH AN EXACT OPTIMAL LAYOUT

The theory of optimal layouts is particularly powerful for *exact* optimal solutions and was first used (e.g. Prager and Rozvany *1977*) for these problems. Exact optimal layouts involve a structural universe consisting of an infinite number of potential members and hence the choice of a suitable method for selecting the optimal layout is particularly important. Whereas we considered displacement constraints in our previous introductory example, with a view to showing the versatility of this approach we shall now consider stress constraints.

5.1 The Underlying Theory of Optimal Truss Layouts for Stress Constraints

(a) Statical determinacy of certain classes of solutions. It was shown by Sved (1954) some forty years ago that a truss optimized for one load condition and stress constraints is always *statically determinate.*

Since possibly not all readers are familiar with this term, we explain its meaning briefly. If a grid contains few enough members and supports so that all member forces can be determined from *equilibrium* (or *static admissibility*) *conditions only*, then it is termed statically determinate. For example, the two-bar truss in Fig. 7b is statically determinate because we can calculate both member forces f^a and f^b from the equilibrium conditions in the horizontal and vertical directions:

$$\frac{f^a}{\sqrt{2}} - 1 = 0, \, f^b + \frac{f^a}{\sqrt{2}} - 2 = 0. \tag{15}$$

However, the three-bar truss in Fig. 7a is *statically indeterminate* or *redundant*, because the two equilibrium equations

$$\frac{f^a - f^c}{\sqrt{2}} - 1 = 0, \qquad \frac{f^a + f^c}{\sqrt{2}} + f^b - 2 = 0 \tag{16}$$

are not sufficient to determine the three forces f^a, f^b and f^c uniquely. The above unknowns can, however, be uniquely determined if we also consider compatibility of the deformations, or briefly *compatibility* or *kinematic admissibility*. For the considered three-bar truss, the compatibility condition becomes (see the bar elongations δ^a, δ^b and δ^c in Fig. 9)

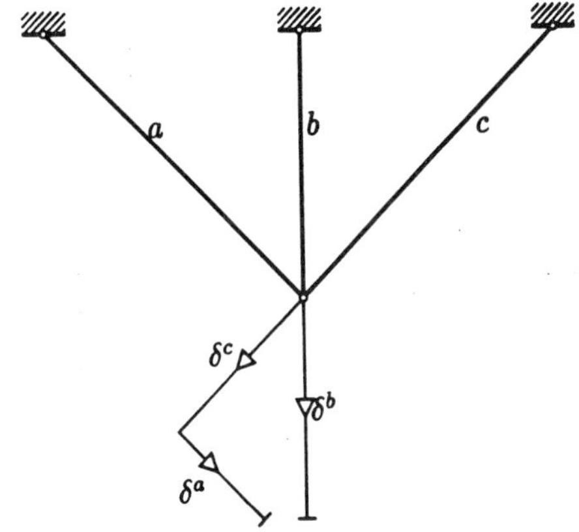

Fig. 9. Bar elongations for a three-bar truss.

$$\frac{\delta^a + \delta^c}{\sqrt{2}} = \delta^b. \tag{17}$$

Using Hooke's law for the above truss (Fig. 8a)

$$\delta^a = \frac{f^a 10\sqrt{2}}{E x^a} \ , \qquad \delta^b = \frac{f^b 10}{E x^b} \ , \qquad \delta^c = \frac{f^c 10\sqrt{2}}{E x^a} \ , \tag{18}$$

the compatibility conditions reduce to

$$\frac{f^a}{x^a} + \frac{f^c}{x^c} = \frac{f^b}{x^b} . \tag{19}$$

The forces f^a, f^b and f^c can then uniquely be determined from (15), (16) and (19).

Note that *for statically determinate structures* (see the two-bar truss above) *the internal forces are not dependent on the cross-sectional areas.*

(b) Optimal elastic vs. optimal plastic design. Structural optimization of statically indeterminate structures can be based on *optimal plastic design* or *on optimal elastic design.* In the former (e.g. Prager and Shield 1967), elastic compatibility conditions can be ignored even for redundant structures and only equilibrium constraints are considered. However, since for *trusses* the optimal solution for *one loading condition and stress constraints* is known to be statically determinate (Sved 1954), the optimal solution for these structures is *equally valid for optimal elastic design and optimal plastic design.* Hence *even for redundant trusses* (with one load condition), *we may consider equilibrium conditions only.*

(c) The layout function. The above conclusion implies that for the problem to be considered, we can use the layout function for plastic design, which for stress constraints can be expressed (e.g. from Michell 1904) as

$$\phi^e = (\sigma_0^e/\varrho^e)|\varepsilon^e|, \tag{20}$$

where σ_0^e is the permissible stress for the member e in both compression and tension:

$$-\sigma_0^e \le \sigma^e \le \sigma_0^e. \tag{21}$$

A modified layout function can be used if the permissible stress is different in compression and tension.

We shall now consider

- a structural universe with members running in all possible directions θ $(0 \le \theta \le \pi)$ at any point $P \in D$, where D is the *available domain* or *structural domain* for the centroidal axes of the truss members; and

- problems for which the values of σ_0^e and ϱ^e are θ-independent, i.e. at a given point P they do not depend on the orientation of the members.

Then it follows from the fundamental conditions of layout optimality [(10) and (11)] that optimal, nonvanishing members may only occur at the directional maxima of the strain at a point. Using the point P and the direction θ for identifying an (infinitesimal) truss element, the above conclusion implies

$$x^P(\theta^*) > 0 \quad \text{only if} \quad |\varepsilon^P(\theta^*)| = \max_\theta |\varepsilon^P(\theta)|, \tag{22}$$

where $x^P(\theta^*)$ and $\varepsilon^P(\theta^*)$ are, respectively, the cross-sectional area of and strain in the element at P in direction θ^* which is some value of θ $(0 \le \theta \le \pi)$.

It is known, however, that the maximum directional value(s) of axial strains can only occur in the so-called principal direction(s), which are at right angles.

5.2 Problem Statement: An Exact Layout Problem

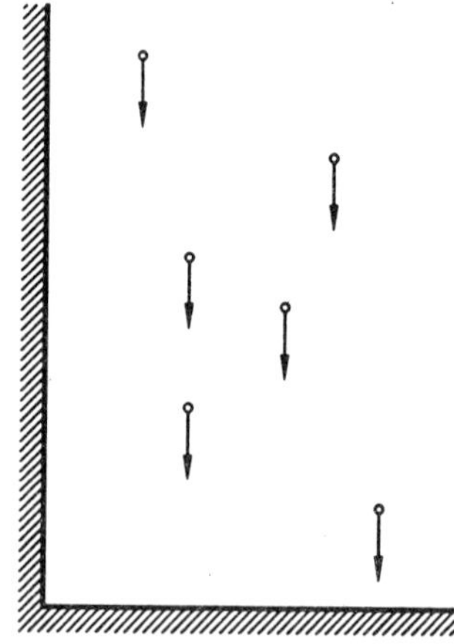

Fig. 10. An exact layout problem.

We consider a structural domain bounded by a horizontal and a vertical support (Fig. 10). Vertical point loads in various locations are to be transmitted to the supports by a plane truss of minimum weight. The specific weight of the truss material is $\varrho = $ const. and the permissible stress in both tension and compression is $\sigma_0 = $ const. in all directions as well as at all points of the structural domain.

5.3 A Demonstration of the Role of the Layout Function ϕ in Deriving Exact Optimal Layouts

Using the layout function (20) introduced in Subsection 5.1 (c), we can use the following procedure for finding the optimal layout:

(i) Divide the structural domain D into so-called *optimal regions* which are characterized by the sign of the strains in the principal directions, in which the layout function takes on a unit value.

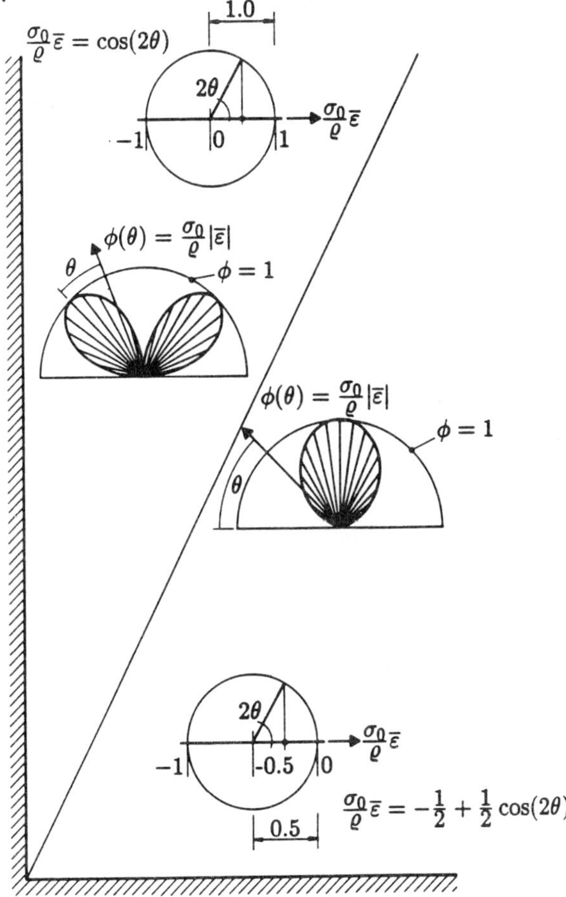

Fig. 11. Values of the layout function in dependence of the orientation at any point of the upper and lower regions.

In the optimal *adjoint* strain field shown in Fig. 11, the solution consists of two such regions. In the upper region, above the line with a slope 2 : 1, the layout function takes on a unit value in *two directions* and the sign of the two principal strains is "+" and "–". This type of region will be termed subsequently a *T region* (Prager **1974**).

In the lower region the layout function takes on a unit value in one direction only and the corresponding strain is negative. This type of region will be termed an R^- *region*.

It can be seen immediately from the variation of the layout function ϕ in Fig. 11 that in the top region optimal members may run in two directions (at $\pm 45°$ to the horizontal support) and in the bottom region nonvanishing members may only run in the vertical direction. This is because the fundamental condition (10) of layout optimality permits members only in directions with a unit value of the layout function [i.e. $\phi(\theta) = 1$].

Moreover, it can also be observed from Fig. 11 that the inequality (11) is satisfied by this adjoint strain field, i.e. $\phi \leq 1$ in all other directions.

(ii) It is still necessary to prove that the strain field in Fig. 11 satisfies

• kinematic boundary conditions along supports, and

• kinematic continuity conditions along region boundaries. This will be done in Section 5.4.

(iii) Finally, a truss must be found in which the statically admissible member forces satisfy the following *optimality conditions*:

$$(\text{for } f^e > 0) \quad \varepsilon^e = 1\,,$$

$$(\text{for } f^e < 0) \quad \varepsilon^e = -1\,. \tag{23}$$

The same relation can also be expressed as

$$(\text{for } f^e \neq 0) \quad \varepsilon^e = \text{sgn} f^e\,, \tag{24}$$

in which the usual sign function "sgn" has the meaning:

$$(\text{for } f^e > 0) \quad \text{sgn} f^e = 1\,, \quad (\text{for } f^e < 0) \quad \text{sgn} f^e = -1\,. \tag{25}$$

5.4 Further Details of the Solution Process

Figure 12b for the case $(\sigma_0/\varrho) = 1$ shows the displacement fields corresponding to the strains and layout function values in Fig. 11. The symbols with the arrows indicate the directions and signs of the principal strains with $\phi = 1$. In the *top (T-type) region*, we have (for $\sigma_0/\varrho = 1$)

$$\overline{\varepsilon}_x = \partial\overline{u}/\partial x = 0\,, \quad \overline{\varepsilon}_y = \partial\overline{v}/\partial y = 0\,, \quad \overline{\gamma}_{xy} = \partial\overline{u}/\partial y + \partial\overline{v}/\partial x = 2\,,$$

$$\overline{\varepsilon}_{I,II} = \frac{\overline{\varepsilon}_x + \overline{\varepsilon}_y}{2} \pm \sqrt{\left(\frac{\overline{\varepsilon}_x - \overline{\varepsilon}_y}{2}\right)^2 + \frac{\overline{\gamma}_{xy}^2}{4}} = \pm 1\,,$$

$$\theta_I = \frac{1}{2}\arctan\frac{\overline{\gamma}_{xy}}{\overline{\varepsilon}_x - \overline{\varepsilon}_y} = \frac{1}{2}\arctan\infty = 45°\,. \tag{26}$$

In the *bottom (R^-) region*, the principal strains become

$$\overline{\varepsilon}_x = \partial u/\partial x = 0\,, \quad \overline{\varepsilon}_y = \partial\overline{v}/\partial y = -1\,, \quad \overline{\gamma}_{xy} = \partial\overline{u}/\partial y + \partial\overline{v}/\partial x = 0\,,$$

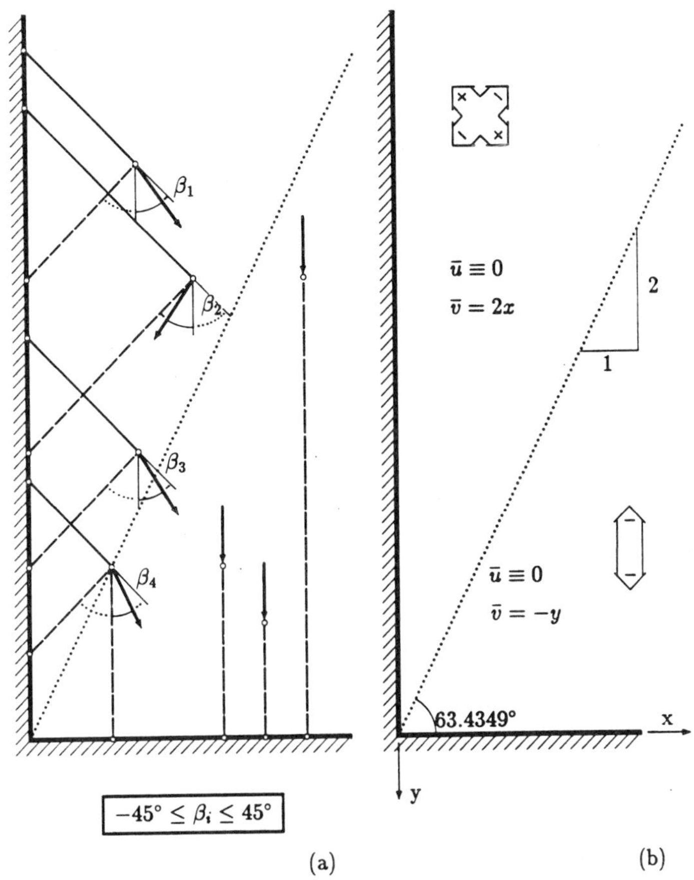

Fig. 12. (a) optimal layout and (b) adjoint strain field for the exact layout problem in Fig. 10.

$$\bar{\varepsilon}_I = -1, \quad \bar{\varepsilon}_{II} = 0, \quad \theta_I = 0, \tag{27}$$

where the angles θ_I of the first principal strain are measured from the vertical axis. The same principal strains are determined graphically (by means of Mohr-circles) in Fig. 11.

It can be checked readily that the displacement field in Fig. 12b satisfies the *kinematic boundary conditions*:

Top Region

$$\left(\bar{u}_t \equiv 0, \bar{v}_t = 2x\right): \quad \text{for} \quad x = 0, \quad \bar{u} = \bar{v} = 0,$$

Bottom Region

$$\left(\bar{u}_b \equiv 0, v_b = -y\right): \quad \text{for} \quad y = 0, \quad \bar{u} = \bar{v} = 0, \tag{28}$$

where the subscripts "b" and "t" refer to the "top" and "bottom" regions.

Moreover, along the *region boundary* $y = -2x$ we have

$$\overline{u}_t = \overline{u}_b = 0, \quad \overline{v}_t = 2x = \overline{v}_b = -y = 2x\,, \tag{29}$$

satisfying *continuity of displacements.*

Finally, we must find a truss in which the members match the directions and sign of principal strains with $\phi = 1$. This is the consequence of the optimality conditions (23) or (24). This means that over the top region members sloping at 45° from left to right must be in tension and those sloping at 45° from right to left must be in compression. Over the bottom region, all members must be vertical and in compression. It turns out that over the top region this solution can be extended from vertical loads to any load enclosing an angle not greater than 45° in either direction (Fig. 12a). This is because all these point loads cause tension (continuous lines in Fig. 12a) and compression (broken lines) in the correct directions. On the other hand, over the bottom region only vertical forces can be transmitted by the vertical bars. For the point load with the angle β_4, an infinite number of optimal solutions of equal weight exists, which include two bars at $\pm 45°$, a vertical bar or any convex combination of the above two. In all the above solutions the bar forces f^e are to be calculated from equilibrium and then the optimal cross-sectional areas are given by

$$x^e = \frac{f^e}{\sigma_0}\,. \tag{30}$$

For the three-bar truss in the corner, all statically admissible forces happen to be also kinematically admissible. This is the consequence of the facts that

- the adjoint displacements and strains $\overline{u}, \overline{v}$ and $\overline{\varepsilon}$ have been shown to be kinematically admissible; and
- for this class of problems, the real and adjoint displacement and strain fields are proportional (so-called *self-adjoint problems*).

5.5 Conclusions Drawn from the Illustrative Examples (Sections 4 and 5)

By studying the above examples carefully, the reader will see that the *layout function* ϕ is highly suitable for

- deciding whether a given *discrete layout* is optimal or not (see Fig. 7); and
- determining directly the *exact optimal layout* by imbedding a grid problem into a continuum (in this case: plane strain field, Fig. 11).

The assumption that the kinematically admissible strains in an infinitely dense truss or trusslike continuum can be replaced with a plane strain field was already inherent in Michell's (1904) theory and was used throughout the work by Prager and the author (e.g. Prager and Rozvany *1977*). However, a more rigorous examination of this assumption would be of considerable theoretical interest.

6. MUTUAL CONFIRMATION OF RESULTS FROM EXACT AND DISCRETIZED LAYOUT OPTIMIZATION AND FROM GENERALIZED SHAPE OPTIMIZATION

It was established by Rozvany, Olhoff, Bendsøe *et al.* (1985, 1987), and more rigorously by Allaire and Kohn (1993a and b) as well as by Bendsøe and Haber (1993) that at low volume fractions the optimal solution for perforated plates in plane stress and bending, respectively, tends to that of least-weight trusses and grillages of given depth. The above conclusion is restricted to one load condition with a compliance constraint or one of its equivalents.

It is clear that an exact optimal truss or grillage layout and an optimal discretized layout for the same problem can be used for a qualitative and quantitative check on the validity of these results. Moreover, due to the conclusion mentioned in the previous paragraph, we may also use optimized perforated plates for qualitatively confirming optimal truss layouts.

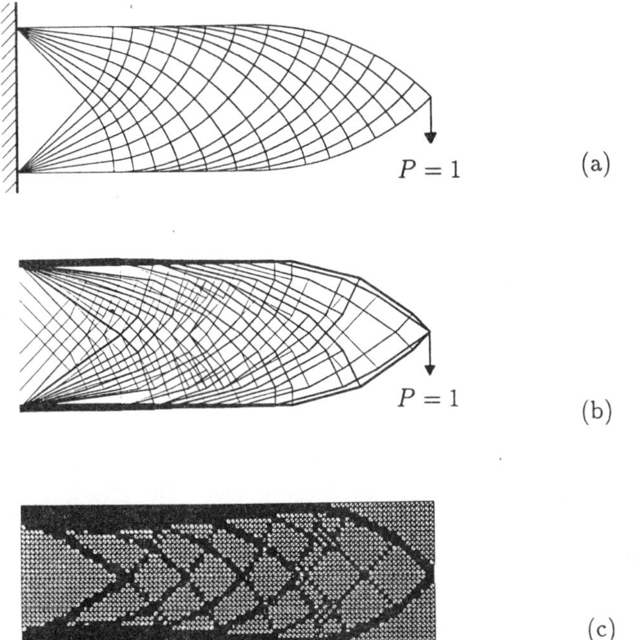

Fig. 13. A comparison of various types of solutions for a layout problem: cantilever truss with a point load (a) exact analytical solution; (b) discretized truss solution by DCOC; (c) discretized plate solution by DCOC.

As an example, Fig. 13a shows the exact optimal layout for a cantilever truss, derived analytically and plotted by computer graphics. The truss is subject to a displacement constraint at the point load. The solution is a truss-like continuum, with an infinite number of members at an infinitesimal spacing but, for obvious reasons, Fig. 13a shows only a finite number of these. The derivation of the exact solution (Lewinski, Zhou and Rozvany 1994) involves a rather lengthy mathematical procedure, using Lommel functions of two variables.

The discretized truss solution in Fig. 13b was derived by Zhou using the DCOC method and 7204 truss elements in the ground structure. The similarity of the layout

                                                                G.I.N. Rozvany

in Figs. 13a and b is obvious. The optimal weight given by the analytical solution is $W = 13.6$ and the one by the discretized solution is $W = 13.7$, a relatively small difference for such a complicated layout. The optimized generalized shape of a perforated plate for the same support and load conditions is shown in Fig. 13c, which also shows a striking similarity with the exact truss solution. The latter was derived by Zhou using 5400 triangular elements.

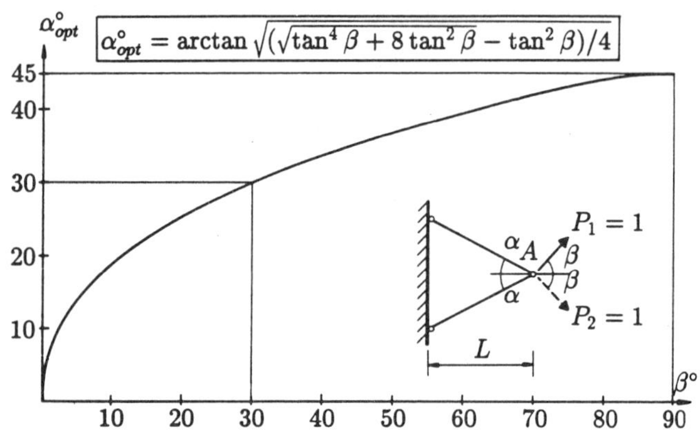

Fig. 14. Exact optimal truss layout for two alternative loads.

The second truss example involves two *alternative* point loads and displacement constraints at and in the direction of these loads. The exact analytical solution for this problem (Rozvany, Zhou and Birker 1993) is shown in Fig. 14. The dependence of the optimal angle of the two truss members on the angle β of the point loads is given by the neat closed form expression at the top of Fig. 14. Figure 15a shows the ground structure used by Zhou for verifying the above solution for $\beta = 5°$ and Fig. 15b indicates the result obtained by the DCOC method. The analytical solution for this problem has bar angles of $\alpha = 13.7613°$. Because the ground structure has no bars in these directions, the discretized solution consists of four bars, but the weighted combination of the corresponding angles agrees well with the analytical solution. The weight of the discretized solution was found $W = 11.316581$ which differs only by 0.007 from the weight of the analytical solution ($W = 11.315835$). Perforated plates of optimized generalized shape (Fig. 16b and c) also show a good agreement with the analytical solutions for trusses (Fig. 14). These plate solutions were obtained by Birker who used $36 \times 72 = 2592$ square elements (Fig. 16a).

Finally, the optimal analytical solution for a square grillage with clamped (built-in) edges is shown in Fig. 17 (Lowe and Melchers 1972/73). The symbols with arrows indicate the optimal direction of the beams, together with the sign of the beam moments. In the central region with the circle, all beam directions are equally optimal, but the beam moments must be positive. Figure 18a shows the ground structure (624 linearly

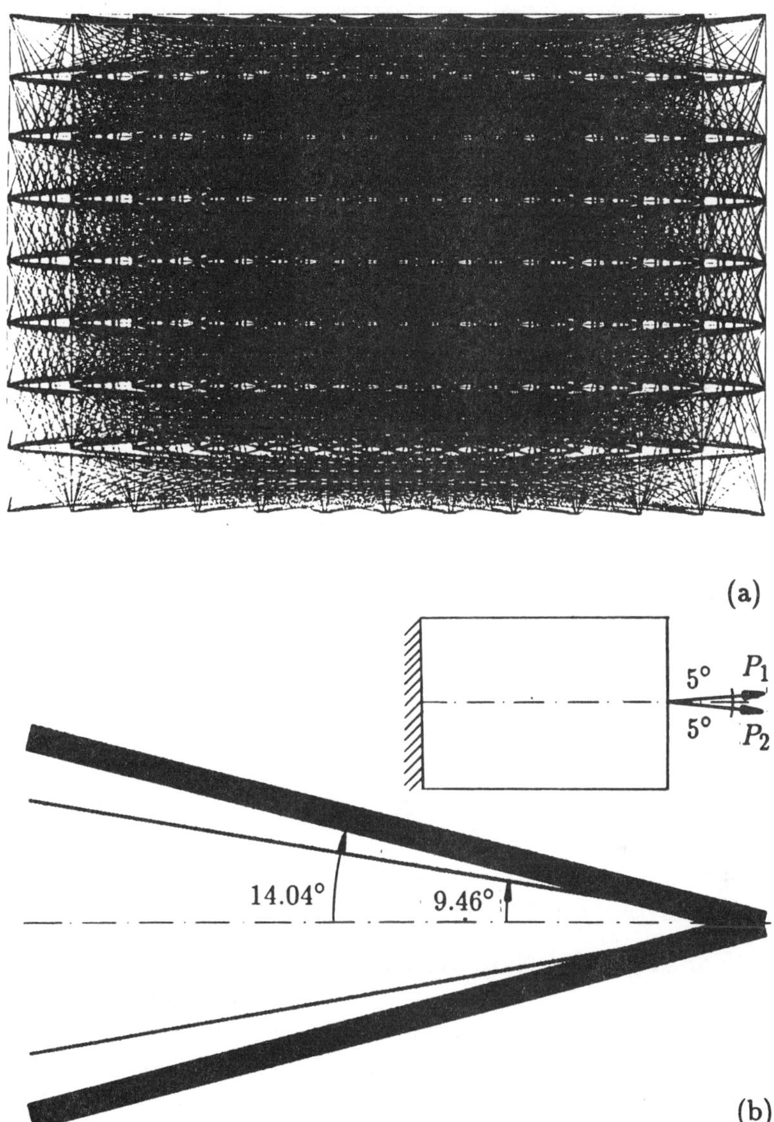

Fig. 15. Discretized truss solution for the problem in Fig. 14 (after Zhou).

varying beam elements) used by Sigmund (*et al.* 1993) for an optimal discretized grillage layout, which is shown in Fig. 18b–c. The perfect agreement with the analytical solution is obvious. More recently, Birker (1996) carried out a generalized shape optimization for a discretized perforated plate with 6400 square elements and for a loading in Fig. 19a obtained the result in Fig. 19b. The similarity with the analytical solution can again clearly be observed.

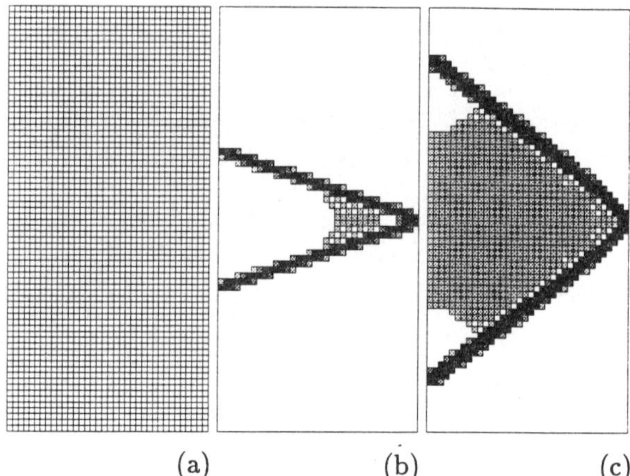

$$\text{(a)} \qquad\qquad \text{(b)} \qquad\qquad \text{(c)}$$

Fig. 16. Discretized plate solutions for the problem in Fig. 14 (after Birker).

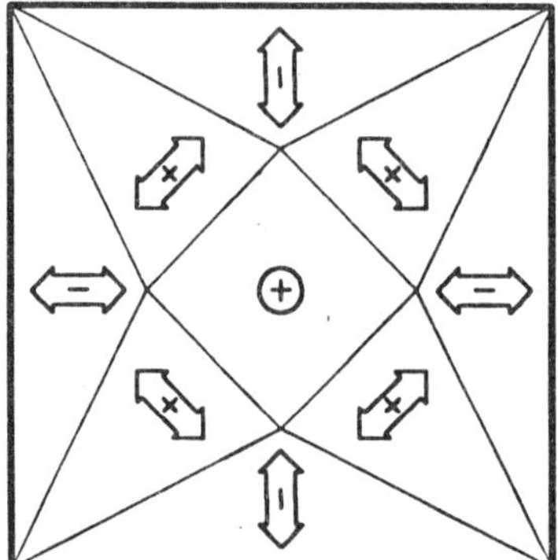

Fig. 17. Exact analytical solution for a square grillage with clamped edges.

7. WHY TRUSSES, WHY NOT GRILLAGES?

Most publications on discretized layout optimization use *trusses* in examples. We have followed this trend in the present chapter. Possible reasons for preferring trusses to grillages are as follows.

- This custom has been established for several decades.
- Particularly mathematicians may find it easier to visualize trusses, which can be regarded as a discretized version of a plane stress field.

However, optimized grillage layouts have the following advantages for demonstration purposes.

- Closed form analytical solutions are available for most boundary and load conditions.
- The optimal grillage layout is often independent of the load distribution.
- Optimal grillage theory has been extended to a number of design conditions.

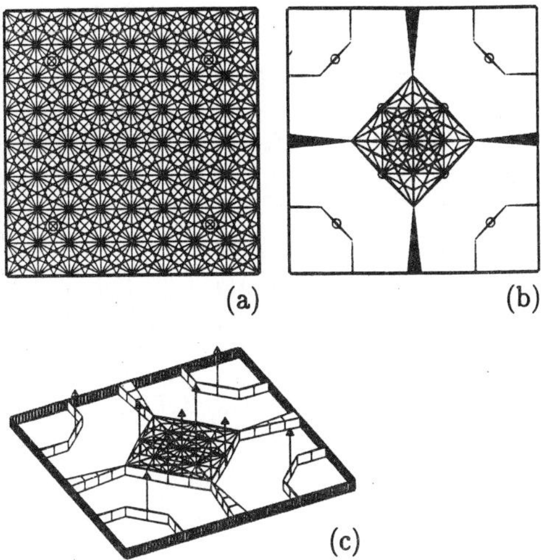

Fig. 18. Discretized grillage solution by DCOC for the problem in Fig. 17 (after Sigmund).

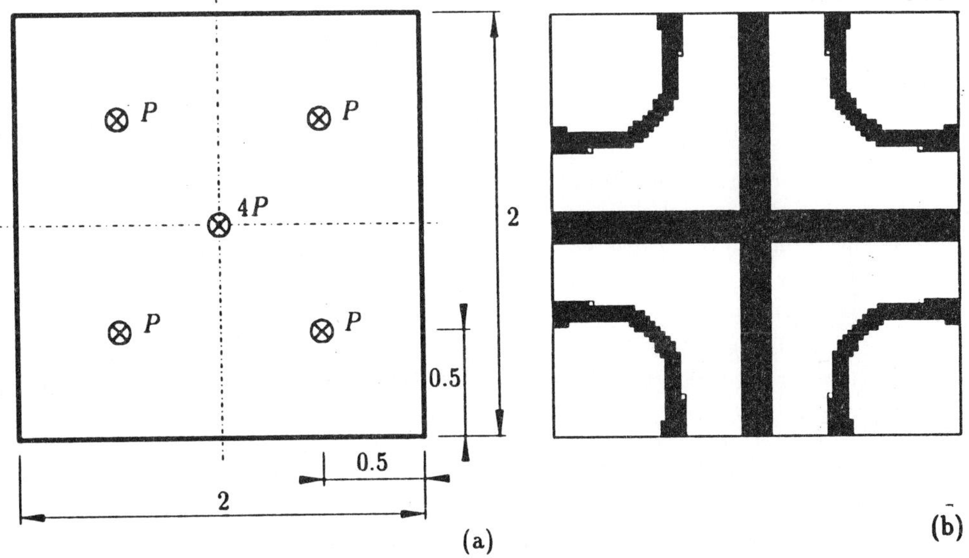

Fig. 19. Discretized plate solution for the problem in Fig. 17 (after Birker).

- They are more practical than trusses if member buckling is ignored in the latter.
- Since the main action (beam moment) varies along its members, whilst usual trusses have a constant axial force, grillages can be used for demonstrating a wider range of problems.

It will be seen in Chapters 3 and 4 that the optimization of truss and grillage layouts is mathematically similar.

8. ON MESH-DEPENDENCE OF THE OPTIMAL TOPOLOGY

Since the exact optimal layout often contains an infinite number of members or holes (voids) at an infinitesimal spacing, no ground structure with a finite number of elements can yield such an exact solution. It follows that with increasing numbers of elements in the ground structure, we obtain optimal layouts with increasing numbers of members or holes. Hence the optimal solution is often strongly "mesh-dependent", where "mesh" refers to the system of member axes in a grid or boundaries of elements in a discretized continuum.

The effect of this mesh-dependence on the optimal weight of the structure (within a given ground structure) can vary considerably depending on the type of problem. Figure 20 shows three grillage examples to demonstrate this. In Figs. 20a and b, respectively, a point load P and a line load p is to be transmitted to two clamped supports (thick lines at right angles). In Fig. 20c, the grillage has two simple supports (double lines), two free edges (single lines) and a point load P. The optimal layouts are also shown in Figs. 20a–c. It is assumed that in Figs. 20a and b the ground structure includes beam elements along the polygons $ABCD$ and $EFGH$ (Fig. 120b) and some additional ones in between the two, with a spacing of $a/2n$. In Fig. 20c the ground structure includes beams running in the directions PR and ST, with a spacing of $a/2n$ and a/n, respectively.

It is shown in Section A.2 that the weight of the above grillages is proportional to the "moment volumes" V_M given in Fig. 20. The dependence of the moment volumes V_M on the number n of divisions within the critical part of the ground structure is shown in Fig. 21, in which the vertical axis represents $\Delta\%$, i.e. the percentage difference between the moment volume V_{Mn}, a discretized solution and the moment volume of the exact solution $V_{M\infty}$ (with an infinite number of divisions in the ground structure). The following conclusions can be drawn from Figs. 20 and 21.

- In some cases the solution is not at all mesh dependent, as long as the considered ground structures contain certain elements (Fig. 20a).
- In other cases, the optimal weight is very weakly mesh-dependent, because the member spacings in the ground structure only influence the distribution of material over a certain region ($ABCDHGFE$ in Fig. 20b). In the quoted example, the percentage difference between the exact solution ($n = \infty$) and a single interval ($n = 1$) is only 7.1% and for five intervals ($n = 5$) it is only 0.29% [see curve (b) in Fig. 21]. The cost of transmission of the load p along its line of action is not taken into consideration when it is discretized into point loads.
- For some layout problem, the structural weight is strongly mesh dependent. In Fig. 20c, for example, the solution consists of only two members along PR and ST for $n = 1$, whilst the optimal layout contains narrow bands of "beam weaves" (Prager) along the free edges SP and TP for larger n values. The percentage difference between the weights for $n = 1$ and $n = \infty$ is 42.9% and even for $n = 5$ and $n = \infty$ it is 8.6%.

Considering continua, most of the above problems are "ill-posed" from a mathematical point of view, because a piece-wise continuous optimal solution (with a finite number of discontinuities) does not exist, resulting in mesh-dependence. Mathematicians have overcome this problem since the early eighties by using "relaxation" or "homogenization", by which they meant that they replace the solution having an infinite number of discontinuities with a smoothed-out version of the same problem, using equivalent stiffness or strength values. To engineers this is not a new concept, although it is now done more rigorously by the mathematician's yardstick. The solution in Fig. 20c, for example, was obtained in the seventies by the author (see e.g. Prager and Rozvany 1977) and his "engineering" method yielded the same result for the exact solution as homogenization methods would for plates with a zero volume fraction.

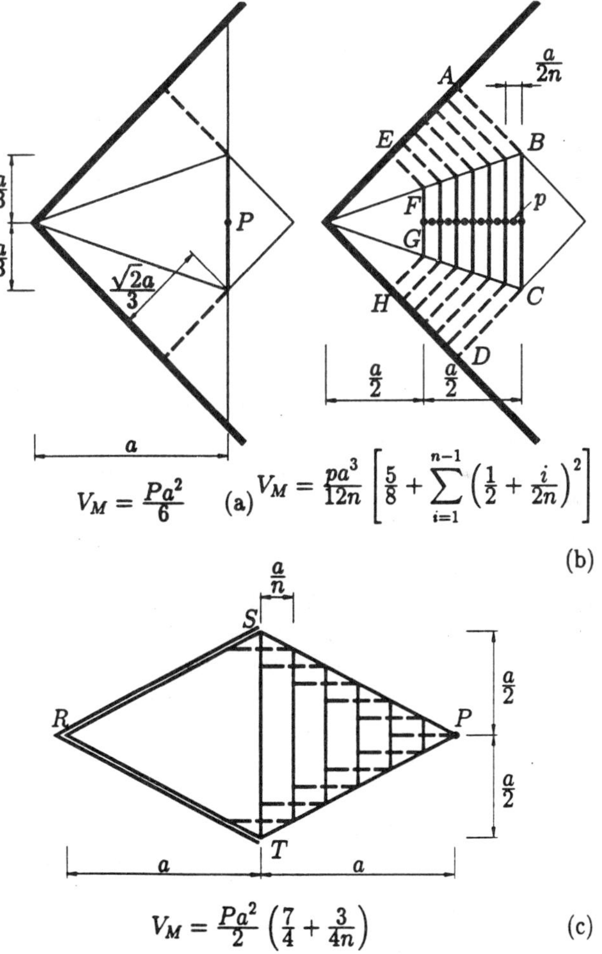

Fig. 20. Examples showing various degrees of mesh-dependence of the optimal layout.

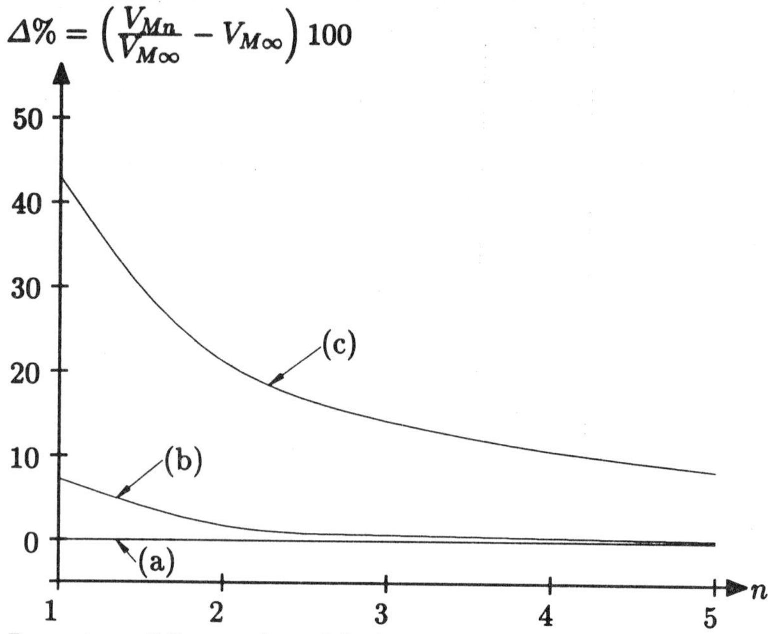

Fig. 21. Percentage difference in weight between solutions with finite mesh size and the exact solution (with $n \to \infty$).

There exist various possibilities for forcing a solution with a finite number of members or holes. One of these was introduced by the author and Prager (Rozvany and Prager 1976) and consists of adding a fixed cost value for each *additional* member or hole as a type of "cost of handling". The above paper shows that this method leads to a finite number of members in optimal grillages, which depends on the magnitude of the handling costs.

Another method was suggested by Niordson (1983a and b), in which the maximum spatial slope of the plate thickness is constrained from above. Although this method does not result in $0 - 1$ plate solutions with either zero thickness or full thickness, it does give continuous functions for the plate thickness in optimal solutions. A detailed treatment of "Niordson-constraints" was given in the author's second book (Rozvany **1989**, Chapt. 5).

A very powerful numerical method for obtaining a finite number of internal boundaries in perforated continua was suggested by Haber *et al.* (1996). In their "perimeter method" the total length of internal boundaries is constrained from above. A practical drawback of this method is the fact that the engineer does not necessarily know which perimeter value to choose for a given problem, whereas in the method by Rozvany and Prager (1976) one may have some idea as to the costs of producing an extra hole. However, a series of mesh-independent solutions could be produced for various given values of the perimeter and then the designer could select the solution he prefers. It is to be remarked that even the perimeter method becomes mesh-dependent if the size of elements is greater than some critical value.

9. NONUNIQUENESS OF THE OPTIMAL TOPOLOGY

There exist three types of nonuniqueness in exact topology optimization, which are explained in the next three subsections.

9.1 Nonconvex Problems with Multiple Local Minima

Almost all *real-world* topology optimization problems are nonconvex, either due to non-linear cost functions or constraints (e.g. displacement constraints for elastic structures) or due to variables with discrete values (e.g. commercially available cross-sections). A simple nonconvex weight function $W(x)$ depending on the design variable x is shown in Fig. 22a. The local minima A and B would represent two different topologies (and two different adjoint fields), which in general would give different weight values. The global optimum (in this case, B) could only be determined by weight comparison.

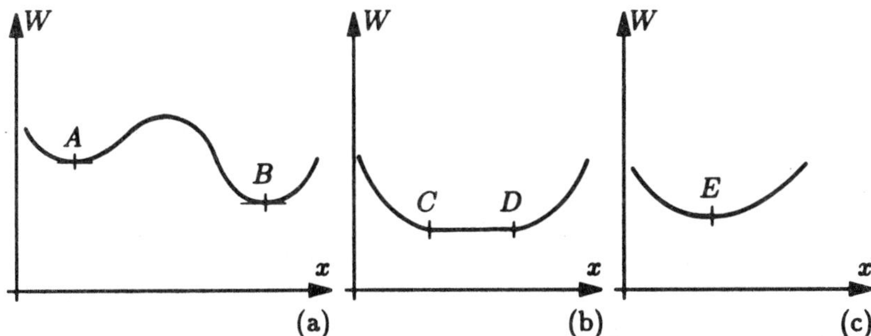

Fig. 22. Types of nonunique solutions in topology optimization.

9.2 Nonstrictly Convex Problems with Multiple Minima of the Same Weight

Most of the *idealized* topology optimization problems in the literature on exact solutions are convex but not strictly convex. This is shown conceptually in Fig. 22b, in which only one minimum (CD) exists but it is over a range of values of the design variable x. Such nonstrictly convex topology optimization problems are trusses and grillages of given depth, considering one load condition and a stress or a compliance constraint.

To illustrate this case with an example, we consider a square grillage with simple supports (double lines in Fig. 23). On the basis of results by Morley (1966) or Rozvany (1966), the optimal adjoint field for this problem (Fig. 23a) admits an infinite number of equally optimal solutions. In the central region (marked with a "+" sign in a circle), beams under positive moment can run in any arbitrary direction. In the corner regions, beams must run parallel to the two diagonals and must have, respectively, positive and negative moments (as an example, see the crossed arrows). Surprisingly to those unfamiliar with the grillage layout theory, all layouts in Figs. 23a–d give exactly the

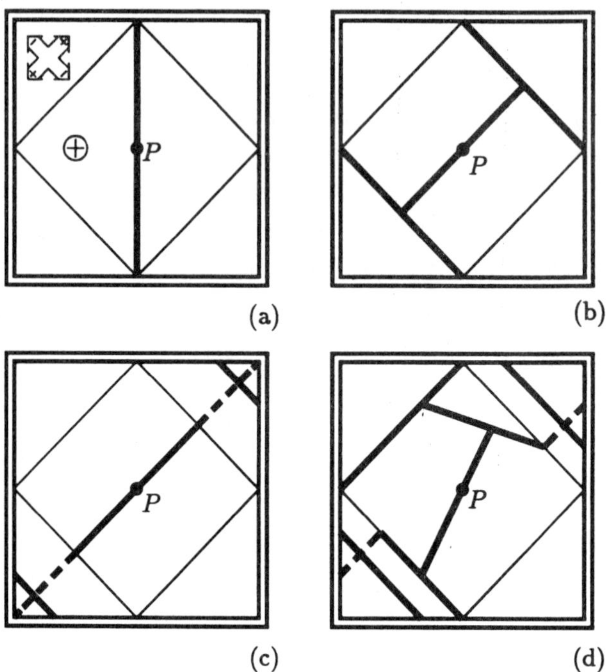

(a)   (b)

(c)   (d)

Fig. 23. Multiple minima of the same weight: grillage example.

same, optimal structural weight.

An important aspect of this example is that *the adjoint displacement field is unique and hence the same for all these alternative optimal layouts.* This is always the case if

- the structural domain is fully loaded; or
- it is sufficiently supported (as in Fig. 23).

An example of a nonunique truss topology is given in Section 2.4.4.

9.3 Unique Optimal Topology with a Nonunique Adjoint Displacement Field

In nonstrictly convex topology optimization problems, such as in grillage or truss topology optimization, the optimal layout may be unique (Fig. 22c) and yet the adjoint displacement field nonunique, if the structural domain is not fully loaded and not sufficiently supported.

This is illustrated in Fig. 24, for which the solution (Fig. 24a) was derived earlier (see Fig. 12, top region and Eq. 26). However, the adjoint field in Fig. 24b is equally admissible and gives the same solution, because along line AB the strain is zero and hence to the right of that line we may use a rigid translation of 2a in the vertical direction. If we had some other load or support on the right hand side, then this solution would not satisfy kinematic admissibility or the optimality conditions.

10. WHAT IS MEANINGFUL IN TOPOLOGY DESIGN FROM THE ENGINEER'S VIEWPOINT?

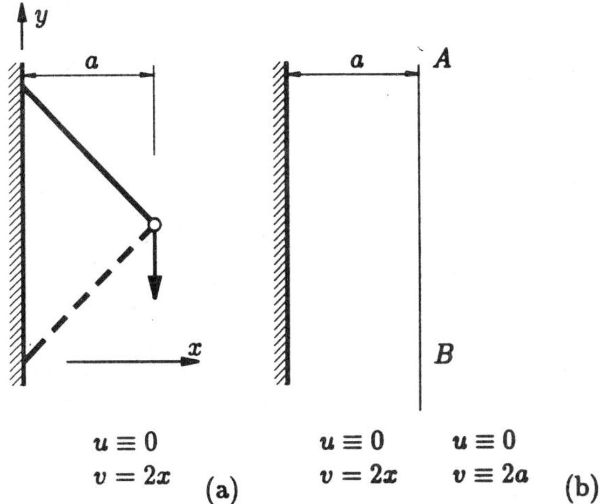

Fig. 24. Nonunique adjoint fields associated with a unique optimal layout: truss example.

Both topology optimization problems and methods for solving them fall into two broad categories which, together with their practical importance, are shown in Table 2.

Table 2 Basic classes of problems and types of solutions in topology optimization.

	Exact (explicit, closed form, analytical) solutions	Approximate (discretized, iterative, numerical) solutions
Complex (real-world, practical) problems	(a) relatively rare	(b) directly useful
Simple (idealized or artificial) problems	(c) indirectly useful	(d) less meaningful

The difference between exact and approximate solutions was explained in detail in Section 2 (Table 1).

The difference between real-world and artificial-idealized problems should be obvious to engineers, but it is explained here for the benefit of more theoretically inclined readers. In most structural problems in the practice,

- *a number of alternative load conditions* must be considered; and
- national (e.g. DIN) or international (e.g. Eurocode) design codes or standards *prescribe limiting values* on stresses, displacements, natural frequencies, ultimate collapse loads, service life expectancy, or alternatively, probabilities of various limit states.

Then a solution is sought whose weight, cost (in a financial sense), some nonstructural (e.g. aerodynamic) property or some weighted combination of the above quantities is as favourable as possible. As an alternative, for multiple objective functions pareto-optimal solutions may be generated.

Idealized problems are usually selected with a view to their mathematical simplicity, ensured through such properties as convexity or selfadjointedness. An idealized problem can be termed "artificial" if it has no physical meaning in engineering applications. An example of such formulation is minimizing the weighted combination of compliances (= total external work) for several load conditions or minimizing the elementwise maximum values of the compliances for a structure. There is no rational justifications for designing a bridge, aeroplane, car or any other structure in a real-world situation for such objective functions, even though some clients may be persuaded to accept them. For a single load condition and for very simple structures (e.g. trusses), a compliance constraint can be shown to be equivalent to a stress constraint (see Cox 1958, Hegemier and Prager 1969, Bendsøe *et al. 1994*), but similar equivalences cannot be extended to

- more complicated structures,
- several load conditions,
- more realistic (e.g. combined stress and displacement) constraints,
- unequal permissible stresses in tension and compression (which is the case in practical problems)

The value judgements in Table 2 can be justified as follows.

(a) In practical, real-world problems, the boundary and load conditions, as well as design constraints are rather complex and therefore closed form, explicit solutions are extremely rare, although recent advances in exact topology optimization of multi-purpose trusses are very promising.

(b) Discretized solutions for real-world topology design problems are *directly* useful for obvious reasons, particularly in view of the unavailability of exact solutions for these problems. However, certain special difficulties must be overcome in the case of stress and stablity constraints.

(c) Exact analytical solutions of idealized problems are *indirectly* useful, as was explained in Section 2.1.

(d) Discretized solutions of artificial problems occupy much of the extensive literature on topology optimization. These however cannot be used directly because of their technological irrelevance, nor necessarily indirectly, because of their fuzzy resolution and often low accuracy in detecting fundamental features of optimal layouts. The solution becomes particularly blurred, if the optimal directions are nonunique in certain regions.

11. METHODS OF PROOF USED BY THE AUTHOR

In the proofs of optimality criteria, the author's method depends on the type of elements in the structure.

If the cross-sectional dimensions and stress resultants vary along an element – in dependence on the spatial coordinates – then the calculus of variations (with Euler-Lagrange equations) is employed. An example of this type of structure is a grillage.

If the cross-sectional area and stress resultant are constant along a member – as in a truss element – then the proof consists of the following three steps.

- Derive necessary conditions of cost minimality on the basis of Kuhn-Tucker conditions for a structure with a finite number of elements, having prescribed minimum cross-sectional dimensions. This step is essentially sizing optimization.
- Allow the prescribed minimum cross-section to tend to zero.
- Increase the number of elements in the ground structure to infinity (structural universe).

In the third step, a strain field in a dense network of members is replaced by a continuum-type strain field. For example, the kinematically admissible axial strains in a truss having an infinitesimal member spacing becomes a plane strain field. No proof of convergence is given. Whilst a more rigorous examination of this step by mathematicians would be most welcome, the requirement of kinematic continuity by the optimality criteria should prevent lack of convergence. Until now, no case has been found where this "engineering" method would break down.

The above procedure for deriving optimality criteria is illustrated with an example (trusses with several load conditions and displacement constraints) in the Appendix (Section A.2).

12. GENERALIZED (VARIABLE TOPOLOGY) SHAPE OPTIMIZATION OR "ADVANCED" LAYOUT OPTIMIZATION

In *classical layout optimization*, we deal with grids whose volume fraction – from a mathematical viewpoint – tends to zero. The optimal microstructure of grids consists of intersecting members having a width of the same (usually first) order infinitesimal, and the effect of member intersections on strength, stiffness or weight is neglected.

In *generalized* (variable topology) *shape optimization*, which was also termed by the author (e.g. Rozvany **1989**) "advanced" layout optimization, the volume fraction is greater than zero and hence the optimization procedure consists of the following two steps:

- determination of the optimal microstructures, which usually contain some free parameters (corresponding to cross-sectional dimensions in classical layout optimization), and
- optimization of the *layout* of these microstructures, i.e. their optimal orientation and the optimal values of their free parameters.

It was shown in the early eighties (Lurie, Cherkaev and Fedorov 1982; Gibiansky and Cherkaev 1984) and confirmed later by others (e.g. Kohn and Strang 1986) that for a minimum compliance (= total external work) of a two-material composite plate we have an orthogonal rank-2 layered microstructure, having layers of first- and second-order infinitesimal width in the two principal directions. For a perforated plate, the weaker material in the above microstructure is replaced by voids.

The solution for the above problem is nonunique. Another (rank-1) optimal microstructure was derived by Vigdergauz (1994).

Whereas the compliance problem for plates of fixed thickness is "selfadjoint" (having proportional real and adjoint displacement fields), Lurie (1994, 1995a and b) showed more recently that for nonselfadjoint composite plate problems the optimal microstructure consists of rank-2 layering, which is in general nonorthogonal. Again, the microstructure for perforated plates is a limiting case of the above class of solutions.

Using optimal microstructures, one may obtain exact-analytical or approximate-discretized solutions, which will be discussed in the next two subsections.

12.1 Exact-Analytical Solutions in Generalized Shape Optimization

Using the orthogonal rank-2 microstructures described above for compliance problems, Rozvany, Olhoff, Bendsøe *et al.* (1985, 1987)

- derived the correct elastic constants (in the homogenization literature termed "rigidity tensor" $E_{ijk\ell}$) of the equivalent anisotropic plate; and
- obtained, by means of optimality criteria methods, closed-form analytical solutions for *axisymmetric perforated plates* with various load and support conditions.

Whilst the above results were for perforated plates with a *Poisson's ratio of zero value*, Ong, Rozvany and Szeto (1988) extended the same results to *nonzero Poisson's ratios*, and the author's doctoral students, Ong (1987) and Szeto (1989) to *two-material composite plates*. A review of this work was given by the author (Rozvany **1989**, pp. 349–351).

Considering nonselfadjoint problems, exact solutions for perforated plates under plane stress with two displacement constraints were obtained quite recently by Károlyi and Rozvany (1997).

12.2 Approximate Discretized Solutions of Generalized Shape Optimization

The basic philosophy of discretized solutions for porous or perforated structures varies considerably in dependence on the basic aim of the exercise. If we want to get a good approximation of the *exact* solution – which is not practical but is indirectly very useful – then we allow three types of elements in the solution, namely

- solid elements – filled with material,
- empty elements – without material,
- porous regions – some material, with cavities of inifinitesimal size.

The above solutions are termed *SEP (Solid–Empty–Porous) topologies*. From an engineering point of view, it is more practical to aim at solutions with only solid and empty elements at the macro-level (*SE topologies*).

(a) Discretized SE topologies (0–1 type problem). It has been demonstrated by the author and his associates that for SE topologies a powerful method is the combination of the discretized continuum-type optimality criteria (*DCOC*) method and *solid isotropic microstructures* with *penalty (SIMP)* for intermediate densities. Compelling evidence of the effectiveness of this approach was given in Figs. 13–19.

The basic idea of this method was first mentioned – together with some arguments against it – by Bendsøe (1989). The author arrived at the SIMP method independently – considering the combined cost of material and the labour involved in manufacturing perforated plates (Rozvany and Zhou 1991a, presented at a meeting in 1990; for a review, see the paper by Zhou and Rozvany 1991).

Before discussing discretized SEP topologies, we shall compare the specific cost (weight) – stiffness (rigidity) relation for various types of microstructures in Fig. 25, considering equal stiffnesses in the two principal directions.

The broken line in Fig. 25 corresponds to the optimal rank-2 microstructure – originally derived by Rozvany, Olhoff, Bendsøe *et al.* (1985/87) and later confirmed by homogenization studies (see for a review e.g. Bendsøe **1995**).

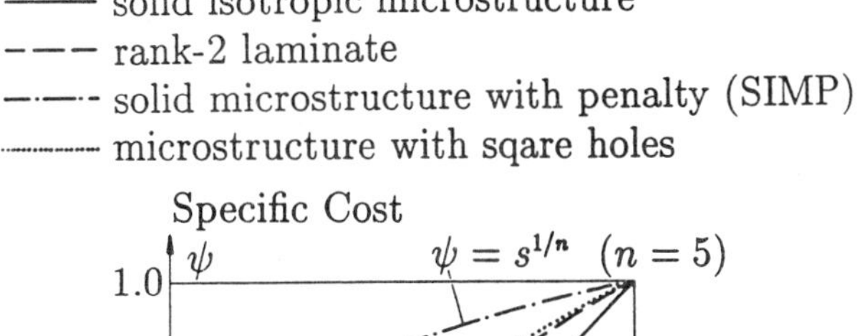

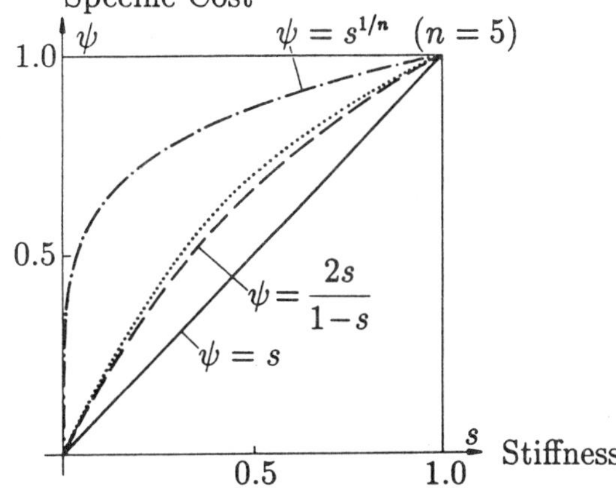

Fig. 25. Stiffness ("rigidity") – specific cost (weight) relations for various microstructures.

The continuous line with the linear relation represents a solid isotropic microstructure without penalty. For a plate in plane stress, this would represent an isotropic, unperforated plate of varying thickness (which is *not* a solution of the original perforated plate problem but can be used as a first step in deriving the optimal SE topology for the latter).

The dotted line gives the numerically calculated stiffness/weight relation for square holes in a rank-1 microstructure (Rozvany **1989**, p. 347).

Finally, the dash-dot line indicates the stiffness–weight relation for a solid, isotropic microstructure, in which the intermediate stiffnesses are penalized (SIMP method).

Three fundamental advantages of the SIMP approach (also termed the "engineering method" in the mathematical literature) are as follows:

• simplicity of analysis and optimization (only one variable per element);

• selective suppression of porous regions by adjustable penalty (see the constant n in Fig. 25); and

- capability of handling a combination of several design constraints for several load conditions by using the DCOC algorithm.

(b) Discretized SEP topologies. In this approach one uses essentially the exact stiffness–weight relation for the optimal microstructures (e.g. the broken line in Fig. 25). The resulting solutions are suitable for approximating the theoretically exact solution but, in general, unsuitable for SE topologies owing to the too small penalization of intermediate densities (in between 0 and 1), see Fig. 25. In a number of generalized shape optimization studies, the Bendsøe group (starting with Bendsøe and Kikuchi 1988; for a review, see Bendsøe **1995**) has been using nonoptimal microstructures, such as square or rectangular holes in the cell. As can be seen from Fig. 25, these have had – possibly inadvertently – a beneficial effect for SE topologies, in so far as they penalize stronger the intermediate densities than the optimal microstructure (compare the dotted and broken lines in Fig. 25). However, this penalization is still too weak and the microstructure with rectangular holes (similarly to the optimal ones) requires three free variables per element: two rib widths and the rib orientation. The author, therefore, believes that the SIMP method – owing to its variable penalization and its simplicity – is preferable. Another possibility is to use the optimal microstructure with penalization (e.g. Allaire and Kohn 1993c). As yet, there is no evidence that the extra effort of this more complicated procedure gives better results than the SIMP algorithm.

Generalized shape optimization by the homogenization method, including some excellent results by Bendsøe and his associates, are not discussed in greater detail by this author, because it is being covered in great detail by another lecturer (by G. Allaire).

13. SPECIAL DIFFICULTIES ASSOCIATED WITH STRESS AND STABILITY CONSTRAINTS

Most publications on topology optimization consider selfadjoint problems with some global (compliance or natural frequency) criterion. In all real-world problems, however, additional stress constraints must be imposed on the elements.

One of the difficulties with stress constraints is the fact that they are *conditional constraints*, that is, they are only valid if a cross-sectional (A) area is nonzero $(A \neq 0)$ and they become invalid if a member vanishes $(A = 0)$ in the optimal solution. This discontinuity in the design procedure was first pointed out by Sved and Ginos (1968). It was first stated by Cheng and Jiang (1992) that for such problems the feasible set is not disjoint, but the optimal solution is connected to the rest of the feasible set by a line segment. This can be shown on a simple example, which was introduced by Kirsch (1990) and is reproduced in Fig. 26. The feasible set for this problem consists of the shaded area $BCDE$ and of the line segment FD, the optimal solution being at F. It can be seen that the usual successive iteration methods would not find the "narrow passage" FD leading to the optimal solution.

The problem of singular solutions was also discussed extensively by the author, providing some deeper insight into their causes and outlining the classes of problems in which they may occur (Rozvany and Birker 1994). In more recent publications, Zhou

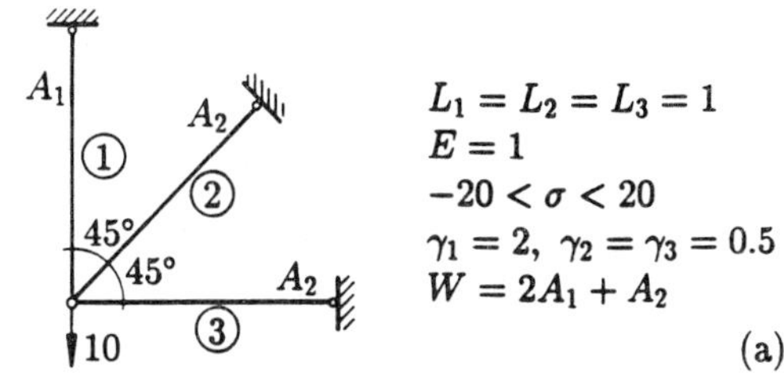

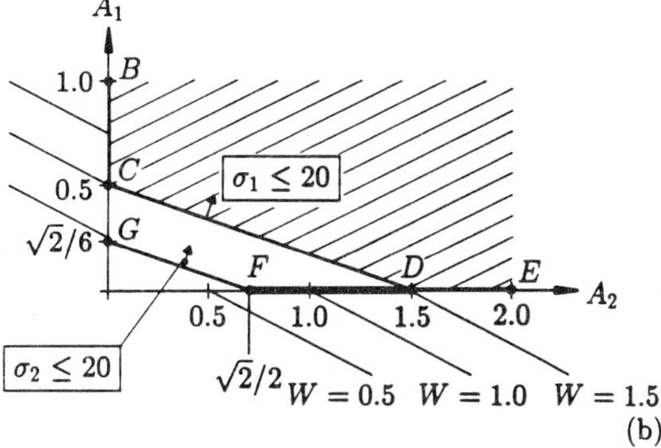

Fig. 26. Kirsch's (1990) example of a singular optimal topology.

(1996) and Rozvany (1996) pointed out further difficulties associated with local and global buckling constraints.

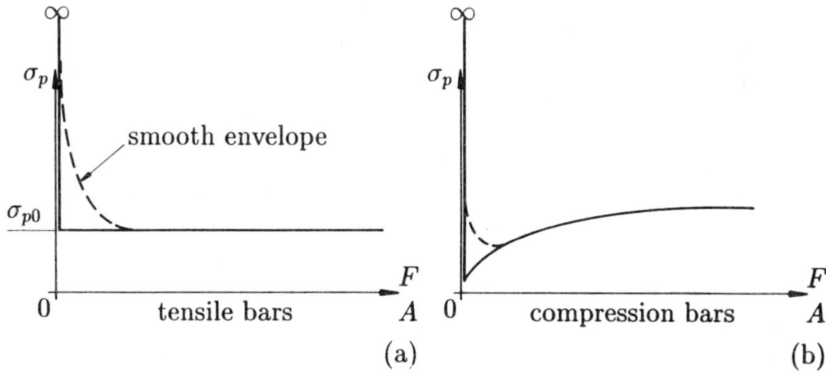

Fig. 27. Permissible stress (σ_p) – member force F (or cross-sectional area A) relations for tension and compression bars.

The problems caused by stress and local buckling constraints are summarized, and a method for overcoming them is given in Fig. 27 (after Rozvany 1996). In Fig. 27a, the continuous line shows that the permissible stress for a *tension bar* is constant for a

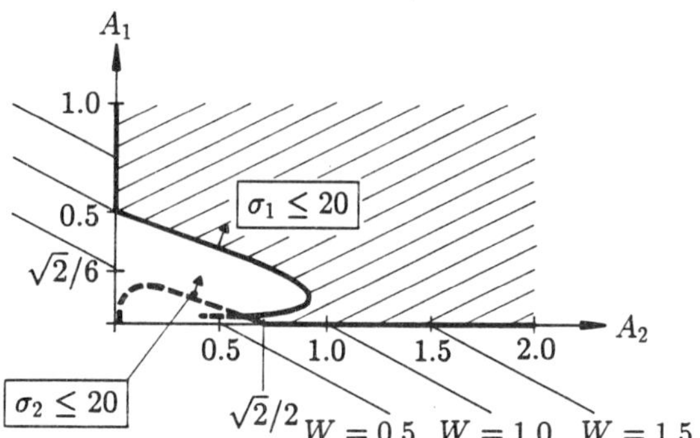

Fig. 28. The effect of using smooth envelope functions on the feasible set in Kirsch's (1990) example.

nonzero member force (F) or cross-sectional area (A), but suddenly changes to infinity for $A = F = 0$. For a *compression bar* this singularity is even more severe, because the permissible stress for small cross-sectional areas (with high slenderness values) tends to zero, but it again changes to infinity at $A = F = 0$. One method of overcoming this problem is also shown in Fig. 27 in broken lines, which represent "smooth envelope functions" (e.g. KS functions, e.g. Rozvany and Sobieski 1992). As a result of this modification, the "narrow passage" FD in Fig. 27 widens out considerably (see the modified design space in Fig. 28) and hence the optimal solution F can be reached in principle by iterative methods. A similar approach was also suggested more recently by Cheng and Guo (1997).

14. "PRACTICAL" TOPOLOGY DESIGN IN ACCORDANCE WITH NATIONAL OR INTERNATIONAL DESIGN STANDARDS

This topic was investigated by the author's doctoral students at Essen University more recently (Gerdes 1994; Birker 1996). Figure 29a, for example, shows two alternative loads and potential supports, Fig. 29b the ground structure with 1036 truss elements and Fig. 29c the optimal layout using a ground structure with 7750 truss elements, considering stress and displacement constraints (after Gerdes).

Figure 30a shows the area – member force relation used by Birker (1996) for tubular members, based on the Eurocode 3 for steel structures. Figure 30b indicates the step function representing the available cross sections specified in the German standard DIN 2448 and Fig. 30c the corresponding smooth envelope (KS) function. Birker used the combination of Figs. 30a and c for optimizing trusses. All these practical results do not exclude the possibility of a local minimum, particularly in view of the problems mentioned in Section 13.

CONCLUSIONS

(1) Topology optimization problems may be solved by *exact-analytical* or *approximate-*

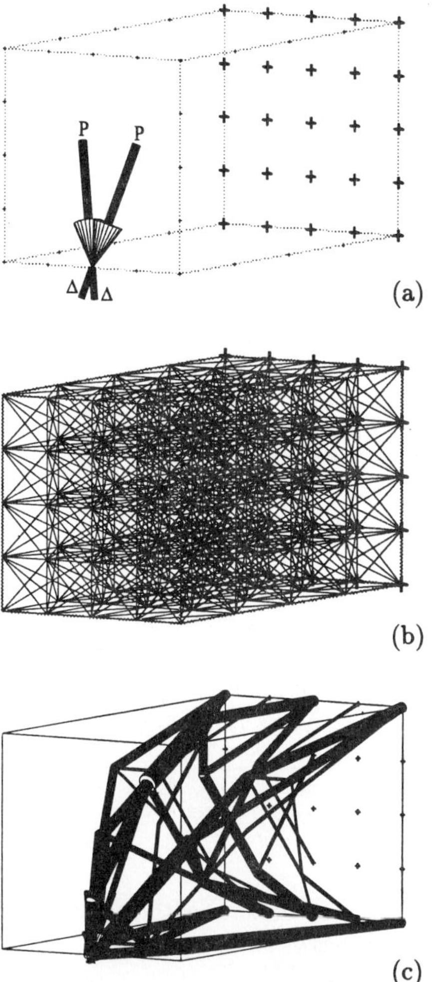

Fig. 29. A discretized multipurpose optimal truss layout: (a) loads, supports and displacement constraints; (b) ground structure; (c) optimal layout (after Gerdes).

discretized formulations. The ground structure for these, respectively, consists of an infinite and a finite number of elements.

(2) *Exact-analytical* type optimality criteria and optimal solutions can now be derived for *multipurpose, multiload* trusses and grillages, which represent nonselfadjoint problems.

(3) Using the optimality criteria method DCOC, *discretized optimal layouts* can be obtained in principle for any grid-type structure subject to a combination of *real-world design constraints* based on national or international design standards.

(4) The combination of the SIMP approach with DCOC enables us to derive dis-

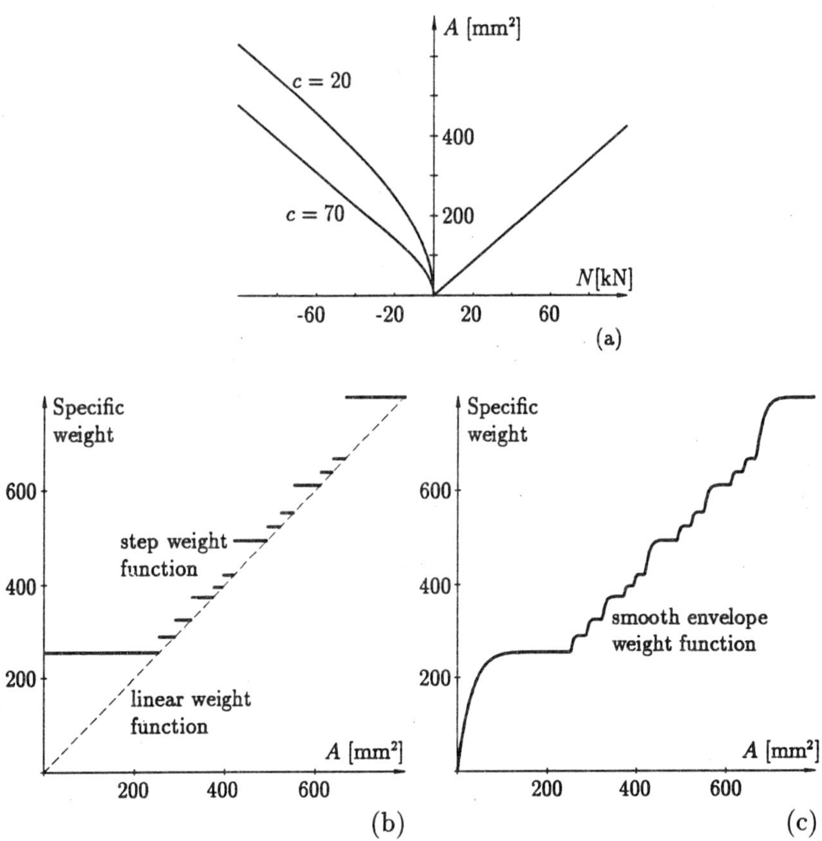

Fig. 30. Truss optimization in accordance with Eurocode 5 (steel structures) and DIN 2448 (standardized cross sections).

cretized *generalized shape optimization* solutions for *any combination of the usual* (e.g. stress, displacement, natural frequency) *constraints*.

(5) However, the solution of problems under (3) and (4) is still subject to *special difficulties* caused by singularity of the solution in the case of stress and/or local stability constraints. Topology optimization cannot yield realistic results, unless and until these pitfalls are overcome. Some initial progress in this direction has been made recently.

ACKNOWLEDGEMENTS

The author is indebted to the Deutsche Forschungsgemeinschaft (Project Ro 744/6) and to the NATO Scientific and Environmental Affairs Division for financial support; to Anne Fischer (text processing) and Elke Becker (drafting) for their help in preparing the manuscript.

REFERENCES

Note. In the text, the year of publication of *books* is given in boldface (e.g. Bendsøe **1995**), for *review papers* in italic (e.g. Bendsøe, Ben-Tal and Zowe *1994*), and for research papers, doctoral theses and research reports in roman typeface (e.g. Bendsøe and Haber 1993).

R.1 Books

Argyris, J.H. and Kelsey, S. 1960: *Energy Theorems and Structural Analysis*. Butterwork, London.

Bendsøe, M.P. 1991: *Methods for the optimization of structural topology, shape and material*. Rep. Math. Inst. Tech. Univ. Denmark, Lyngby, Denmark (also to be published as a book by Springer-Verlag).

Prager, W. 1974: *Introduction to Structural Optimization*. (Course held at Int. Centre for Mech. Sci. Udine, CISM) **212**. Springer-Verlag, Vienna.

Rozvany, G.I.N. 1989: *Structural Design via Optimality Criteria*. Kluwer, Dordrecht.

R.2 Review Papers

Bendsøe, M.P.; Ben-Tal, A. and Zowe, J. 1994: Optimization methods for truss geometry and topology design. *Struct. Optim.* **7**, 141–159.

Prager, W. and Rozvany, G.I.N. 1977: Optimization of structural geometry. In: Bednarek, A.R. and Cesari, L. (Eds.) *Dynamical Systems*, pp. 265–293. Academic Press, New York.

Rozvany, G.I.N. and Ong, T.G. 1986b: Optimal plastic design of plates, shells and shellgrids. In: Bevilacqua, L.; Feijóo, R. and Valid, R. (Eds.) *Inelastic Behaviour of Plates and Shells* (Proc. IUTAM Symp., Rio de Janeiro, 1985), pp. 357–384. Springer-Verlag, Berlin.

R.3 Research Papers, Doctoral Theses and Research Reports

Allaire, G. and Kohn, R.V. 1993a: Explicit bounds on the elastic energy of a two-phase composite in two space dimensions. *Q. Appl. Math.* **51**, 675–699.

Allaire, G. and Kohn, R.V. 1993b: Optimal bounds on the effective behaviour of a mixture of two well-ordered elastic materials. *Q. Appl. Math.* **51**, 643–674.

Allaire, G. and Kohn, R.V. 1993c: Topology optimization and optimal shape design using homogenization. In: Bendsøe, M.P. and Mota Soares, C.A. (Eds.) *Topology Design of Structures* (Proc. NATO ARW, Sesimbra, Portugal, 1992), pp. 207–218. Kluwer, Dordrecht.

Bendsøe, M.P. 1989: Optimal shape design as a material distribution problem. *Struct. Optim.* **1**, 193–202.

Bendsøe, M.P. and Haber, R.B. 1993: The Michell layout problem as a low volume fraction limit of the perforated plate topology optimization problem: an asymptotic study. *Struct. Optim.* **6**, 263–267.

Bendsøe, M.P. and Kikuchi, N. 1988: Generating optimal topologies in structural design using a homogenization method. *Comp. Meth. Appl. Mech. Eng.* **71**, 197–224.

Birker, T. 1996: *New developments in structural optimization using optimality criteria.* Doctoral Thesis, Essen University. Also: Series 18, No. 199, VDI-Verlag, Düsseldorf.

Cheng, G.D. and Guo, X. 1997: ε-relaxed approach in structural topology. *Struct. Optim.* **13** (in press).

Cheng, K.-T. and Jiang, Z. 1992: Study on topology optimization with stress constraints. *Eng. Optim.* **20**, 129–148.

Cox, H.L. 1958: The theory of design. *Rep. Aeronautical Res. Council*, No. 2037.

Dorn, W.S.; Gomory, R.E. and Greenberg, H.J. 1964: Automatic design of optimal structures. *J. de Mécanique* **3**, 25–52.

Gibiansky, L.V. and Cherkaev, A.V. 1984: Design of composite plates of optimal rigidity. *Rept. No. 914*, A.F. Ioffe Physical Technical Institute, Academy of Sciences of the USSR, Leningrad (in Russian).

Hegemier, G.A. and Prager, W. 1969: On Michell trusses. *Int. J. Mech. Sci.* **11**, 209–215.

Huang, N.C. 1971: On principle of stationary mutual complementary energy and its application to structural design. *ZAMP* **22**, 608–620.

Károlyi, G. and Rozvany, G.I.N. 1997: Exact analytical solutions for nonselfadjoint variable-topology shape optimization problems: perforated cantilever plates in plane stress subject to displacement constraints. *Struct. Optim.* **13** (in press).

Kirsch, U. 1990: On singular topologies in structural design. *Struct. Optim.* **2**, 133-142.

Kirsch, U. 1995: Layout optimization using reduction and expansion processes. In: Olhoff, N. and Rozvany, G.I.N. (Eds.) *Proc. First World Congr. Struct. Multidisc. Optim.* pp. 95–102. Pergamon, Oxford.

Kohn, R.V. and Strang, G. 1986a: Optimal design and relaxation of variational problems, I, II, and III. *Comm. Pure Appl. Math.* **39**, 113–137, 139–182, 353–377.

Lewiński, T.; Zhou, M. and Rozvany, G.I.N. 1994: Extended exact solutions for least-weight truss layouts – Part I: cantilever with a horizontal axis of symmetry. *Int. J. Mech. Sci.* **36**, 5, 375–398.

Lowe, P.G. and Melchers, R.E. 1972-1973: On the theory of optimal constant thickness fibre-reinforced plates. I, II, III. *Int. J. Mech. Sci.* **14**, 311–324, **15**, 157–170, **15**, 711–726.

Lurie, K.A. 1994: Direct relaxation of optimal layout problems for plates. *JOTA* **80**, 93–116.

Lurie, K.A. 1995a: On direct relaxation of optimal material design problems for plates. *WSSIAA* **5**, 271–296.

Lurie, K.A. 1995b: An optimal plate. In: Olhoff, N.; Rozvany, G.I.N. (Eds.) *Proc. First World Cong. Struct. Multidisc. Optim.* pp. 169–176. Pergamon, Oxford.

Lurie, K.A.; Cherkaev, A.V. and Fedorov, A.V. 1982: Regularization of optimal design problems for bars and plates. JOTA, **37**, 499–522 (Part I); **37**, 523–543 (Part II) and **42**, 247–282 (Part III).

Michell, A.G.M. 1904: The limits of economy of material in frame-structures. *Phil. Mag.* **8**, 589–597.

Morley, C.T. 1966: The minimum reinforcement of concrete slabs. *Int. J. Mech. Sci.* **8**, 305–319.

Niordson, F.I. 1983a: Optimal design of elastic plates with constraint on the slope of the thickness function. *Int. J. Solids Struct.* **19**, 141–151.

Niordson, F.I. 1983b: Some new results regarding optimal design of elastic plates. In: Eschenauer, H.; Olhoff, N. (Eds.) *Optimization Methods in Structural Design*, pp. 380–286. Wissenschaftsverlag, Mannheim.

Ong, T.G. 1987: Structural optimization via static-kinematic optimality criteria. *Ph.D. Thesis*, Monash University, Melbourne.

Ong, T.G.; Rozvany, G.I.N. and Szeto, W.T. 1988: Least-weight design of perforated elastic plates for given compliance: non-zero Poisson's ratio. *Comp. Meth. Appl. Mech. Eng.* **66**, 301–322.

Prager, W. 1978a: Nearly optimal design of trusses. *Comp. Struct.* **8**, 451–454.

Prager, W. 1978b: Optimal layout of trusses with a finite number of joints. *J. Mech. Phys. Solids* **26**, 241–250.

Prager, W. and Shield, R.T. 1967: A general theory of optimal plastic design. *J. Appl. Mech.* **34**, 1, 184–186.

Rozvany, G.I.N. 1966: A rational approach to plate design. *J. Amer. Conc. Inst.* **63**, 1077–1094.

Rozvany, G.I.N. 1992: Optimal layout theory: analytical solutions for elastic structures with several deflection constraints and load conditions. *Struct. Optim.* **4**, 247–249.

Rozvany, G.I.N. 1996: Difficulties in truss topology optimization with stress, local buckling and system buckling constraints. *Struct. Optim.* **11**, 213–217.

Rozvany, G.I.N. and Birker, T. 1994: On singular topologies in exact layout optimization. *Struct. Optim.* **8**, 228–235.

Rozvany, G.I.N.; Gerdes, D. and Gollub, W. 1996: Ein analytischer Computeralgorith-mus zum Auffinden optimaler Tragstrukturen bei Beigebeanspruchung. *Research Rep.* No. 69. Civil Eng. Dept., Essen University.

Rozvany, G.I.N. and Hill, R.H. 1978: A computer algorithm for deriving analytically and plotting optimal structural layout. *Comp. Struct.* **10**, 295–300.

Rozvany, G.I.N.; Olhoff, N.; Bendsøe, M.P.; Ong, T.G. and Szeto, W.T. 1985: Least-weight design of perforated elastic plates. *DCAMM Report* No. 306. Techn. Univ. Denmark, Lyngby.

Rozvany, G.I.N.; Olhoff, N.; Bendsøe, M.P.; Ong, T.G.; Sandler, R. and Szeto, W.T. 1987: Least-weight design of perforated elastic plates I, II. *Int. J. Solids Struct.* **23**, 521–536, 537–550.

Rozvany, G.I.N. and Prager, W. 1976: Optimal design of partially discretized grillages. *J. Mech. Phys. Solids* **24**, 125–136.

Rozvany, G.I.N. and Sobieszczanski-Sobieski, J. 1992: New optimality criteria methods: forcing uniqueness of the adjoint strains by corner-rounding at constraint intersections. *Struct. Optim.* **4**, 244–246.

Rozvany, G.I.N. and Zhou, M. 1991a: Applications of the COC method in layout optimization. In: Eschenauer, H.; Matteck, C. and Olhoff, N. (Eds.) *Proc. Int. Conf. "Engineering Optimization in Design Processes"* (Karlsruhe 1990), pp. 59–70. Springer-Verlag, Berlin.

Rozvany, G.I.N. and Zhou, M. 1996: Advances in overcoming computational pitfalls in topology optimization. *Proc. 6th AIAA/NASA/ISSMO Symp. on Multidisc. Anal. Optim.* A.I.A.A., Reston, VA.

Rozvany, G.I.N.; Zhou, M. and Birker, T. 1993: Why multiload topology designs based on orthogonal microstructures are in general nonoptimal. *Struct. Optim.* **6**, 200–204.

Shield, R.T. and Prager, W. 1970: Optimum structural design for given deflection. *ZAMP* **21**, 513–523.

Sigmund, O.; Zhou, M. and Rozvany, G.I.N. 1993: Layout optimization of large FE-systems by new optimality criteria methods: applications to beam systems. In: Haug, E.J. (Ed.) *Proc. NATO ASI Concurrent Engineering Tools and Technologies for Mechanical System Design* (Iowa, 1992). Springer-Verlag, Berlin.

Sved, G. 1954: The minimum weight of certain redundant structures. *Austral. J. Appl. Sci.* **5**, 1–8.

Szeto, W.T. 1989: Microstructure studies and structural optimization. *Ph.D. Thesis*, Monash University, Melbourne.

Vigdergauz, S. 1986: Effective elastic parameters of a plate with a regular system of equal-strength holes. *Mechanics of Solids* (transl. of Mekh. Tver. Tela.) **21**, 162–166.

Zhou, M. 1996: Difficulties in truss topology optimization with stress and local buckling constraints. *Struct. Optim.* **11**, 134–136.

Zhou, M. and Rozvany, G.I.N. 1992/1993: DCOC: an optimality criteria method for large systems. Part I: theory. Part II: algorithm. *Struct. Optim.* **5**, 12–25; **6**, 250–262.

APPENDIX

A.1 LAYOUT PROBLEM WITH A THREE-BAR GROUND STRUCTURE

A.1.1 Optimal Layout (a, b)

By (3) and (7) with $\varrho^e = 1$ and $E^e = E$ (for all e) we have

$$x^e = \sqrt{\frac{f^e \left(\nu_1 \hat{f}_1^e + \nu_2 \hat{f}_2^e \right)}{E}} . \tag{A.1}$$

Substituting the force values from Fig. 8b-d, (A.1) implies

$$x^a = \sqrt{\frac{\sqrt{2}\nu_1\sqrt{2}}{E}} = \sqrt{\frac{2\nu_1}{E}} , \tag{A.2}$$

$$x^b = \sqrt{\frac{1[\nu_1(-1) + \nu_2(1)]}{E}} . \tag{A.3}$$

The values of the Lagrange multipliers will be determined from the displacement constraints, assuming that they are both active. The displacement constraints are expressed by means of virtual work equations:

$$u_d = \sum_d \sum_e \frac{f^e \hat{f}_d^e L^e}{E x^e} = \Delta_d , \tag{A.4}$$

implying together with Fig. 8a, (A.2) and (A.3)

$$E\Delta_1 = 20 = \left[\frac{\sqrt{2}\sqrt{2}\sqrt{2}(10)}{\sqrt{2\nu_1}} + \frac{(1)(-1)(10)}{\sqrt{\nu_2 - \nu_1}} \right] \sqrt{E} , \tag{A.5}$$

$$E\Delta_2 = 10 = \frac{(1)(1)(10)}{\sqrt{\nu_2 - \nu_1}} \sqrt{E} \tag{A.6}$$

Then we have

$$(A.6) \Rightarrow \sqrt{\nu_2 - \nu_1} = \sqrt{E} , \tag{A.7}$$

$$(A.5)(A.7) \Rightarrow 2 = \frac{2\sqrt{E}}{\sqrt{\nu_1}} - 1 \Rightarrow \sqrt{\nu_1} = \frac{2}{3}\sqrt{E} , \Rightarrow \nu_1 = \frac{4}{9}E , \tag{A.8}$$

$$(A.7)(A.8) \Rightarrow \sqrt{\nu_2 - \frac{4}{9}E} = \sqrt{E} \Rightarrow \nu_2 = \frac{13}{9}E , \tag{A.9}$$

$$(A.2)(A.8) \Rightarrow x^a = \frac{2}{3}\sqrt{2} , \tag{A.10}$$

$$(A.3)(A.8)(A.9) \Rightarrow x^b = \sqrt{\frac{13}{9} - \frac{4}{9}} = 1 . \tag{A.11}$$

The optimal total weight W_{opt} of the truss then becomes

$$W_{\mathrm{opt}} = \sum_e x^e L^e = \frac{2}{3}\sqrt{2}\sqrt{2}(10) + (1)(10) = \frac{70}{3} = 23.333\,. \tag{A.12}$$

The real strains are given by (13), Fig. 8b, (A.10) and (A.11)

$$\varepsilon^a = \frac{\sqrt{2}}{\frac{2}{3}\sqrt{2}E} = \frac{3}{2E}\,, \; \varepsilon^b = \frac{1}{E}\,, \tag{A.13}$$

and the adjoint strains by (9), Figs. 8c-d, (A.8), (A.9), (A.10) and (A.11)

$$\overline{\varepsilon}^a = \frac{\nu_1 \hat{f}_1^a + \nu_2 \hat{f}_2^a}{x^a E} = \frac{\frac{4}{9}\sqrt{2}}{\frac{2}{3}\sqrt{2}} = \frac{2}{3}\,,$$

$$\overline{\varepsilon}^b = \frac{\nu_1 \hat{f}_1^b + \nu_2 \hat{f}_2^b}{x^b E} = \frac{\frac{4}{9}(-1) + \frac{13}{9}(1)}{1} = 1\,, \tag{A.14}$$

In order to calculate the strains along the vanishing bar "c", we first determine the displacements in the direction of that bar (Fig. 25a). These are denoted Δ_c (for the real strains) and $\overline{\Delta}_c$ (for the adjoint strains). These displacements are calculated from virtual work equations, using the forces $\hat{f}_c^e$ caused by the unit "dummy" load in Fig. 25a and the results in Figs. 8e-g or in (A.10), (A.11) (A.13) and (A.14).

$$\Delta_c = \sum_e \hat{f}_c^e \varepsilon^e L^e = (1)\left(\frac{3}{2E}\right)\sqrt{2}(10) + \left(-\sqrt{2}\right)\left(\frac{1}{E}\right)(10) = \frac{1}{2E}\sqrt{2}(10)\,, \tag{A.15}$$

$$\overline{\Delta}_c = \sum_e \hat{f}_c^e \overline{\varepsilon}^e L^e = (1)\left(\frac{2}{3}\right)\sqrt{2}(10) + \left(-\sqrt{2}\right)(1)(10) = \frac{1}{3}\sqrt{2}(10)\,. \tag{A.16}$$

The corresponding strains can be calculated by dividing the above displacements by the negative length of the bar "c", i.e. by $-L^c = -\sqrt{2}(10)$. The neative sign is due to the fact that a positive displacement Δ_c or $\overline{\Delta}_c$ in Fig. 25a corresponds to a compressive strain in the bar "c" which is negative. This means that (A.15) and (A.16) imply

$$\varepsilon^c = -\frac{\Delta_c}{L^c} = -\frac{\frac{1}{2}\sqrt{2}(10)}{\sqrt{2}(10)} = -\frac{1}{2E}\,, \tag{A.17}$$

$$\overline{\varepsilon}^c = -\frac{\overline{\Delta}_c}{L^c} = -\frac{\left(\frac{-1}{3}\right)\sqrt{2}(10)}{\sqrt{2}(10)} = \frac{1}{3}\,. \tag{A.18}$$

The layout function ϕ can then be calculated by multiplying the product of the real and adjoint strains in Figs. 8f and g by E (see Fig. 8h).

A.1.2 Nonoptimal Layout (b, c)

The real and virtual forces f^e, $\hat{f}_1^e$ and $\hat{f}_2^e$ for this layout are shown in Figs. 25b-d and the internal forces $\hat{f}_a^e$ caused by a unit load in the direction of the member "a" in Fig. 25e. First we assume that both displacement constraints are active and then we show that the corresponding solution is nonoptimal. Then we investigate the solution with one active displacement constraint.

(a) Solution with two active constraints nonoptimal. By (3) and (7) we have

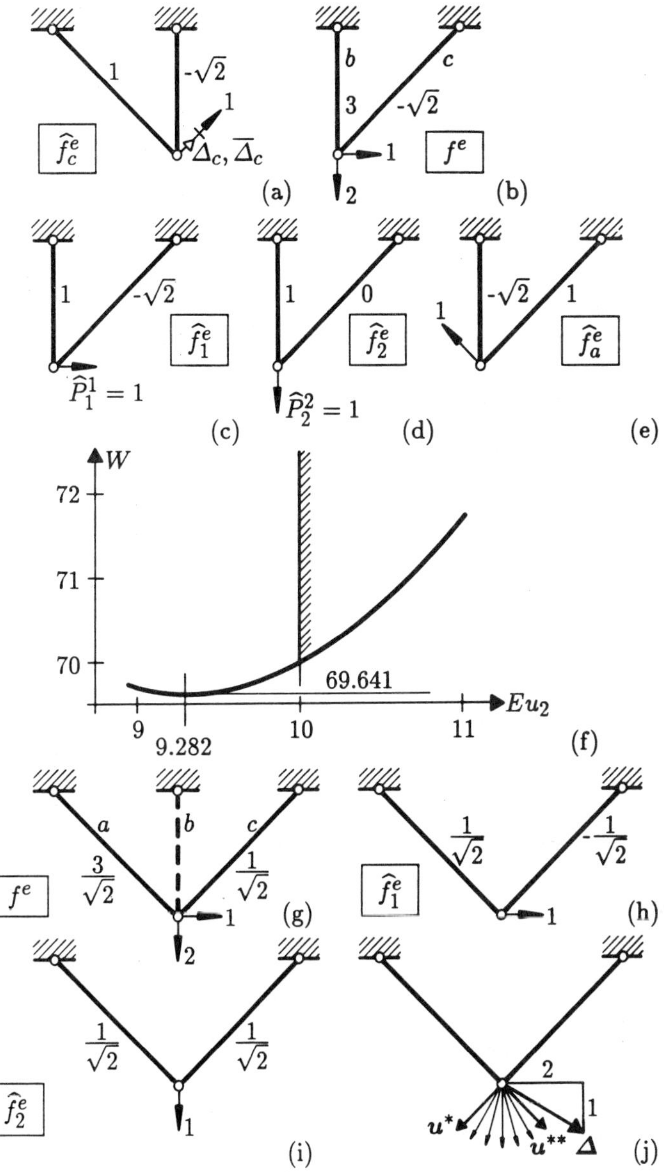

Fig. 31. Derivation of the optimal solution for a three-bar truss problem in Fig. 7a.

$$x^b = \sqrt{\frac{3(\nu_1 + \nu_2)}{E}}\,, \quad x^c = \sqrt{\frac{(-\sqrt{2})\nu_1(-\sqrt{2})}{E}} = \sqrt{\frac{2\nu_1}{E}}\,. \tag{A.19}$$

Then the displacement constraints yield the following Lagrange multipliers.

$$EΔ_1 = 20 = \frac{(3)(1)(10)}{\sqrt{3(\nu_1 + \nu_2)}}\sqrt{E} + \frac{(-\sqrt{2})(-\sqrt{2})(\sqrt{2})10}{\sqrt{2\nu_1}}, \tag{A.20}$$

$$EΔ_2 = 10 = \frac{(3)(1)(10)}{\sqrt{3(\nu_1 + \nu_2)}}\sqrt{E}. \tag{A.21}$$

Then we have

$$(A.21) \Rightarrow \sqrt{\nu_1 + \nu_2} = \sqrt{3E}, \ \nu_1 + \nu_2 = 3E, \tag{A.22}$$

$$(A.20)(A.22) \Rightarrow 20 = \frac{30}{3} + \frac{20}{\sqrt{\nu_1}}\sqrt{E} \Rightarrow \sqrt{\nu_1} = 2\sqrt{E}, \ \nu_1 = 4E, \tag{A.23}$$

$$(A.22)(A.23) \Rightarrow \nu_2 = -E. \tag{A.24}$$

Since in the Kuhn-Tucker conditions all Lagrange multipliers must be positive for an inequality constraint, the considered solution cannot be optimal.

(b) Independent proof of nonoptimality of the solution with two active constraints. To convince the reader about the above conclusion, we offer here an independent proof. For the considered layout, by Figs. 25b and d the displacement u_2 in the vertical direction can be expressed from the virtual work equation as

$$Eu_2 = \frac{(3)(1)(10)}{x^b}, \tag{A.25}$$

implying

$$x^b = \frac{30}{Eu_2}. \tag{A.26}$$

Assuming that the first displacement constraint is active, we have the work equation with (A.26)

$$Eu_1 = EΔ_1 = 20 = \frac{(3)(1)(19)u_2}{30} + \frac{(-\sqrt{2})(-\sqrt{2})\sqrt{2}(10)}{x^c}, \tag{A.27}$$

implying

$$x^c = \frac{20\sqrt{2}}{20 - u_2}. \tag{A.28}$$

This means that the total weight in terms of u_2 becomes

$$W = \sum_e x^e L^e = \frac{300}{u_2} + \frac{400}{20 - u_2}, \tag{A.29}$$

which is shown graphically in Fig. 25f.

It can be seen from the latter that the minimum weight is *not* at the active constraint $u_2 E = 10$ (with $W = 70$) but at $u_2 E = 9.282$ (with $W_{\text{opt}} = 69.641$). By (A.26) and (A.28) this solution corresponds to

$$x^b = 3.232, \ x^c = 2.639. \tag{A.30}$$

(c) Correct optimal solution with one active constraint. We shall now confirm the optimality of the above solution using optimality criteria.

Assuming that only the first displacement constraint is active, (3) and (7) with $\nu_2 = 0$ Figs. 25b-c imply

$$x^b = \sqrt{\frac{3\nu_1}{E}}\,,\ x^c = \sqrt{\frac{(-\sqrt{2})(-\sqrt{2})\nu_1}{E}} = \sqrt{\frac{2\nu_1}{E}}\,, \tag{A.31}$$

and then the active displacement constraint becomes

$$E\Delta_1 = 20 = \frac{(3)(1)(10)}{\sqrt{3(\nu_1)}}\sqrt{E} + \frac{(-\sqrt{2})(-\sqrt{2})(\sqrt{2})10}{\sqrt{2\nu_1}}\sqrt{E}$$

$$\Rightarrow 20 = \left(\frac{30}{\sqrt{3}} + 20\right)\sqrt{\frac{E}{\nu_1}} \Rightarrow \sqrt{\nu_1} = \left(\frac{\sqrt{3}}{2} + 1\right)\sqrt{E}\,, \tag{A.32}$$

Then by (A.31) and (A.32) we have

$$x^b = \sqrt{3}\left(\frac{\sqrt{3}}{2} + 1\right) = 3.232\,,\ x^c = \sqrt{2}\left(\frac{\sqrt{3}}{2} + 1\right) = 2.639\,, \tag{A.33}$$

$$W = (2 + \sqrt{3})\left(\frac{\sqrt{3}}{2} + 1\right)10 = (35 + 20\sqrt{3}) = 69.641\,. \tag{A.34}$$

The vertical displacement for this solution is given by the virtual work equation with Figs. 25b and d

$$Eu_2 = \frac{(3)(1)(10)}{\sqrt{3}\left(\frac{\sqrt{3}}{2} + 1\right)} = 9.282\,, \tag{A.35}$$

which agrees with the result under Subsection (b) and is within the limit set by the second displacement constraint.

The real and adjoint strains for the above solution by Fig. 25b, (A.32) and (A.33) become

$$\varepsilon^b = \frac{3}{\sqrt{3}\left(\frac{\sqrt{3}}{2} + 1\right)E} = \frac{\sqrt{3}}{\left(\frac{\sqrt{3}}{2} + 1\right)E}\,,\ \varepsilon^c = \frac{-\sqrt{2}}{\sqrt{2}\left(\frac{\sqrt{3}}{2} + 1\right)E} = \frac{-1}{\left(\frac{\sqrt{3}}{2} + 1\right)E}\,,$$

$$\bar{\varepsilon}^b = \left(\frac{\sqrt{3}}{2} + 1\right)^2 \frac{1}{\sqrt{3}\left(\frac{\sqrt{3}}{2} + 1\right)} = \frac{\frac{\sqrt{3}}{2} + 1}{\sqrt{3}}\,,\ \bar{\varepsilon}^c = \left(\frac{\sqrt{3}}{2} + 1\right)^2 \frac{-\sqrt{2}}{\sqrt{2}\left(\frac{\sqrt{3}}{2} + 1\right)} = -\left(\frac{\sqrt{3}}{2} + 1\right)\,,$$

$$\tag{A.36}$$

implying

$$\phi^b = \phi^c = 1\,, \tag{A.37}$$

The strains in the vanishing member "a" can be calculated using the virtual forces $\hat{f}_a^e$ caused by the unit "dummy" load in Fig. 25e.

The real and adjoint displacements u_a and $\bar{u}_a$ in the direction of member "a" can be calculated from the work equation:

$$u_a = \sum_{e=b,c} \hat{f}_a^e \varepsilon_a^e L^e = \frac{10}{E}\left(-\sqrt{2}\frac{\frac{\sqrt{3}}{2}}{\frac{\sqrt{3}}{2}+1} - \frac{\sqrt{2}}{\frac{\sqrt{3}}{2}+1}\right) = -\frac{\sqrt{3}+1}{\frac{\sqrt{3}}{2}+1}\frac{\sqrt{210}}{E},$$

$$\bar{u}_a = \sum_{e=a,b} \hat{f}_a^e \bar{\varepsilon}_a^e L^e = 10\left[-\sqrt{2}\frac{\frac{\sqrt{3}}{2}+1}{\sqrt{3}} - \sqrt{2}\left(\frac{\sqrt{3}}{2}+1\right)\right] = -\left(\frac{\sqrt{3}}{2}+1\right)\left(\frac{1}{\sqrt{3}}+1\right)\sqrt{210},$$

$$(A.38)$$

The above displacements must be divided by $-L^a = (-10\sqrt{2})$ to get the appropriate strains because positive displacements in (A.38) correspond to negative (compressive) strains:

$$\varepsilon^a = \frac{\sqrt{3}+1}{\left(\frac{\sqrt{3}}{2}+1\right)E}, \; \bar{\varepsilon}^a = \left(\frac{\sqrt{3}}{2}+1\right)\left(\frac{1}{\sqrt{3}}+1\right), \tag{A.39}$$

implying

$$\phi^a = E\varepsilon^a\bar{\varepsilon}^a = (\sqrt{3}\ 1)\left(\frac{1}{\sqrt{3}}+1\right) = 2 = \frac{1}{\sqrt{3}}+\sqrt{3} = 4.309, \tag{A.40}$$

A.1.3 Another Nonoptimal Layout (a, c)

For the sake of eliminating any other possible optimum in this nonconvex problem, we shall also investigate the layout (a, c).

(a) Solution with two active constraints (nonoptimal). The real internal forces f^e and virtual internal forces $\hat{f}_1^e$ and $\hat{f}_2^e$ for this layout are shown in Figs. 25g-i. The two active displacement constraints for this problem imply the virtual work equations

$$E\Delta_1 = 20 = \left[\frac{\frac{3}{\sqrt{2}}\frac{1}{\sqrt{2}}}{x^a} + \frac{\frac{1}{\sqrt{2}}\left(-\frac{1}{\sqrt{2}}\right)}{x^c}\right]\sqrt{210}, \tag{A.41}$$

$$E\Delta_2 = 10 = \left(\frac{\frac{3}{\sqrt{2}}\frac{1}{\sqrt{2}}}{x^a} + \frac{\frac{1}{\sqrt{2}}\frac{1}{\sqrt{2}}}{x^c}\right)\sqrt{210}. \tag{A.42}$$

Subtracting (A.42) from (A.41), we have

$$10 = -\frac{\sqrt{2}}{x^c}10, \tag{A.43}$$

which would imply $x^c < 0$. This shows that for the considered layout two active displacement constraints are infeasible.

The same conclusion can be confirmed by simple inspection. If a member "a" had a finite cross-sectional area and that of member "b" tended to zero, the resultant displacement vector $\underline{u}^*$ would have the direction shown in Fig. 25j. If member "a" was vanishing and "b" did not, the resultant displacement vector $\underline{u}^{**}$ would have the direction shown in the same diagram. For finite cross-sectional areas of both bars, the displacement vector would be some convex combination of the above two, shown by the cone of vectors in Fig. 25j. None of these agree with the direction of $\underline{\Delta}$ in Fig. 25j, which would be the resultant vector if both displacement constraints were active. This simple agreement confirms the nonfeasibility of two active constraints for the considered layout.

(b) Optimal solution with one active constraint. The above discussion proves that for the layout (a, c) only the second displacement constraint can be active. Then making use again of (3) and (7) with Figs. 25g and 25i, we have

$$x^a = \sqrt{\frac{3}{\sqrt{2}}\frac{\nu_2}{\sqrt{2}}\frac{1}{E}} = \sqrt{\frac{3\nu_2}{2E}},$$

$$x^c = \sqrt{\frac{1}{\sqrt{2}}\frac{\nu_2}{\sqrt{2}}\frac{1}{E}} = \sqrt{\frac{\nu_2}{2E}}. \tag{A.44}$$

Moreover, the second displacement constraint becomes

$$E\Delta_2 = 10 = \left(\frac{\frac{3}{\sqrt{2}}\frac{1}{\sqrt{2}}}{\sqrt{\frac{3\nu_2}{2E}}} + \frac{\frac{1}{\sqrt{2}}\frac{1}{\sqrt{2}}}{\sqrt{\frac{\nu_2}{2E}}}\right) 10\sqrt{2} \Rightarrow \sqrt{\frac{\nu_2}{E}} = \sqrt{3}+1. \tag{A.45}$$

Then (A.44) and (A.45) imply

$$x^a = \sqrt{\frac{3}{2}}(\sqrt{3}+1)\,, \; x^c = \sqrt{\frac{1}{2}}(\sqrt{3}+1)\,,$$

$$W = (x^a + x^c)\, 10\sqrt{2} = 10(\sqrt{3}+1)^2 = 74.641\,. \tag{A.46}$$

The real and adjoint strains and the layout function for the nonvanishing bars are as follows:

$$E\varepsilon^a = \frac{\frac{3}{\sqrt{2}}}{\sqrt{\frac{3}{2}}(\sqrt{3}+1)} = \frac{\sqrt{3}}{\sqrt{3}+1}\,,$$

$$\bar{\varepsilon}^a = \nu_2\hat{\varepsilon}_2^a = \frac{\frac{1}{\sqrt{2}}(\sqrt{3}+1)^2}{\sqrt{\frac{3}{2}}(\sqrt{3}+1)} = \frac{\sqrt{3}+1}{\sqrt{3}}\,,$$

$$\phi^a = E\varepsilon^a\bar{\varepsilon}^a = 1\,, \tag{A.47}$$

$$E\varepsilon^c = \frac{\frac{1}{\sqrt{2}}}{\sqrt{\frac{1}{2}}(\sqrt{3}+1)} = \frac{1}{\sqrt{3}+1}\,,$$

$$\bar{\varepsilon}^c = \nu_2\hat{\varepsilon}_2^c = \frac{\frac{1}{\sqrt{2}}(\sqrt{3}+1)^2}{\sqrt{\frac{1}{2}}(\sqrt{3}+1)} = \sqrt{3}+1\,,$$

$$\phi^c = E\varepsilon^c\bar{\varepsilon}^c = 1\,. \tag{A.48}$$

The real strain in the vanishing bar "b" is given by the known vertical displacement $(E\Delta_2 = 10$, since active constraint), divided by $L^b = 10$:

$$\varepsilon^b = \frac{10}{10} = 1 . \tag{A.49}$$

The vertical displacement in the joint of the adjoint truss is given by

$$\bar{u}_{\text{vert}} = \sum_e \hat{\varepsilon}^e_2 \bar{\varepsilon}^e_2 L^e = \left(\frac{\sqrt{3}+1}{\sqrt{3}} \frac{1}{\sqrt{2}} + \frac{\sqrt{3}+1}{\sqrt{2}} \right) 10\sqrt{2} = \frac{(\sqrt{3}+1)^2}{\sqrt{3}} 10 . \tag{A.50}$$

The adjoint strain in member "b" is given by the above displacement $\bar{u}_{\text{vert}}$ divided by the member length (10)

$$\bar{\varepsilon}^b = \frac{(\sqrt{3}+1)^2}{\sqrt{3}} . \tag{A.51}$$

It follows from (A.50) and (A.51) that the layout function for the vanishing bar "b" becomes

$$\phi^b = E\varepsilon^b\bar{\varepsilon}^b = \frac{(\sqrt{3}+1)^2}{\sqrt{3}} = 4.309 . \tag{A.52}$$

A.1.4 Check on the Weight Values Using a Dual Formula

The optimal weight for a given topology can be calculated from either of the following expressions:

$$W = \sum_e x^e L^e , \quad \overline{W} = \sum_d \Delta_d \nu_d , \tag{A.53}$$

where Δ_d is the prescribed limit in a displacement constraint. We have for all layouts:

$$L^a = L^c = 10\sqrt{2} , \quad L^b = 10 , \tag{A.54}$$

$$\Delta_1 = 20/E , \quad \Delta_2 = 10/E . \tag{A.55}$$

Moreover, for particular layouts we have

Layout (a, b), optimal

$$\nu_1 = \frac{4}{9}E \quad , \quad \nu_2 = \frac{13}{9}E \quad ,$$

$$x^a = \frac{2}{3}\sqrt{2} \quad , \quad x^b = 1 \quad ,$$

$$W = \sqrt{2}(10)\frac{2}{3}\sqrt{2} + 10 = 10\left(\frac{7}{3}\right) \quad ,$$

$$\overline{W} = 20\left(\frac{4}{9}\right) + 10\left(\frac{13}{9}\right) = 10\left(\frac{21}{9}\right) = 10\left(\frac{7}{3}\right) . \tag{A.56}$$

Layout (b, c)

$$\nu_1 = \left(\frac{\sqrt{3}}{2}+1\right)^2 E \quad , \quad \nu_2 = 0 \quad ,$$

$$x^b = \sqrt{3}\left(\frac{\sqrt{3}}{2} + 1\right) \quad , \quad x^c = \sqrt{2}\left(\frac{\sqrt{3}}{2} + 1\right) \quad ,$$

$$W = 10\sqrt{3}\left(\frac{\sqrt{3}}{2} + 1\right) + \sqrt{2}10\sqrt{2}\left(\frac{\sqrt{3}}{2} + 1\right) = 10(2 + \sqrt{3})\left(\frac{\sqrt{3}}{2} + 1\right) = 20\left(\frac{\sqrt{3}}{2} + 1\right)^2 ,$$

$$\overline{W} = 20\left(\frac{\sqrt{3}}{2} + 1\right)^2 . \tag{A.57}$$

Layout (a, c)

$$\nu_1 = 0 \quad , \quad \nu_2 = (\sqrt{3} + 1)^2 E \quad ,$$

$$x^a = \sqrt{\frac{3}{2}}(\sqrt{3} + 1) \quad , \quad x^b = \sqrt{\frac{1}{2}}(\sqrt{3} + 1) \quad ,$$

$$W = \sqrt{2}10\frac{1}{\sqrt{2}}(\sqrt{3} + 1)^2 = 10(\sqrt{3} + 1)^2 \quad ,$$

$$\overline{W} = 10(\sqrt{3} + 1)^2 . \tag{A.58}$$

It will be seen that all three solutions have been confirmed by the dual formula.

A.2 EXAMPLE ILLUSTRATING THE DERIVATION OF OPTIMALITY CRITERIA

To illustrate the author's method for deriving optimality criteria, we consider a *plane truss subject to several load conditions and displacement constraints.*

A.2.1 Derivation for a Finite Number of Truss Elements with Lower Side Constraints

This problem can be formulated as follows:

$$\min W = \sum_e \rho^e A^e L^e , \tag{A.59}$$

subject to displacement and lower side constraints as well as equilibrium constraints. It will be shown subsequently that *the elastic compatibility conditions are fulfilled automatically as optimality criteria.*

The *displacement constraints* can be expressed as

$$\sum_e \frac{f_k^e \hat{f}_{kd}^e L^e}{A^e E^e} - \Delta_{kd} \leq 0 \ (k = 1, \ldots, K; d = 1, \ldots, D_k) , \tag{A.60}$$

where k denotes a load condition, d a displacement constraint, f_k^e is the force in the member e under the load k, $\hat{f}_{kd}^e$ is the force in the member e under the virtual load (*e.g.* "unit dummy" load) for the d-th displacement constraint under the load k and Δ_{kd} is the limiting value of the corresponding displacement. Other symbols were defined earlier.

54   G.I.N. Rozvany

The *lower side constraints* take the form

$$\underline{A}^e - A^e \leq 0 \quad (e = 1, \ldots, E), \tag{A.61}$$

where $\underline{A}^e$ is the prescribed minimum cross-sectional area for the member e.

Finally, the equilibrium conditions can be expressed by means of the theorem of virtual displacements (e.g. Argyris and Kelsey **1960**) as

$$\sum_\ell \overline{u}_k^\ell P_k^\ell - \sum_e \overline{\varepsilon}_k^e L^e f_k^e = 0 \quad (k = 1, \ldots, K), \tag{A.62}$$

$$\sum_\ell u_{kd}^\ell \hat{P}_{kd}^\ell - \sum_e \varepsilon_{kd}^e L^e f_{kd}^e = 0 \quad (k = 1, \ldots, K; d = 1, \ldots, D_k), \tag{A.63}$$

where $P_k^\ell(\ell = 1, \ldots, L)$ constitute the k-th load condition, $\hat{P}_{kd}^\ell(\ell = 1, \ldots, L)$ the virtual load for the d-th displacement condition under the k-th load condition and $(\overline{u}_k, \overline{\varepsilon}_k)$ and $(u_{kd}, \varepsilon_{kd})$ represent, at this stage, *any* kinematically admissible displacements and strains. Introducing the Lagrange multipliers $\nu_{kd}, \beta^e, \overline{\alpha}_k$ and α_{kd} for the constraints in (A.60)-(A.63), the Kuhn-Tucker conditions are as follows for the variables indicated:

$$A^e : \rho^e - \beta^e - \sum_k \sum_d \nu_{kd} \frac{f_k^e \hat{f}_{kd}^e}{A^{e2} E^e} = 0, \tag{A.64}$$

$$f_k^e : \overline{\alpha}_k \overline{\varepsilon}_k^e = \sum_d \nu_{kd} \frac{\hat{f}_{kd}^e}{A^e E^e}, \tag{A.65}$$

$$\hat{f}_{kd}^e : \alpha_{kd} \varepsilon_{kd}^e = \nu_{kd} \frac{f_k^e}{A^e E^e}. \tag{A.66}$$

It can be seen that ε_{kd}^e in (A.66) is a factored (ν_{kd}/α_{kd}) value of the elastic strain for the load condition k. On the other hand, the principle of virtual displacements required kinematical admissibility of ε_{kd}. It follows that *kinematical admissibility of the member strains follows from the optimality conditions.* This was the reason for not incorporating the compatibility conditions in the formulation above. Independent proofs of the kinematical admissibility of the optimal solution were given by Huang (1971) as well as Shield and Prager (1970), who used the principle of stationary mutual energy and by Zhou and Rozvany (1992/1993), who used matrix notation with statics and kinematics matrices. For the Lagrange multipliers we have from the Kuhn-Tucker conditions

$$\nu_{kd} \geq 0 \text{ and } \nu_{kd} \neq 0 \text{ only if } \sum_e \frac{f_k^e \hat{f}_{kd}^e L^e}{A^e E^e} = \Delta_{kd}, \; \beta^e \geq \text{ and } \beta^e \neq 0 \text{ only if } A^e = \underline{A}^e. \tag{A.67}$$

Since linear scaling of the strains does not affect the solution, we can adopt $\overline{\alpha} = 1$. It can then be seen from (A.65) that the quantity $\overline{\varepsilon}_k^e$ for the k-th load condition can be interpreted as the *strain in a fictitious truss which is subject to the sum of the products of Lagrange multipliers* ν_{kd} *and virtual loads* $\hat{P}_{kd}$ *representing the displacement constraints* $(d = 1, \ldots, D_k)$ *for the load condition* k. This fictitious truss is termed an *ad joint truss.* Due to the theorem of virtual displacements, the adjoint strains $\overline{\varepsilon}_k^e$ must be *kinematically admissible.*

The optimal cross-sectional areas A^e can be determined from (A.64) as

$$A^e = \sqrt{\frac{\sum_k f_k^e \sum_d \nu_{kd}\hat{f}_{kd}^e}{E^e \rho^e (1 - \beta^e)}} = \sqrt{\frac{\sum_k f_k^e \overline{f}_k^e}{E^e \rho^e (1 - \beta^e)}}, \tag{A.68}$$

where $\overline{f}^e = \sum_{d=1}^{D_k} \nu_{kd}\hat{f}_{kd}^e$. Then (A.67) and (A.68) imply

$$(\text{for } A^e > \underline{A}^e) \quad A^e = \sqrt{\frac{\sum_k f_k^e \overline{f}_k^e}{E^e \rho}}, \quad (\text{for } A^e = \underline{A}^e) \quad A^e \geq \sqrt{\frac{\sum_k f_k^e \overline{f}_k^e}{E^e \rho}}. \tag{A.69}$$

Substituting (A.69) into the expressions for strains

$$\varepsilon_k^e = \frac{f_k^e}{A^e E^e}, \quad \overline{\varepsilon}_k^e = \frac{\sum_d \nu_{kd}\hat{f}_{kd}^e}{A^e E^e} = \frac{\overline{f}_k^e}{A^e E^e}, \tag{A.70}$$

we obtain the optimality criteria

$$(E^e/\rho^e) \sum_k (\varepsilon_k^e \overline{\varepsilon}_k^e) \leq 1 \ (\text{for } A^e = \underline{A}^e), \quad (E^e/\rho^e) \sum_k (\varepsilon_k^e \overline{\varepsilon}_k^e) = 1 \ (\text{for } A^e > \underline{A}^e). \tag{A.71}$$

If we define the layout function for this class of problems as

$$\phi = (E^e/\rho^e) \sum_k (\varepsilon_k^e \overline{\varepsilon}_k^e), \tag{A.72}$$

then (A.71) and (A.72) reduce to

$$\phi \leq 1 \ (\text{for } A^e = \underline{A}^e), \quad \phi = 1 \ (\text{for } A > \underline{A}^e). \tag{A.73}$$

A.2.2 Extension to Vanishing Cross Sections

If $\underline{A}^e \to 0$, then (A.73) reduces to

$$\phi \leq 1 \ (\text{for } A^e = 0), \quad \phi = 1 \ (\text{for } A^e > 0, \tag{A.74}$$

where ϕ is defined in (A.72).

A.2.3 Extension to an Infinite Number of Members

As explained in Section 11, for exact optimal layouts the structural universe consists of an infinite number of members and, due to kinematic admissibility of the real and adjoint strains in the members, the strains can be represented in a continuum fashion. For plane trusses, for example, (A.74) can be replaced by

$$\phi(x,y,\theta) \leq 1 \ [\text{for } A(x,y,\theta) = 0], \quad \phi(x,y,\theta) = 1 \ [\text{for } A(x,y,\theta) > 0], \tag{A.75}$$

where x and y denote the location and θ the orientation of an infinitesimal truss element.

TOPOLOGY OPTIMIZATION OF DISCRETE STRUCTURES
AN INTRODUCTION IN VIEW OF COMPUTATIONAL AND NONSMOOTH ASPECTS

W. Achtziger

University of Erlangen-Nuremberg, Erlangen, Germany

Abstract

We discuss standard problems of topology optimization of discrete structures. This paper is an attempt to provide an (almost) self-contained introduction and stresses the techniques of reformulating problems and mathematical tools needed for a successful numerical treatment.

First, relations between several classical formulations of single load problems are shown in order to illustrate the mathematical techniques used, such as minimax-Theorems, duality etc. Numerical approaches are discussed. The outlined concept is then generalized to the multiple load case where minimization of compliance in the sense of worst case design is considered. We end up with displacement based nonsmooth optimization formulations. In the fourth section the problem of simultaneous optimization of topology and geometry is considered. We illustrate its mathematical treatment as a bilevel problem which again uses nonsmooth optimization. In each section a few numerical examples show the applicability of the proposed approaches.

Since nonsmooth analysis is non-standard in structural optimization, we close with a short introduction to important terms up to the concept of an algorithm.

1 Introduction, Problem and Notations

1.1 Topology Optimization

One of the classical problems in structural optimization is optimal design of trusses, i.e., given support conditions and external forces, how should a pin-jointed framework look like? When Linear Programming and computers came up in the fifties and sixties, Dorn, Gomory and Greenberg (1964, [1]) were among the first who developed a basis for a computational approach. Though at that time capabilities of computers were limited, a relation of trusses and continuum structures could already be seen: For analogous load- and support conditions the trusses obtained coincide (in a "discretized sense") with the level lines of principal stresses of an optimal continuum structure.

The last decade reviewed interest in topology optimization of trusses. There may be several reasons:

First, since the topology problem is typically large-scaled and/or possesses a difficult mathematical structure of variables dependent on each other, it is quite natural that new classes of problems can be tackled nowadays due to fast development of computer capabilities. Thus difficult but realistic problem formulations can be treated in a way which leads to extensive computational procedures (see e.g. [2]).

Second, new theories in mechanics automatically lead to consideration of classical problems from a new point of view (e.g. [3]) as well as to new problem formulations which closer and closer model reality (e.g. [4, 5, 6]). This goes hand in hand with progress made in the numerical treatment of problems of structural optimization. For the first time new techniques allow the treatment of thousands of potential members even coupled by difficult constraints (e.g. [7]). A new class of methods applied in topology optimization is based on nonsmooth optimization techniques, i.e., methods dealing with functions that are not differentiable at each point of their domain (e.g. [8, 9]).

A third reason why to investigate topology problems has appeared in the recent past: Topology and material optimization of continuum structures lead to truss-like structures, provided the volume fraction is low. Moreover, optimization problems considered in this field are (after finite element discretization) closely related to truss topology problems (see e.g. [9]). Thus new solution techniques and formulations of truss problems may well be carried over to that new area.

Introductions to truss topology problems can be found, e.g., in [6, 10, 11], and [12]. The latter reference also discusses continuum problems.

1.2   Introduction to the Problem

For simplicity we first concentrate on the well-known problem of optimizing trusses. Below we will then state general requirements such that the terms explained also cover more general types of discrete structures/problems.

1.2.1   The Ground Structure Approach

We use the common and widely used truss model formulated by the so-called *ground structure approach* (see, e.g., [1, 13]). That means, we start with a set of N *nodal points* (or *nodes*) in space (dim = 3) or in the plane (dim = 2). At some of these nodes given ("external") forces apply, at some nodes support conditions are defined. Moreover, a grid of m *potential bars* is given where each bar connects two of the nodal points. This grid (including forces and boundary conditions) is called *ground structure*. The task is to find a good (optimal) truss structure that satisfies all load- and support conditions, and whose bar connections are given by a subset of bars and nodes from the given ground structure.

In this approach one has in mind that bars forming a good structure are contained as a "substructure" in the *structural universe* pre-defined by the ground structure. Usually, the positions of the nodal points are fixed to keep mathematics of the optimization problem simple. In this case the ground structure should be very "dense" to obtain a good real structure. Therefore, m is usually chosen as large as possible, i.e., $m \approx \frac{1}{2}N(N-1)$ (long bars overlapping shorter ones are usually not considered). The ground structure approach allows for inclusion of geometrical constraints (like "holes" in the structure), or for implicit restriction on the spectrum of possible member lengths [14]. Figure 1 shows two ground structures for a 8 by 6 nodal layout where only two different

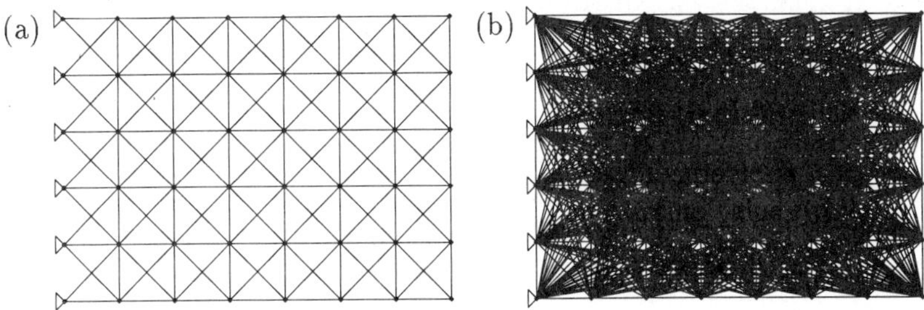

Figure 1 Ground structures of plane truss; only neighbours connected (only two bar lengths) (a), and all possible connections (b)

bar lengths appear in (a) resulting in $m = 147$ potential bars while in (b) all possible connections (except overlapping ones) are shown ($m = 730$).

In an (uncommon) approach in Sec. 4 we will also allow for variations of nodal positions. In this sense we mix the question of topology with the problem of geometry. However, the ground structure approach as such is not touched: The decision which bar should be included in an optimal structure has still to be made. However, if nodal positions are free then we can allow for smaller number N of nodes (and thus for much less potential bars) since we do not have to mimic variable nodal positions by a dense grid of nodes. The disadvantage of simultaneous geometry and topology optimization is, however, that the underlying mathematical model becomes much more complicated.

In the following we assume that the nodal positions are fixed, and that the ground structure is chosen dense enough such that the underlying real life problem can be properly modelled.

1.2.2 *Problem Modelling versus Numerical Tractability*

When modelling problems arising in structural optimization, the choice of objective and constraints is essential for a successful numerical treatment of the underlying mathematical optimization problem. It is clear that more complicated (but more realistic) modelling will lead to a more difficult numerical treatment. Of course, in most cases, it is possible (at least for very small problem sizes) to treat difficult models in the sense that a known structure can (only) be *improved* by extensive numerical computations if it is not yet (locally) optimal. However, this will not help in a topology context where optimal structures are generated in a genuine way, and usually the result is not known. Therefore, during modelling the engineer should always have in mind which constraints are ruling the design process, and which other constraints can be neglected (at least at an early stage of optimization) since otherwise numerics will not be able to find a (local or global) solution.

We stress that *convexity* (see, e.g., [15]) is one of the mathematical properties that gives numerical algorithms a chance to find good solutions. Since optimization algorithms are usually able to find only *local* optima, we will succeed in finding a *global* optimum in case of convexity: In a convex problem each local optimum is automatically a global one (cf. e.g. [15] Th. 3.4.2). Algorithms for convex problems are well investigated [15], and trickily programmed routines (for the differentiable case) are presently available in standard software packages. The meaning of convexity in structural optimization problems is discussed, e.g., in [16].

In many applied problems, however, convexity will not be automatically given in its natural mathematical formulation using the usual terms from modelling (like stresses, forces, displacements, strains, etc.). As we will see below, in certain situations convexity can be "enforced" by accepting other disadvantages (like, e.g., nonsmoothness; see Sec. 5).

Another popular method of tackling a non-convex problem is to mimic convexity via approximation by a sequence of convex problems. This leads to optimization techniques of sequential convex programming (SCP) [17]. For problems of structural optimization, convex linearizations (see, e.g., [18, 19]) or the frequently used Method of Moving Asymptotes (MMA) [20, 21] proved to be successful tools.

Optimality of structures is related to a particularly given objective function and constraints. Moreover, these functions determine the types of variables which have to be considered in addition to the design variables. For trusses, minimization of weight subject to equilibrium (of external and member forces) and stress constraints is mainly considered since this problem only possesses linear constraints [11]. If a displacement based formulation is used then a kind of elastic equilibrium must be stated as a constraint in order to link displacements with the design variables. This will result in non-linear terms. For the most formulations of single load problems it is known that there exists an optimal solution which is statically determinate. Thus geometric compatibility constraints (linking displacements with bar elongations, see below) can be neglected in the problem formulation. This, however, is not possible for the multiple load case (cf. [6, 10, 12]).

1.2.3 Basic Notations, Relations and Assumptions

We collect basic definitions for ground structures which can be found, e.g., in [13]. Throughout the paper we work with column vectors, and "z^T" denotes the transpose of a vector z. Inequality sings between vectors are meant component-wise.

The *number of degrees of freedom* of a ground structure consisting of N nodes is defined by

$$n := \dim \cdot N - s \qquad (1.1)$$

where "dim" is the dimension of space (2 for planar and 3 for spatial trusses), and s is the number of support conditions, i.e., the number of fixed nodal coordinate directions.

Due to the truss model, given external loads apply only at nodal points. In global reduced coordinates we collect these forces in a vector $f \in \mathbb{R}^n$. To exclude pathological cases we require $f \neq 0$.

For the i-th bar, $i = 1, \ldots, m$, we define a vector $\gamma_i \in \mathbb{R}^n$ that contains the cosines and sines of the angles between bar and spatial coordinate axes in its components corresponding to the end nodes of this bar. In this way, any vector $q \in \mathbb{R}^m$ of member forces (i.e., forces *along* the bars) can be transformed into global reduced coordinates, i.e., into forces applied at the nodes. Collecting the vectors γ_i in the *geometry (compatibility, statics) matrix* $B := (\gamma_1 \cdots \gamma_m) \in \mathbb{R}^{n \times m}$ we require *equilibrium of forces*,

$$Bq = f, \tag{1.2}$$

to be satisfied for (unknown) member forces $q \in \mathbb{R}^m$ and (given) external loads $f \in \mathbb{R}^n$ (Note that the coordinates belonging to support conditions are already removed from γ_i (i.e., from B), and thus boundary conditions are already incorporated ("global *reduced* coordinates")).

Moreover, when working with nodal displacements, denoted by $u \in \mathbb{R}^n$, we want to satisfy *(geometric) compatibility*,

$$B^T u = \Delta \ell \in \mathbb{R}^m. \tag{1.3}$$

This means that nodal displacements u geometrically "fit" with all bar elongations $\Delta \ell_i$, $i = 1, \ldots, m$.

Usually, assumption of *linear elasticity* links nodal displacements u with member forces q, i.e., for the i-th bar with cross-sectional area a_i and length ℓ_i,

$$\sigma_i = E_i \epsilon_i, \qquad \text{where} \qquad \sigma_i := \frac{q_i}{a_i}, \qquad \epsilon_i := \frac{\Delta \ell_i}{\ell_i}. \tag{1.4}$$

Here σ_i denotes stress, ϵ_i strain, and E_i is Young's modulus for the material of the i-th bar.

For $i = 1, \ldots, m$ we denote by $x_i K_i$ the *element stiffness matrix of the i-th bar in global (reduced) coordinates* where $x_i := a_i \ell_i \geq 0$ denotes the *volume* of this bar, and

$$K_i := \frac{E_i}{\ell_i^2} \gamma_i \gamma_i^T \in \mathbb{R}^{n \times n} \tag{1.5}$$

(In mathematics, such a (matrix-)product of a row and a column vector is called "dyadic product"). Note that, in opposite to definitions stated elsewhere, K_i does *not* contain information on the cross-sectional area a_i of the i-th bar; this area is hidden in $x_i = a_i \ell_i$ (= volume(!) of the i-th bar). Moreover, note that K_i is defined in *global* (reduced) coordinates (i.e., K_i is an $n \times n$ matrix which does *not* contain any rows/columns which belong to fixed nodal coordinate directions).

Vector $x \in \mathbb{R}^m$ containing values $x_1, \ldots, x_m \geq 0$ of bar volumes is called *(vector of) design variable(s)*. Analogously, vector u of nodal displacements is the *(vector of) state variable(s)*.

For a structure $\boldsymbol{x} \in \mathbb{R}^m$, $\boldsymbol{x} \geq 0$, the *global stiffness matrix (in global reduced coordinates)* is defined by

$$\boldsymbol{K}(\boldsymbol{x}) := \sum_{i=1}^{m} x_i \boldsymbol{K}_i \in \mathbb{R}^{n \times n}. \tag{1.6}$$

Assuming *linearly elastic behaviour of the bars* (1.4), *geometric compatibility* (1.3), and *equilibrium of forces* (1.2) we end up with the well-known *conditions of elastic equilibrium*

$$\boldsymbol{K}(\boldsymbol{x})\boldsymbol{u} = \boldsymbol{f}. \tag{1.7}$$

In the following we mainly concentrate on *minimization of compliance*. This objective function has been considered in many publications (e.g. [9, 12, 22, 23, 24, 25, 26]) and is a common practice in structural optimization. If $(\boldsymbol{x}, \boldsymbol{u})$ is a couple of bar volumes $\boldsymbol{x}$ and the corresponding nodal displacements $\boldsymbol{u}$ under load $\boldsymbol{f}$, then the *compliance* is defined as the the number $\boldsymbol{f}^T \boldsymbol{u}$. Besides the interpretation of $\boldsymbol{f}^T \boldsymbol{u}$ as external work we mention a purely mathematical description: Since the euclidean product $\boldsymbol{f}^T \boldsymbol{u}$ measures the length of $\boldsymbol{u}$ along the direction of $\boldsymbol{f}$, compliance is the sum of nodal movements along $\boldsymbol{f}$ when $\boldsymbol{x}$ is under load $\boldsymbol{f}$. Thus minimizing compliance amounts to minimizing the "movement of the structure" along the direction of force $\boldsymbol{f}$. Later on we will link minimization of compliance and the classical optimization of weight/volume.

Until now we have only concentrated on optimization of trusses. The ground structure approach, however, may also be applied (in a different sense) to other systems consisting of a discrete number of "potential members". We think, e.g., of a discrete problem which arises from finite element discretization of a continuum shape optimization problem or in material optimization. Examples for such formulations can be found, e.g., in [9]. Discrete structures in this generalized sense lack of a dyadic structure of the element stiffness matrices like in (1.5) for the truss problem. Sometimes $\boldsymbol{K}_i$ is a *sum* of dyadic products.

Thus for the development below we only assume that

$$\boldsymbol{K}_i \in \mathbb{R}^{n \times n} \quad \text{is symmetric and positive semi-definite for all } i = 1, \ldots, m.$$

When minimizing compliance we have to work with a pre-stated volume bound V limiting the total volume of the structure. Clearly, V has to satisfy $V > 0$, and, for a fixed ground structure, we denote the set of structures whose volume is limited by V by

$$X_V := \left\{ \boldsymbol{x} \in \mathbb{R}^m \mid \boldsymbol{x} \geq 0, \ \sum_{i=1}^{m} x_i \leq V \right\}. \tag{1.8}$$

Note that optimization over the structures in X_V describes the feasible set of a "pure" (volume-constrained) topology and sizing problem: Each element is allowed to get volume between zero and V.

We mention that in the mathematical development below we have to deal with infinite numbers. Moreover, we do not discuss in detail the solvability of all considered problems, i.e., the question whether an infimum value of the objective function is attained at a feasible point or not. Therefore, we simplify the strictly mathematical notation using "inf" and "sup" a little bit to the more relaxed notation using only "min" and "max".

2 Equivalences in Truss Topology Optimization

2.1 Equivalent Basic Formulations

In this section we illustrate the techniques of transforming different formulations of single load topology problems into equivalent problems. Our intention is to provide a mathematical framework which can be applied also to other questions than those discussed here. In this way we want to stress the methodologies used. The mathematical tools used are duality, minimax-Theorems, or simple substitutions. We mention that convexity is needed for equivalent reformulations via duality or minimax.

Some of the results below can be considered as classical ones and are known for many years (see, e.g., [1, 25, 27, 28]). Some of them are known intuitively and can be illustrated by mechanical principles. However, we want to stay inside a mathematical framework, and thus concentrate on mathematical proofs. In this way we collect results, e.g., from [8, 9, 22, 28, 29]. A brief summary including some practical aspects can be found in [30].

We start with consideration of the *maximum-stiffness design problem*. That means, we seek a structure that minimizes compliance. As mentioned above in Sec. 1.2.3, compliance is an important and common criterion in structural optimization.

Since compliance is considered, we have to deal with displacements $u \in \mathbb{R}^n$ as state variables which are linked to the design variables $x \in \mathbb{R}^m$, $x \geq 0$. This link is done, as outlined in Secs. 1.2.2 and 1.2.3, by equations of elastic equilibrium (1.7). In order to obtain a well-posed problem, we have to bound the volume of the truss. For later purposes we scale the objective function by the factor $\frac{1}{2}$. All in all we end up with the following optimization problem:

$$(\mathrm{P_{compl}}) \qquad \min_{x \in \mathbb{R}^m,\, u \in \mathbb{R}^n} \frac{1}{2} f^T u$$
$$\text{s.t.} \quad K(x)u = f,$$
$$x \in X_V$$

Note that we only assume positive semi-definiteness of the K_i's but not necessarily the dyadic structure of K_i (i.e., that K_i is given by (1.5)). Thus $(\mathrm{P_{compl}})$ can also be considered for discretized structures/problems that are more general than trusses.

Our goal is to show reformulation techniques transforming $(\mathrm{P_{compl}})$ also for large-scaled topology problems. For this reason, and for keeping the presentation on a general level, we do not take more constraints into account (such as stress bounds, local or global stability, bounds on displacements, etc.). More specific formulations and approaches can be found, e.g., in [2, 7, 31, 32, 33]. Computation of *globally* optimal structures can only be expected for convex formulations. Sometimes, additional knowledge on optima is used, like preliminary selection of active constraints or knowledge on statical determinacy. The latter can be proved for a minimum weight problem w.r.t. local stability constraints, and thus methods can be applied which proceed from one statically determinate design to another [5, 34, 35, 36].

However, we mention that the development below partly can be carried out for more general settings. For trusses, the extension to inclusion of self-weight and the problem of reinforcing a given structure is straightforward and is discussed in [9]. Also lower and upper bounds on the design variables can be considered [22]. The extension to different behaviour of the bars under tension and compression is a mathematical exercise [37]. Also extensions to formulations including unilateral contact which are based on the reformulation techniques outlined below have been recently investigated [38, 39, 40].

Let us have a first glance at (P_{compl}) from the practical solution point of view: Due to elastic equilibrium, (P_{compl}) is not a convex problem. Thus we can only expect to compute *local* optima when applying a suitable optimization routine. Moreover, the dimension of the problem is $(n + m)$ which may be very large. This is the case, e.g., when a truss ground structure is considered with almost each of the N nodes pairwise connected by potential bars (then $m \approx \frac{1}{2}N(N - 1)$, and $n \approx \dim \cdot N$; cf. (1.1) and Fig. 1(b)). Thus a direct numerical treatment of (P_{compl}) for finding a (local) optimum by standard methods of nonlinear optimization (cf. [15]) will only work for moderate problem size. This behaviour can indeed be observed in numerical experiments.

Many algorithms tackling (P_{compl}) (or similar displacement based formulations) require the condition that $\sum_{i=1}^{m} \boldsymbol{K}_i$ is positive definite (This condition implies that $\boldsymbol{K}(\boldsymbol{x})$ is positive definite for all structures $\boldsymbol{x}$ with $\boldsymbol{x} > 0$). The reason is that these algorithms (like the well-known stress-rationing-method; e.g. [6, 9, 24]) perform a sensitivity analysis based on the equality

$$\boldsymbol{u} = \boldsymbol{K}(\boldsymbol{x})^{-1}\boldsymbol{f} \tag{2.1}$$

calculating $\boldsymbol{u}$ for fixed $\boldsymbol{x} > 0$. In truss problems positive definiteness of $\sum_{i=1}^{m} \boldsymbol{K}_i$ assures that each structure with $\boldsymbol{x} > 0$ does not represent a mechanism. However, this property of the ground structure may not be satisfied for problems with only few bars ("sparse" ground structure), but it will hold for a well-posed topology problem (many bars, "dense" ground structure).

Nevertheless, elimination of state variables by (2.1) is problematic: In a topology context, the optimal structure will often consist of few bars and nodes, and thus the global stiffness matrix $\boldsymbol{K}(\boldsymbol{x})$ usually becomes singular (at least close to the optimum). This causes numerics to break down. The same argument applies when putting a lower bound $0 < \epsilon \leq x_i$ on the design variables: Solution of (2.1) for fixed $\boldsymbol{x}$ is an ill-conditioned problem if $\boldsymbol{K}(\boldsymbol{x})$ is almost singular (i.e., possesses eigenvalues close to zero) [41]. Thus ϵ plays the role in a fragile balance: If ϵ is chosen too small then numerical errors will destroy the result. If ϵ is too large then it is not clear whether a computed cross-sectional area should be small (and positive) or zero. Once more, this problem shows the difficulty which is *inherently* contained in a topology problem, namely the question which of the design variables are zero at an optimal point. Moreover, eliminating u from (P_{compl}) by (2.1) does not overcome the difficulty of large size of the problem since typically the number m of potential members is much larger than n (see above).

For trusses, problem (P_{compl}) can result in an optimal topology that is a mechanism. But anyway, we expect $K(x)$ to be singular due to the fact that many nodal points of the ground structure are not used in the solution. Also, it may happen that the optimal topology possesses straight bars with inner nodal points. These points should be ignored (cf. Ex. 2.10).

When optimizing discrete structures, minimization of weight is the classical and much more popular objective than minimization of compliance [10]. Hence, we briefly mention a simple relation of (P_{compl}) to a minimum weight formulation. For simplicity we assume that the specific weights of all bars are the same, and thus we may minimize volume. Consider the following problem where $C > 0$ is a given upper bound for (half) compliance:

$$(P_{\text{vol}}) \qquad \min_{x\in\mathbb{R}^m,\, u\in\mathbb{R}^n} \sum_{i=1}^{m} x_i$$

$$\text{s.t.} \quad K(x)u = f,$$
$$\tfrac{1}{2}f^T u \le C,$$
$$x \ge 0$$

Clearly any relation of (P_{compl}) and (P_{vol}) depends on the balance of the bounds C and V. It can be easily shown that for fixed values V and C each solution (x^*, u^*) of (P_{compl}) gives rise to the solution $(\frac{C^*}{C}x^*, \frac{C}{C^*}u^*)$ of (P_{vol}) where $C^* := \tfrac{1}{2}f^T u^*$ is the optimal function value for (P_{compl}). Vice versa a completely analogous statement holds. This shows that (P_{compl}) can be viewed as a minimum volume (weight) problem subject to a compliance constraint, i.e., a displacement constraint in the direction of f (cf. Sec. 1.2.3).

After this first equivalence we prove a second reformulation rewriting (P_{compl}) as a minimax-problem. We use the well-known step expressing compliance by minimum potential energy. Mathematically speaking, this relation is nothing else than a necessary and sufficient optimality condition applied to the quadratic function $q_x : \mathbb{R}^n \longrightarrow \mathbb{R}$,

$$q_x(u) := \tfrac{1}{2}u^T K(x)u - f^T u.$$

For fixed structure x, this function represents potential energy in displacements u.

Proposition 2.1 (Principle of min. pot. energy) *Let $x \in \mathbb{R}^m$, $x \ge 0$ be fixed. Then*

$$\min_{u\in\mathbb{R}^n}\left\{\tfrac{1}{2}f^T u \mid K(x)u = f\right\} = \max_{u\in\mathbb{R}^n}\left\{f^T u - \tfrac{1}{2}u^T K(x)u\right\}$$

$$= \begin{cases} \tfrac{1}{2}f^T u^* & \text{for all } u^* \text{ with } K(x)u^* = f \quad \text{if } f \in \text{range}(K(x)), \\ +\infty & \text{if } f \notin \text{range}(K(x)), \end{cases}$$

where $\text{range}(K(x)) := \{K(x)u \mid u \in \mathbb{R}^n\}$*, and the convention* $\min \emptyset = +\infty$ *is used.*

Proof: Let $f \in \text{range}(K(x))$. Since the Hessian $K(x)$ of the function q_x is positive semi-definite by general assumption on the K_i's, q_x is a convex function. Thus condition $\nabla q_x(u^*) = 0$ is necessary and sufficient for a point $u^* \in \mathbb{R}^n$ to be optimal for the max-problem (where ∇ denotes the gradient). Since $\nabla q_x(u) = K(x)u - f$, we see that u^* is optimal if and only if it satisfies $K(x)u^* = f$. Moreover, elementary calculus shows

$$\tfrac{1}{2}f^T u^1 = \tfrac{1}{2}f^T u^2 = f^T u^2 - \tfrac{1}{2}u^{2T} K(x)u^2$$

for all u^1, u^2 with $K(x)u^1 = K(x)u^2 = f$.

If $f \notin \text{range}(K(x))$ then, by convention, $\min_{u \in \mathbb{R}^n} \left\{ \tfrac{1}{2}f^T u \mid K(x)u = f \right\} = +\infty$.

By means of linear algebra, condition $f \notin \text{range}(K(x))$ shows (use symmetry of $K(x)$) that there exists $\bar{u} \neq 0$ with $K(x)\bar{u} = 0$ and $f^T \bar{u} > 0$. With this $\bar{u}$ we obtain

$$\lim_{\alpha \to +\infty} \left\{ f^T(\alpha\bar{u}) - \tfrac{1}{2}(\alpha\bar{u})^T K(x)(\alpha\bar{u}) \right\} = \lim_{\alpha \to +\infty} \alpha f^T \bar{u} = +\infty$$

which shows the assertion. $\qquad\qquad\qquad\qquad\qquad\qquad\qquad\qquad\qquad\qquad\square$

By this proposition, (P_{compl}) can be equivalently rewritten as

$$(P_{potEn}) \qquad\qquad \min_{x \in \mathbb{R}^m} \max_{u \in \mathbb{R}^n} \left\{ f^T u - \tfrac{1}{2}u^T K(x)u \right\}$$
$$\text{s.t.} \quad x \in X_V.$$

Note that in this formulation uncomfortable elastic equilibrium conditions are avoided, and only linear constraints on x are left. The price for this is the additional max-term. We mention a slight mathematical difference between (P_{compl}) and (P_{potEn}): A structure $x \in X_V$ which cannot carry load f is not feasible for (P_{compl}) due to equilibrium while *it is* feasible for (P_{potEn}). However, since such structures result in an infinite function value in (P_{potEn}) (cf. Prop. 2.1), they are not important for minimization. In this way, feasibility w.r.t. elastic equilibrium is handled in (P_{potEn}) automatically by optimization.

Direct solution of (P_{potEn}) by minimization of $c : \mathbb{R}^n \longrightarrow \mathbb{R} \cup \{+\infty\}$,

$$c(x) := \max_{u \in \mathbb{R}^n} \left\{ f^T u - \tfrac{1}{2}u^T K(x)u \right\}, \qquad\qquad (2.2)$$

over X_V can be performed by nonsmooth optimization methods. However, this will only work for modest problem size m. Nevertheless, a brief investigation of c shows an interesting insight in the problem structure of (P_{potEn}):

Proposition 2.2   *(a) Function c is convex.*

(b) If $x \geq 0$ and $K(x)$ is nonsingular then c is differentiable at x, and

$$\nabla c(x) = -\tfrac{1}{2}\left(u(x)^T K_1 u(x), \ldots, u(x)^T K_m u(x) \right)^T \in \mathbb{R}^m$$

where $u(x) := K(x)^{-1} f$.

Proof: Ad (a): By definition, c is the pointwise supremum function (cf. (5.1)) of linear (and, thus, convex) functions in $\boldsymbol{x}$:

$$c(\boldsymbol{x}) = \sup\{c_u(\boldsymbol{x}) \mid \boldsymbol{u} \in \mathbb{R}^n\} \quad \text{where } c_u(\boldsymbol{x}) := \boldsymbol{f}^T\boldsymbol{u} - \sum_{i=1}^{m} x_i(\tfrac{1}{2}\boldsymbol{u}^T\boldsymbol{K}_i\boldsymbol{u}).$$

Hence, c is convex on $\mathbb{R}^n$ (easy to prove, or cf. [42] Th. 5.5).

Ad (b): Since $\boldsymbol{K}(\boldsymbol{x})$ is nonsingular, $\boldsymbol{K}(\boldsymbol{y})$ is nonsingular for $\boldsymbol{y}$ in a small neighbourhood of $\boldsymbol{x}$. This shows that $\boldsymbol{u}(\,.\,)$ is well-defined as a function on this neighbourhood. The formula for the gradient then follows from the chain rule. $\square$

Item (a) of this proposition re-states a result of Svanberg [16] who showed convexity under the assumptions of (b) by proving that the Hessian of c is positive semi-definite (see also [23]). Under the assumptions of (b), c reduces to the function $(\boldsymbol{x} \mapsto \tfrac{1}{2}\boldsymbol{f}^T\boldsymbol{K}(\boldsymbol{x})^{-1}\boldsymbol{f})$ (cf. (2.1) and Prop. 2.1). One can prove that c is even infinitely many times differentiable at points $\boldsymbol{x}$ where $\boldsymbol{K}(\boldsymbol{x})$ is nonsingular, and explicit formulas similar to that in (b) can be derived for any derivative [29, 43].

As already mentioned, in our context the number m is large, and thus we do not focus on minimization of c. Before tackling the problem of large number m, we mention another problem which is equivalent to $(\mathrm{P_{compl}})$ (via $(\mathrm{P_{potEn}})$) for trusses, i.e., where $\boldsymbol{K}_i$ is given by (1.5). Consider the following stress-based formulation which is known as the *dual formulation* of $(\mathrm{P_{potEn}})$:

$$(\mathrm{P}_\sigma) \qquad \min_{x,q\in\mathbb{R}^m} \frac{1}{2} \sum_{i:x_i>0} \frac{\ell_i^2}{E_i} \frac{(q_i)^2}{x_i}$$

$$\text{s.t.} \quad \boldsymbol{B}\boldsymbol{q} = \boldsymbol{f},$$

$$q_i = 0 \quad \text{for all } i \text{ with } x_i = 0,$$

$$\boldsymbol{x} \in X_V$$

The objective function of this problem is convex (i.e., jointly in $(\boldsymbol{x}, \boldsymbol{q})$) which can be seen from the positive semi-definiteness of the Hessian of the objective function. The following theorem states that this problem is equivalent to problem $(\mathrm{P_{potEn}})$ in the sense that each structure that is optimal for one of the problems is automatically optimal for the other one. This is astonishing at a first glance since (P_σ) is a formulation which does not include linear elasticity (1.4) a priori (opposite to formulation $(\mathrm{P_{compl}})$).

Theorem 2.3   *(a) Let $(\boldsymbol{x}, \boldsymbol{q})$ be optimal for (P_σ). Then there exists $\boldsymbol{u} \in \mathbb{R}^n$ such that*

$$q_i = \frac{E_i x_i \boldsymbol{\gamma}_i^T \boldsymbol{u}}{\ell_i^2} \quad \text{for all } i = 1,\ldots,m, \tag{2.3}$$

and $(\boldsymbol{x}, \boldsymbol{u})$ is optimal for $(\mathrm{P_{potEn}})$.

(b) Let $(\boldsymbol{x}, \boldsymbol{u})$ be optimal for $(\mathrm{P_{potEn}})$. Then $(\boldsymbol{x}, \boldsymbol{q})$ is optimal for (P_σ) where $\boldsymbol{q}$ is given by (2.3).

Proof: In fact, the only relation we use is duality of the min-problem in q and the max-problem in u, respectively:

For fixed $x \geq 0$ consider minimization in q in problem (P_σ) (where $I_x := \{i \mid x_i = 0\}$)

$$\min_{q \in \mathbb{R}^m} \left\{ \tfrac{1}{2} \sum_{i \notin I_x} \frac{\ell_i^2}{E_i} \frac{(q_i)^2}{x_i} \,\middle|\, Bq = f, \; q_i = 0 \text{ for } i \in I_x \right\} =$$

(use Lagrangian duality [15] for constraints expressing equilibrium of forces with multipliers $u \in \mathbb{R}^n$; note that the function to be minimized is convex (quadratic) in q)

$$= \max_{u \in \mathbb{R}^n} \min_{q \in \mathbb{R}^m} \left\{ \tfrac{1}{2} \sum_{i \notin I_x} \frac{\ell_i^2}{E_i} \frac{(q_i)^2}{x_i} + (f - Bq)^T u \,\middle|\, q_i = 0 \text{ for } i \in I_x \right\}$$

$$= \max_{u \in \mathbb{R}^n} \min_{q \in \mathbb{R}^m} \left\{ \tfrac{1}{2} \sum_{i \notin I_x} \frac{\ell_i^2}{E_i} \frac{(q_i)^2}{x_i} - \sum_{i \notin I_x} q_i \gamma_i^T u + f^T u \,\middle|\, q_i = 0 \text{ for } i \in I_x \right\}$$

$\Big($Now we solve the inside min-problem in q for fixed u which is a convex quadratic problem, and thus the necessary and sufficient optimality conditions are

$$\frac{\ell_i^2}{E_i} \frac{q_i}{x_i} = \gamma_i^T u \quad \text{for all } i \notin I_x,$$

which is the equation in (2.3) for $i \notin I_x$. For $i \in I_x$ the equation in (2.3) trivially holds since $q_i = 0$ for $i \in I_x$ by the explicit constraint. Thus we may plug in this optimal q_i for fixed u, and continue the above with:$\Big)$

$$= \max_{u \in \mathbb{R}^n} \left\{ \tfrac{1}{2} \sum_{i \notin I_x} \frac{E_i x_i}{\ell_i^2} (\gamma_i^T u)^2 - \sum_{i \notin I_x} \frac{E_i x_i}{\ell_i^2} (\gamma_i^T u)^2 + f^T u \right\}$$

$$= \max_{u \in \mathbb{R}^n} \left\{ f^T u - \tfrac{1}{2} \sum_{i \notin I_x} x_i u^T K_i u \right\} = \max_{u \in \mathbb{R}^n} \left\{ f^T u - \tfrac{1}{2} \sum_{i=1}^m x_i u^T K_i u \right\}$$

$$= \max_{u \in \mathbb{R}^n} \{ f^T u - \tfrac{1}{2} u^T K(x) u \}$$

Thus, if (x, q) is optimal for (P_σ) then there exists a vector u of multipliers that satisfies (2.3), and that gives rise to the same objective function value in (P_{potEn}) as in (P_σ). By the above, (x, u) is optimal for (P_{potEn}). Completely analogous arguments prove (b). $\qquad\square$

We now go on with problem (P_{potEn}) and focus on the technique how we can get rid of the design-variables x_i. Note that due to large number m such a step is a necessity in order to end up with a problem formulation that can be numerically treated. The function

$$(x, u) \mapsto f^T u - \tfrac{1}{2} u^T K(x) u \tag{2.4}$$

is convex in x (cf. (1.6): $K(.)$ even linearly depends on x), and concave in u (the Hessian $(-K(x))$ is negative semi-definite due to general assumption on the K_i's). Since, finally, X_V is a compact (i.e., bounded and closed) set, we may interchange "min" and "max" in (P_{potEn}) by the well-known minimax-Theorem (cf. e.g. [42] Cor. 37.3.2).

We end up with the problem

$$\max_{u\in\mathbb{R}^n}\ \min_{x\in X_V}\{\ \boldsymbol{f}^T\boldsymbol{u} - \tfrac{1}{2}\boldsymbol{u}^T\boldsymbol{K}(\boldsymbol{x})\boldsymbol{u}\}. \tag{2.5}$$

Since $\boldsymbol{K}(\boldsymbol{x}) = \sum x_i\boldsymbol{K}_i$ linearly depends on $\boldsymbol{x}$, the inside min-problem in $\boldsymbol{x}$ is a linear programming problem with the simple feasible set X_V, namely the unit-simplex scaled by V (cf. (1.8)). We may rewrite this inside LP-problem for fixed $\boldsymbol{u}$ as

$$\min_{x\in X_V}\{\ \boldsymbol{f}^T\boldsymbol{u} - \tfrac{1}{2}\boldsymbol{u}^T\boldsymbol{K}(\boldsymbol{x})\boldsymbol{u}\} = \boldsymbol{f}^T\boldsymbol{u} - \max_{x\in X_V}\Big\{\sum_{i=1}^{m} x_i\cdot(\tfrac{1}{2}\boldsymbol{u}^T\boldsymbol{K}_i\boldsymbol{u})\Big\}. \tag{2.6}$$

Due to simplicity of X_V (cf. (1.8)), it is immediately clear that $\boldsymbol{x} := V\cdot\boldsymbol{e}_{i_0}$ solves the latter max-problem where $\boldsymbol{e}_{i_0}\in\mathbb{R}^m$ denotes the i_0-th unit vector in $\mathbb{R}^m$, and where i_0 is an index i such that $\tfrac{1}{2}\boldsymbol{u}^T\boldsymbol{K}_i\boldsymbol{u}$ is maximal. This optimal $\boldsymbol{x}$ is plugged in (2.6), and problem (2.5) (and thus $(\mathrm{P_{potEn}})$) finally becomes

$$(\mathrm{P_{strEn}}) \qquad\qquad \max_{u\in\mathbb{R}^n}\ \min_{1\le i\le m}\{\ \boldsymbol{f}^T\boldsymbol{u} - \tfrac{V}{2}\boldsymbol{u}^T\boldsymbol{K}_i\boldsymbol{u}\}. \tag{2.7}$$

Note that $(\mathrm{P_{strEn}})$ is a formulation in the displacements only, and therefore possesses a much lower dimension than $(\mathrm{P_{potEn}})$ since the number n of variables is the degree of freedom. For trusses and a "dense" ground structure with $m\approx\tfrac{1}{2}N(N-1)$ we obtain that the number N of nodal points appears quadratically in m but only linearly in n (cf. (1.1)). The m design variables have vanished and are hidden in the min-term as multipliers (see Th. 2.4 below). The price for this dramatic reduction of the number of variables is the more complicated mathematical structure of the remaining function in $\boldsymbol{u}$: Due to minimization over $1\le i\le m$, this new objective function is nonsmooth. For $\boldsymbol{u}\in\mathbb{R}^n$ we put

$$\phi(\boldsymbol{u}) := \min_{1\le i\le m}\{\ \boldsymbol{f}^T\boldsymbol{u} - \tfrac{V}{2}\boldsymbol{u}^T\boldsymbol{K}_i\boldsymbol{u}\}.$$

With the notations from Sec. 1.2.3 we may interpret each term $\boldsymbol{u}^T\boldsymbol{K}_i\boldsymbol{u}$ in the truss case as the *strain energy density* of the i-th bar,

$$\boldsymbol{u}^T\boldsymbol{K}_i\boldsymbol{u} = E_i\frac{\boldsymbol{u}^T\boldsymbol{\gamma}_i}{\ell_i}\cdot\frac{\boldsymbol{\gamma}_i^T\boldsymbol{u}}{\ell_i} = E_i\epsilon_i\epsilon_i = \sigma_i\epsilon_i. \tag{2.8}$$

Formulation $(\mathrm{P_{strEn}})$ can be viewed as a nonsmooth optimization problem maximizing ϕ. To stay inside the framework of *minimizing* functions, we may consider

$$\min_{u\in\mathbb{R}^n}\psi(\boldsymbol{u}) \qquad \text{where } \psi(\boldsymbol{u}) := \max_{1\le i\le m}\{\tfrac{V}{2}\boldsymbol{u}^T\boldsymbol{K}_i\boldsymbol{u} - \boldsymbol{f}^T\boldsymbol{u}\} = -\phi(\boldsymbol{u}) \tag{2.9}$$

instead of maximizing $\phi\equiv-\psi$ in $(\mathrm{P_{strEn}})$ (clearly, optima $\boldsymbol{u}^*$ are the same). Illustrating how optimality conditions for nonsmooth functions enter this particular problem, we state the following theorem which gives back a structure $\boldsymbol{x}^*$ optimal for $(\mathrm{P_{compl}})$ from an optimum $\boldsymbol{u}^*$ for $(\mathrm{P_{strEn}})$.

Theorem 2.4   *(a) Function ψ is convex (i.e., ϕ is concave) on $\mathbb{R}^n$.*

(b) A point $\boldsymbol{u}^* \in \mathbb{R}^n$ is optimal for (P_{strEn}) if and only if there exists $\boldsymbol{x}^* \in X_V$ such that

$$K(\boldsymbol{x}^*)\boldsymbol{u}^* = \boldsymbol{f}, \tag{2.10}$$

$$\boldsymbol{x}^* \in X_V, \tag{2.11}$$

$$x_i^* \cdot \tfrac{1}{2}\boldsymbol{u}^{*T}\boldsymbol{K}_i\boldsymbol{u}^* = x_i^* \cdot \max_{1 \le j \le m}\{\tfrac{1}{2}\boldsymbol{u}^{*T}\boldsymbol{K}_j\boldsymbol{u}^*\} \quad \text{for all } i = 1,\ldots,m. \tag{2.12}$$

Moreover, $(\boldsymbol{x}^*, \boldsymbol{u}^*)$ is optimal for (P_{compl}).

Proof: Ad (a): Since each $\boldsymbol{K}_i$ is positive semi-definite by general assumption, each of the quadratic functions $(\boldsymbol{u} \mapsto \tfrac{V}{2}\boldsymbol{u}^T\boldsymbol{K}_i\boldsymbol{u} - \boldsymbol{f}^T\boldsymbol{u})$, $i = 1,\ldots,m$, is convex. Thus ψ is the pointwise sup-function (cf. (5.1)) of convex functions, and hence convex itself.
Ad (b): Since $\psi \equiv -\phi$, a point $\boldsymbol{u}^*$ is optimal for (P_{strEn}) if and only if it is a global minimizer of ψ. Since ψ is convex by (a), we know (cf. Th. 5.3) that optimality of $\boldsymbol{u}^*$ is equivalent to

$$0 \in \partial\psi(\boldsymbol{u}^*) \tag{2.13}$$

where the subdifferential $\partial\psi$ of ψ is given by (cf. Th. 5.4(a))

$$\partial\psi(\boldsymbol{u}) = \mathrm{conv}\Big\{ V\boldsymbol{K}_i\boldsymbol{u} - \boldsymbol{f} \,\Big|\, i \text{ such that } \tfrac{V}{2}\boldsymbol{u}^T\boldsymbol{K}_i\boldsymbol{u} = \max_{1 \le j \le m}\{\tfrac{V}{2}\boldsymbol{u}^T\boldsymbol{K}_j\boldsymbol{u}\}\Big\}. \tag{2.14}$$

Therefore, (2.13) is satisfied if and only if there exists $\boldsymbol{\lambda} \in \mathbb{R}^m$ such that

$$\boldsymbol{\lambda} \ge 0, \ \sum_{i=1}^{m} \lambda_i = 1, \tag{2.15}$$

$$0 = \sum_{i=1}^{m} \lambda_i(V\boldsymbol{K}_i\boldsymbol{u} - \boldsymbol{f}), \tag{2.16}$$

$$\lambda_i\Big(\tfrac{V}{2}\boldsymbol{u}^T\boldsymbol{K}_i\boldsymbol{u} - \boldsymbol{f}^T\boldsymbol{u} - \max_{1 \le j \le m}\{\tfrac{V}{2}\boldsymbol{u}^T\boldsymbol{K}_j\boldsymbol{u} - \boldsymbol{f}^T\boldsymbol{u}\}\Big) = 0 \quad \text{for all } i = 1,\ldots,m. \tag{2.17}$$

By putting $\boldsymbol{x}^* := V\boldsymbol{\lambda}$, (2.15) is exactly (2.11). Eq. (2.16) becomes (2.10), and (2.17) is (2.12) since $\boldsymbol{f}^T\boldsymbol{u}$ is independent from i and j.
Easy calculations which use (2.10) to (2.12) show that $\phi(\boldsymbol{u}^*) = \tfrac{1}{2}\boldsymbol{f}^T\boldsymbol{u}^*$. By the above development (equivalence of (P_{compl}), (P_{potEn}), and (P_{strEn})) this shows optimality of $(\boldsymbol{x}^*, \boldsymbol{u}^*)$ for (P_{compl}). $\square$

This theorem shows how an optimal structure $\boldsymbol{x}^*$ can be redetected from a solution $\boldsymbol{u}^*$ of (P_{strEn}): Usually, the λ_i's (i.e., the x_i's) are obtained as a direct output of the optimization algorithm. Alternatively, direct calculation of $\boldsymbol{x}^*$ from $\boldsymbol{u}^*$ is straightforward by solution of a least square problem.

Note that each solution $(\boldsymbol{x}^*, \boldsymbol{u}^*)$ computed via problem (P_{strEn}) is a *global* solution of the nonconvex(!) problem (P_{compl}). As mentioned above, the number n of variables is much less than $(m + n)$ in (P_{compl}).

Theorem 2.4 also shows that each solution $(\boldsymbol{x}^*, \boldsymbol{u}^*)$ is fully-stressed in the sense of (2.12): Multiplying (2.8) by E_i we get (use (1.4))

$$E_i \boldsymbol{u}^T \boldsymbol{K}_i \boldsymbol{u} = \sigma_i^2.$$

Hence, eq. (2.12) means that if the i-th bar is contained in the optimal structure (i.e., if $x_i^* > 0$) then the absolute value of its stress is maximal among all bars in the structure (provided all bars consist of the same material). This fact is exploited, e.g., in the classical stress-rationing-method (cf., e.g., [6, 9, 24]) which by a simple update scheme (like a fixpoint iteration; see also Sec. 2.2) tries to satisfy (2.12)).

Problem (P_{strEn}) can be numerically solved by nonsmooth optimization methods (cf. Sec. 5) simply minimizing ψ. In particular, this is possible for extremely large m. A second possibility is rewriting (P_{strEn}) as a smooth optimization problem by introduction of an auxiliary variable $\alpha \in \mathbb{R}$:

$$(P_{\text{strEn}}^{\text{smooth}}) \qquad \max_{u \in \mathbb{R}^n,\, \alpha \in \mathbb{R}} \boldsymbol{f}^T \boldsymbol{u} - \frac{V}{2}\alpha$$
$$\text{s.t.} \quad \boldsymbol{u}^T \boldsymbol{K}_i \boldsymbol{u} \le \alpha \quad \text{for all } i = 1, \ldots, m.$$

This problem has a linear objective function with convex quadratic constraints, and hence is convex. Nowadays there are algorithms which can deal with such problems even for a very large number of quadratic constraints (note: m is large), see, e.g., [44, 45, 46, 47].

Since $(P_{\text{strEn}}^{\text{smooth}})$ is convex (and a constraint qualification holds [15]), optimality conditions ("Karush-Kuhn-Tucker(KKT)-conditions" [15]) are necessary and sufficient. By these conditions it is an easy exercise left to the reader to show the following result which parallels Th. 2.4 in terms of smooth analysis:

Theorem 2.5 *A point $(\boldsymbol{u}^*, \alpha^*)$ is optimal for $(P_{\text{strEn}}^{\text{smooth}})$ if and only if there exists $\boldsymbol{x}^* \in X_V$ such that (2.10) to (2.12) in Th. 2.4 are satisfied. Moreover, $(\boldsymbol{x}^*, \boldsymbol{u}^*)$ is optimal for (P_{compl}).*

In the following we go on transforming problem $(P_{\text{strEn}}^{\text{smooth}})$ for trusses, i.e., when each $\boldsymbol{K}_i$ is given by a dyadic product, cf. (1.5). Since optimal α represents the maximum of the terms $\boldsymbol{u}^T \boldsymbol{K}_i \boldsymbol{u}$, it quadratically depends on $\boldsymbol{u}$. Hence, it plays the role of a scaling factor which can be eliminated from the problem. Moreover, the constraints in $(P_{\text{strEn}}^{\text{smooth}})$ show that α must be non-negative. This indicates that we may take square roots on both sides of each of the constraints, ending up with *linear* constraints and with the following formulation which is a *linear programming problem* (LP):

$$(P_{\text{strEn}}^{\text{LP}}) \qquad \max_{\tilde{u} \in \mathbb{R}^n} \boldsymbol{f}^T \tilde{\boldsymbol{u}}$$
$$\text{s.t.} \quad -1 \le (\sqrt{E_i}/\ell_i)\boldsymbol{\gamma}_i^T \tilde{\boldsymbol{u}} \le 1 \quad \text{for all } i = 1, \ldots, m.$$

The following theorem parallels Th. 2.5 and can be easily proved by means of optimality conditions.

Theorem 2.6 *Let $\tilde{\boldsymbol{u}}^*$ be optimal for $(\mathrm{P}^{\mathrm{LP}}_{\mathrm{strEn}})$, and let $\boldsymbol{\rho}^{*-}, \boldsymbol{\rho}^{*+} \in \mathbb{R}^m$, $\boldsymbol{\rho}^{*-}, \boldsymbol{\rho}^{*+} \geq 0$, be the Lagrangian multipliers ($\rho_i^{*\mp}$ corresponds to inequality $\mp(\sqrt{E_i}/\ell_i)\boldsymbol{\gamma}_i^T\tilde{\boldsymbol{u}} - 1 \leq 0$). Then $(\boldsymbol{u}^*, \alpha^*)$ is optimal for $(\mathrm{P}^{\mathrm{smooth}}_{\mathrm{strEn}})$, $\boldsymbol{u}^*$ is optimal for $(\mathrm{P}_{\mathrm{strEn}})$, and $(\boldsymbol{x}^*, \boldsymbol{u}^*)$ is optimal for $(\mathrm{P}_{\mathrm{compl}})$ where*

$$\alpha^* := \left(\tfrac{1}{V}\sum_{i=1}^{m}(\rho_i^{-*} + \rho_i^{+*})\right)^2, \quad \boldsymbol{u}^* := \sqrt{\alpha^*}\,\tilde{\boldsymbol{u}}^*, \quad \boldsymbol{x}^* := \frac{1}{\sqrt{\alpha^*}}(\boldsymbol{\rho}^{-*} + \boldsymbol{\rho}^{+*}). \qquad (2.18)$$

In this problem formulation, fully-stressed design is hidden in the definition of $\boldsymbol{x}^*$: Volume x_i^* is positive if and only if $\rho_i^{-*} > 0$ or $\rho_i^{+*} > 0$ (note $\boldsymbol{\rho}^{\pm*} \geq 0$). Since these values are the multipliers of the constraints in $(\mathrm{P}^{\mathrm{LP}}_{\mathrm{strEn}})$, one of the two corresponding "twin constraints" must be satisfied with equality. This means that (after scaling by α^*)

$$\boldsymbol{u}^{*T}\boldsymbol{K}_i\boldsymbol{u}^* = (E_i/\ell_i^2)(\boldsymbol{\gamma}_i^T\boldsymbol{u}^*)^2$$

is at its maximum.

Optimality of fully-stressed designs is the reason why there are even more equivalent well-known single load problem formulations (see below). However, one should always have in mind that this property is simply an optimality condition which holds only in special situations. That means, computation of a fully-stressed design for any other setting reduces to pure heuristics if a corresponding optimality condition cannot be shown.

We close this section by sketching equivalence of $(\mathrm{P}_{\mathrm{compl}})$ to minimization of weight under stress constraints. This equivalence is again based on duality.

We consider the problem of minimization of volume w.r.t. stress constraints where for each i we have the bound $\bar{\sigma}_i > 0$ which is the same for tension and compression, respectively:

$$(\mathrm{P}_{\mathrm{weight}}) \qquad \min_{x,q\in\mathbb{R}^m} \sum_{i=1}^{m} x_i$$
$$\text{s.t.} \quad \boldsymbol{Bq} = \boldsymbol{f},$$
$$\boldsymbol{x} \geq 0,$$
$$-\bar{\sigma}_i x_i \leq \ell_i q_i \leq \bar{\sigma}_i x_i \quad \text{for all } i = 1,\ldots,m.$$

If for each i some specific weight κ_i of the material is given then the (formal) substitutions $x_i \rightarrow \kappa_i x_i$, $\bar{\sigma}_i \rightarrow \bar{\sigma}_i/\kappa_i$ (and the appropriate changes of units) make this formulation work as a minimum weight problem (and therefore we call it $(\mathrm{P}_{\mathrm{weight}})$). One can easily see from the form of the stress constraints and from the objective function that each structure optimal for $(\mathrm{P}_{\mathrm{weight}})$ is automatically fully-stressed. By the substitutions

$$x_i = (\ell_i/\bar{\sigma}_i)(q_i^+ + q_i^-), \qquad (2.19)$$
$$q_i = q_i^+ - q_i^-,$$

for $i = 1,\ldots,m$, we obtain the following problem which is equivalent to $(\mathrm{P}_{\mathrm{weight}})$ (having the same optimal objective function value).

$(\mathrm{P}^{q^+,q^-}_{\mathrm{weight}})$

$$\min_{q^+,q^-\in\mathbb{R}^m} \sum_{i=1}^{m} \frac{\ell_i}{\bar{\sigma}_i}(q_i^+ + q_i^-)$$
$$\text{s.t.} \quad B(q^+ - q^-) = f,$$
$$q^+, q^- \geq 0.$$

Here, we may interpret $q_i^{\pm}$ as the absolute values of member forces q_i under tension and compression, respectively.

Due to the fact that $(\mathrm{P}^{q^+,q^-}_{\mathrm{weight}})$ is a linear programming problem (LP), we may apply LP-duality (cf. e.g. [48]), and we get the dual problem of $(\mathrm{P}^{q^+,q^-}_{\mathrm{weight}})$:

$(\mathrm{P}^{\mathrm{dual}}_{\mathrm{weight}})$

$$\max_{\tilde{u}\in\mathbb{R}^n} f^T \tilde{u}$$
$$\text{s.t.} \quad -\frac{\ell_i}{\bar{\sigma}_i} \leq \gamma_i^T \tilde{u} \leq \frac{\ell_i}{\bar{\sigma}_i} \quad \text{for all } i = 1,\ldots,m.$$

LP-duality guarantees that the optimal function values for $(\mathrm{P}^{q^+,q^-}_{\mathrm{weight}})$ and $(\mathrm{P}^{\mathrm{dual}}_{\mathrm{weight}})$ are the same (provided one of the problems has a solution). We immediately see the relation to $(\mathrm{P}^{\mathrm{LP}}_{\mathrm{strEn}})$. LP-Theory shows that at an optimal point u^* for $(\mathrm{P}^{\mathrm{dual}}_{\mathrm{weight}})$ the corresponding multipliers form an optimal point (q^{+*}, q^{-*}) for $(\mathrm{P}^{q^+,q^-}_{\mathrm{weight}})$. By substitution (2.19) we can recognize the formula for x^* in (2.18). A more precise consideration shows the following relation between solutions for $(\mathrm{P}_{\mathrm{weight}})$ and for $(\mathrm{P}_{\mathrm{compl}})$.

Theorem 2.7 *Consider problems* $(\mathrm{P}_{\mathrm{compl}})$ *and* $(\mathrm{P}_{\mathrm{weight}})$ *with*

$$\sqrt{E_i} = \bar{\sigma}_i \quad \text{for all } i = 1,\ldots,m. \tag{2.20}$$

If (x^*, q^*) *is a solution for* $(\mathrm{P}_{\mathrm{weight}})$*, and* u^* *is the corresponding vector of multipliers for the force equilibrium constraints, then* $(\frac{V}{V^*}x^*, \frac{V^*}{V}u^*)$ *is a solution for* $(\mathrm{P}_{\mathrm{compl}})$ *where* $V^* := \sum x_i^*$.
An analogous relation holds which transforms solutions for $(\mathrm{P}_{\mathrm{compl}})$ *into solutions for* $(\mathrm{P}_{\mathrm{weight}})$*.*

The assertion of this theorem is somehow obvious after the above development: Since the solution x^* for $(\mathrm{P}_{\mathrm{weight}})$ is fully-stressed, the settings in (2.20) assure that this property is preserved in problem $(\mathrm{P}^{\mathrm{LP}}_{\mathrm{strEn}})$ where u^* plays the role of a scaled optimum. As we know (e.g., by (2.18) or by (2.12)), fully-stressed designs immediately lead to optimal designs for $(\mathrm{P}_{\mathrm{compl}})$. A generalization of this theorem to different stress bounds for tension and compression can be found in [37].

For the sake of completeness we mention the following limit-load-problem where for given volume $V > 0$ a structure is sought which can carry a maximal load in a given direction f:

$(\mathrm{P_{liml}})$
$$\max_{x,q\in\mathbb{R}^m,\,\beta\in\mathbb{R}} \beta$$
$$\text{s.t.}\quad \boldsymbol{Bq} = \beta\boldsymbol{f},$$
$$\boldsymbol{x} \in X_V,$$
$$-\bar{\sigma}_i x_i \leq \ell_i q_i \leq \bar{\sigma}_i x_i \quad \text{for all } i = 1,\ldots,m.$$

It is easy to prove that this is again an equivalent reformulation of $(\mathrm{P_{weight}})$ (and each solution of $(\mathrm{P_{liml}})$ is a scaled solution of $(\mathrm{P_{weight}})$ and vice versa).

Summarizing the whole section, we get the following Theorem:

Theorem 2.8 *(a) For positive semi-definite stiffness matrices, problems $(\mathrm{P_{compl}})$, $(\mathrm{P_{vol}})$, $(\mathrm{P_{potEn}})$, $(\mathrm{P_{strEn}})$ and $(\mathrm{P_{strEn}^{smooth}})$ are equivalent (i.e., any solution of one of these problems leads – by duality or rescaling – to solutions of the others). For trusses (i.e., where (1.5) holds), these formulations are also equivalent to $(\mathrm{P_\sigma})$ and $(\mathrm{P_{strEn}^{LP}})$.*

(b) In truss problems, formulations $(\mathrm{P_{weight}})$, $(\mathrm{P_{weight}^{q^+,q^-}})$, $(\mathrm{P_{weight}^{dual}})$, $(\mathrm{P_{liml}})$ are equivalent. With settings (2.20) these problems are equivalent to all problems in (a).

We mention that this complete theory only works in the single load case. Some of the equivalences remain valid under certain extensions (see above). However, we have heavily exploited optimality conditions which appear as the property of fully-stressed solutions. Therefore, e.g., in the multiple load case, a lot of the above framework must break down since it is well-known that even simple problems do not possess optimal fully-stressed structures [6, 12].

2.2 Numerical Approaches and Examples

For the class of problems discussed in the previous section a variety of numerical methods and approaches have been considered. We only mention some of them which are challenging by opinion of the author.

We first discuss approaches that do *not* require dyadic element stiffness matrices, i.e., approaches which manage with the more general assumption of positive semi-definite stiffness matrices. Since fully-stressed designs appear also in other problems and applications, the stress-rationing method (see, e.g., [6, 9, 24]) is the most famous one among structural engineers. This method can be interpreted as a fixed point algorithm [49] which tries to satisfy optimality condition (2.12). This is done by computation of strain energy densities $\frac{1}{2}\boldsymbol{u}(\boldsymbol{x})^T \boldsymbol{K}_i \boldsymbol{u}(\boldsymbol{x})$ for all $i = 1,\ldots,m$ (where $\boldsymbol{u}(\boldsymbol{x}) := -\boldsymbol{K}(\boldsymbol{x})^{-1}\boldsymbol{f}$), and updating x_i by

$$x_i^{\mathrm{new}} := x_i \frac{\frac{1}{2}\boldsymbol{u}(\boldsymbol{x})^T \boldsymbol{K}_i \boldsymbol{u}(\boldsymbol{x})}{\max_{1\leq j\leq m}\{\frac{1}{2}\boldsymbol{u}(\boldsymbol{x})^T \boldsymbol{K}_j \boldsymbol{u}(\boldsymbol{x})\}} \tag{2.21}$$

(Finally, $\boldsymbol{x}$ is scaled assuring $\boldsymbol{x} \in X_V$). Since the displacements have to be computed for each iterate, this approach is not appropriate for large degree of freedom. It also requires the positive definiteness of the stiffness matrix which in a topology context is not satisfied (cf. above, comments after eq. (2.1)). This difficulty is circumvented by putting a small positive lower bound on the design variables (modifying (2.21) in a straightforward way). It can be proved [29, 43] that solutions obtained with lower bounds converge to a solution of the problem with zero lower bounds. Note, however, that positive bounds do not circumvent numerical difficulties when computing $u(x)$: If $K(x)$ is almost singular (i.e., possesses eigenvalues close to zero) then computation of $u(x)$ is an ill-conditioned problem [41]. Summarizing, the stress-rationing method is a method which can easily be implemented and which is fast for problems with n not too large. Since it can be interpreted in physical terms (assigning material proportionally to the specific energies of the members), it is interesting from an engineering point of view. However, it does not exploit the fact that $K(.)$ *linearly* depends on x. Svanberg proved convergence of the stress-rationing method [24]. He also proposed a numerical method which uses the Taylor expansion of c (cf. (2.2) and Prop. 2.2) under the condition that $K(x)$ is nonsingular [23].

Formulation ($\mathrm{P}_{\mathrm{strEn}}^{\mathrm{smooth}}$) is a smooth convex formulation with linear objective function and m quadratic constraints. Hence, any optimization routine of constrained convex nonlinear programming could be used. However, large numbers m of constraints will require an active set strategy. Moreover, sparsity of element stiffness matrices K_i should be exploited. Nowadays, effective *interior point methods* exist [15] which are able to solve problems even with many constraints [44, 45, 47]. However, if only standard methods of nonlinear optimization are available then large m may exceed the number of treatable constraints in smooth approach ($\mathrm{P}_{\mathrm{strEn}}^{\mathrm{smooth}}$). Then we propose the direct solution of ($\mathrm{P}_{\mathrm{strEn}}$) via nonsmooth optimization (cf. Sec. 5), i.e., direct minimization of ψ from (2.9). Subgradients of ψ are given by (2.14).

If the element stiffness matrices are dyadic products (cf. (1.5)) then formulation ($\mathrm{P}_{\mathrm{strEn}}^{\mathrm{LP}}$) or ($\mathrm{P}_{\mathrm{weight}}^{q^+,q^-}$) should be used which are simple Linear Programs. Routines solving Linear Problems are contained in any optimization software library. Exploiting sparsity of B (i.e., of the γ_i's) substantially decreases computation time. A solver with this ability is, e.g., contained in the `netlib`-library (URL: `http://www.netlib.org`).

Remark 2.9 We mention that statical determinacy of a structure (a, q) feasible for ($\mathrm{P}_{\mathrm{weight}}$) is reflected by the fact that the columns

$$\{\gamma_i \mid i : a_i > 0\}$$

of B are linearly independent. By this, statically determinate structures can be identified as so-called "basis solutions" which form the potential iteration points for the well-known simplex-algorithm in Linear Programming [48]. Statical determinacy is also used for problems more general than ($\mathrm{P}_{\mathrm{weight}}$) where equilibrium of forces $Bq = f$ is considered. In this way, some algorithms using statically determinate designs (e.g., [5, 30, 34, 35, 36]) can be interpreted as simplex-like algorithms. $\Diamond$

In the following we show three numerical examples dealing with trusses. As we know from Th. 2.8, the resulting structures shown can be viewed as (scaled) solutions for

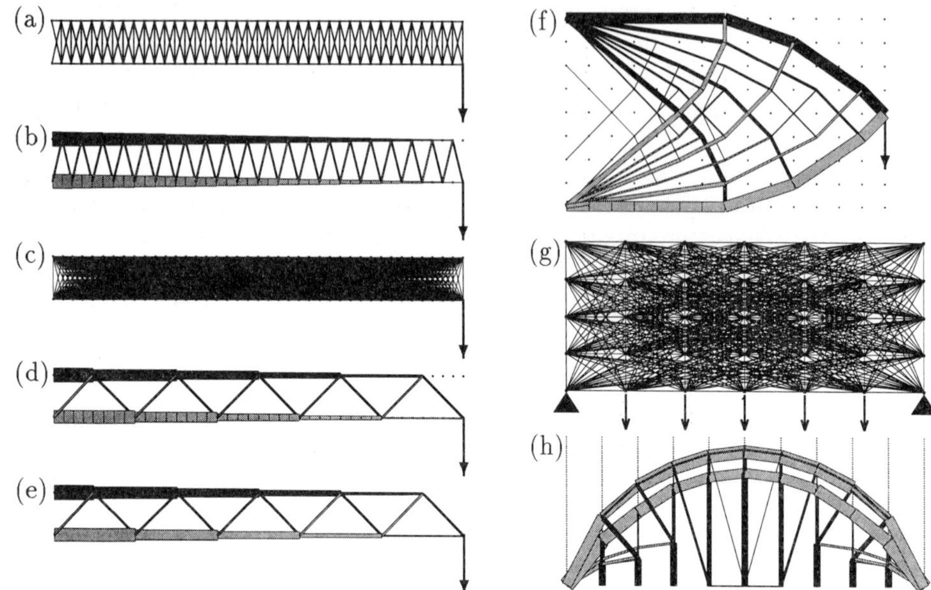

Figure 2 Single load test examples (2D Cantilever, Michell structure, bridge)

all the problems discussed in Sec. 2.1. More examples for single loaded problems are included in Ex. 3.3 and Ex. 3.5 below. In each figure showing a 2D result, black bars indicate bars under tension ($\boldsymbol{\gamma}_i^T \boldsymbol{u} = \Delta\ell_i > 0$) while grey bars illustrate compression.

Example 2.10 This example deals with a 2D cantilever arm modelled by $N = 82$ nodal points. In the first calculation the ground structure (shown in Fig. 2(a)) is used consisting of $m = 200$ potential bars. The load is indicated by the arrow, and both nodes at the left are fixed. The result is shown in Fig. 2(b). The ground structure which uses all possible connections ($m = 1761$ bars; cf. Fig. 2(c)) leads to the optimal structure in Fig. 2(d) (computed via LP ($P_{\text{strEn}}^{\text{LP}}$)). Observe that the interior bars are perpendicular "alternating" in tension and compression. This is also known from theoretical considerations [3]. The value of optimal compliance in (d) is about 30% less than in (b). We mention that the structure in Fig. 2(d) formally represents a mechanism. However, after omitting intermediate nodes we obtain the structure in Fig. 2(e) which is even statically determinate. $\diamond$

Example 2.11 We consider the well-known example of a 2D structure which is fixed at one side, and which is loaded by a vertical force at the mid-point of the opposite side. This example has been studied extensively for the continuum case, and in the famous early works by Maxwell [50] and Michell [51] optimality conditions are stated telling that – in sense of infinitesimal calculus – the lines of principal stresses for tension and compression are perpendicular.

Figure 2(f) shows the result (obtained via $(\mathrm{P}_{\mathrm{strEn}}^{\mathrm{LP}})$) for a ground structure of 15×9 nodal points and all possible potential connections (except overlapping ones), $m = 5614$. We observe that the above mentioned optimality conditions for the continuum setting are almost satisfied up to discretization. Investigations on (so-called) "Michell structures" and their relation to trusses can be found, e.g., in [3, 12, 28]. $\diamond$

Example 2.12 We ask for the optimal design of the side part of a bridge. As an illustration, Fig. 2(g) shows boundary conditions and a ground structure with 7×5 nodal points. Taking an 11×64 grid of $N = 704$ nodal points we get $m = 150031$ potential bars. Here we tested the applicability of an interior point method, i.e., $(\mathrm{P}_{\mathrm{strEn}}^{\mathrm{smooth}})$ has been solved by a method/code by Jarre et al. [44]. The resulting structure is shown in Fig. 2(h). Note that the shape of the bridge is not predetermined, and is a pure outcome of the optimization process. $\diamond$

3 Maximum Stiffness Topology Optimization of Multiply Loaded Discrete Structures

In this section we illustrate how some of the equivalences obtained for the *single* load problem can be generalized to the *multiple* load case. Particularly, we again want to get rid of the (many) design variables and find a formulation which parallels/generalizes formulation $(\mathrm{P}_{\mathrm{strEn}})$. This, as outlined above, is the basis for numerical treatment of large-scaled topology problems. We follow the development in [9, 29, 43, 52, 53].

3.1 Problem Statement and Some Reformulations

Parallel to $(\mathrm{P}_{\mathrm{compl}})$, we start with the formulation of a minimum compliance problem for p load cases $\boldsymbol{f}_1, \ldots, \boldsymbol{f}_p \in \mathbb{R}^n$ (i.e., p given external forces apply at p different points of time). As for the single load case, we assume $\boldsymbol{f}_j \neq 0$ for all $j = 1, \ldots, p$. For each load case $k = 1, \ldots, p$ we have to deal with a different state vector $\boldsymbol{u}_k \in \mathbb{R}^n$ that describes the nodal displacements arising when force vector $\boldsymbol{f}_k$ applies to the structure. For each load case we have *conditions of elastic equilibrium* (compare (1.7))

$$\boldsymbol{K}(\boldsymbol{x})\boldsymbol{u}_k = \boldsymbol{f}_k \quad \text{for } k = 1, \ldots, p. \tag{3.1}$$

To simplify notations, we collect the p load vectors as well as their corresponding state variables in one (column) vector, respectively:

$$\boldsymbol{f} := (\boldsymbol{f}_1^T, \ldots, \boldsymbol{f}_p^T)^T \in \mathbb{R}^{p \cdot n}, \quad \boldsymbol{u} := (\boldsymbol{u}_1^T, \ldots, \boldsymbol{u}_p^T)^T \in \mathbb{R}^{p \cdot n}. \tag{3.2}$$

As in Sec. 2, our main goal is minimization of compliance. However, for multiple load cases a technical problem arises: We have to deal with the p values $\boldsymbol{f}_1^T \boldsymbol{u}_1, \ldots, \boldsymbol{f}_p^T \boldsymbol{u}_p$ in common. Usually, this problem is avoided by minimization of a certain compromise of all values, e.g., a weighted sum [9, 13],

$$\sum_{k=1}^{p} \lambda_k \boldsymbol{f}_k^T \boldsymbol{u}_k, \tag{3.3}$$

where $\boldsymbol{\lambda} \in \mathbb{R}^p$ is a fixed vector of weighting parameters, $\boldsymbol{\lambda} > 0$. In this way we may consider the following problem (which for $p = 1$ becomes $(\mathrm{P_{compl}})$ up to the scaling constant $\lambda_1 > 0$):

$$(\mathrm{P^{mult,\lambda}_{compl}}) \qquad \min_{x \in \mathbb{R}^m,\, u \in \mathbb{R}^{p\cdot n}} \frac{1}{2} \sum_{k=1}^{p} \lambda_k \boldsymbol{f}_k^T \boldsymbol{u}_k$$
$$\text{s.t.} \quad \boldsymbol{K}(\boldsymbol{x})\boldsymbol{u}_k = \boldsymbol{f}_k \quad \text{for all } k = 1, \ldots, p,$$
$$\boldsymbol{x} \in X_V$$

With the choices

$$\bar{\boldsymbol{f}}^{\boldsymbol{\lambda}} := (\lambda_1 \boldsymbol{f}_1^T, \ldots, \lambda_p \boldsymbol{f}_p^T)^T \in \mathbb{R}^{p\cdot n}, \tag{3.4}$$

$$\overline{\boldsymbol{K}}_i^{\boldsymbol{\lambda}} := \begin{pmatrix} \lambda_1 \boldsymbol{K}_i & & \\ & \ddots & \\ & & \lambda_p \boldsymbol{K}_i \end{pmatrix} \in \mathbb{R}^{(p\cdot n)\times(p\cdot n)} \quad \text{for } i = 1, \ldots, m \tag{3.5}$$

problem $(\mathrm{P^{mult,\lambda}_{compl}})$ can formally be rewritten as $(\mathrm{P_{compl}})$ with $p \cdot n$ instead of n. Thus we can repeat all reformulations shown in Sec. 2 where dyadic structure (1.5) of matrices $\boldsymbol{K}_i$ is not needed. That means (cf. Th. 2.8(a)), equivalences of problems $(\mathrm{P^{mult,\lambda}_{compl}}) \equiv$ $(\mathrm{P_{compl}})$, $(\mathrm{P_{vol}})$, $(\mathrm{P_{potEn}})$, $(\mathrm{P_{strEn}})$, and $(\mathrm{P^{smooth}_{strEn}})$ for $\boldsymbol{f} := \bar{\boldsymbol{f}}^{\boldsymbol{\lambda}}$, $\boldsymbol{K}_i := \overline{\boldsymbol{K}}_i^{\boldsymbol{\lambda}}$, and $\boldsymbol{u}$ from (3.2) hold with dimension $p \cdot n$ instead of n. Stating the precise formulations of these problems is left to the reader.

However, the choice of weighting parameters $\lambda_1, \ldots, \lambda_p$ may have big influence on the resulting structure, and hence a good guess for $\boldsymbol{\lambda}$ should be known in advance. This needs deep knowledge of the influence of each loading scenario on the design which is usually not known. Thus we propose a much more conservative strategy which is sometimes referred to as *worst case design* (e.g. [13]). That means, we try to find a feasible structure which is able to carry its worst loading best. In terms of compliance ("good" structure means small "total compliance"; "bad load case k" means large compliance $\boldsymbol{f}_k^T \boldsymbol{u}_k$) we get the generalization of $(\mathrm{P^{mult,\lambda}_{compl}})$:

$$(\mathrm{P^{mult}_{compl}}) \qquad \min_{x \in \mathbb{R}^m,\, u \in \mathbb{R}^{p\cdot n}} \max_{1 \leq k \leq p} \frac{1}{2} \boldsymbol{f}_k^T \boldsymbol{u}_k$$
$$\text{s.t.} \quad \boldsymbol{K}(\boldsymbol{x})\boldsymbol{u}_k = \boldsymbol{f}_k \quad \text{for all } k = 1, \ldots, p,$$
$$\boldsymbol{x} \in X_V$$

For this problem we can also apply a similar mathematical apparatus as for the single load case. This is the topic of the rest of this section.

In order to avoid maximization in k over the discrete index set $\{1, \ldots, p\}$ we rewrite the objective $(\boldsymbol{u} \mapsto \max \frac{1}{2} \boldsymbol{f}_k^T \boldsymbol{u}_k)$ by means of the unit simplex $\Lambda := \{\boldsymbol{\lambda} \in \mathbb{R}^p \,|\, \boldsymbol{\lambda} \geq 0,\, \sum \lambda_k = 1\}$ as a weighted sum over all compliances (note that here $\boldsymbol{\lambda}$ plays the role of a new variable in opposite to (3.3) where $\boldsymbol{\lambda}$ was a fixed chosen parameter). By this, $(\mathrm{P^{mult}_{compl}})$ can be equivalently rewritten as

$(\mathrm{P}^{\mathrm{mult}}_{\mathrm{compl}})$ $\qquad$ $\min\left\{ \max_{\lambda\in\Lambda} \tfrac{1}{2} \sum_{k=1}^{p} \lambda_k \boldsymbol{f}_k^T \boldsymbol{u}_k \mid \boldsymbol{x}\in X_V, \boldsymbol{K}(\boldsymbol{x})\boldsymbol{u}_k = \boldsymbol{f}_k \text{ for all } k\right\}.$

The switch to this purely continuous form is motivated by practical purposes: Computation of an optimal weighting parameter $\boldsymbol{\lambda}$ helps to get deeper insight in the influence of each particular load case on the structure (see Ex. 3.4 below). There are also mathematical reasons: Minimax-Theorems and convex analysis may be applied leading to problem formulations of $(\mathrm{P}^{\mathrm{mult}}_{\mathrm{compl}})$ that can be numerically treated much more efficiently (see below).

Let us have a first view on $(\mathrm{P}^{\mathrm{mult}}_{\mathrm{compl}})$. As in the single load case, $(\mathrm{P}^{\mathrm{mult}}_{\mathrm{compl}})$ is not a convex problem, and the dimension $(p\cdot n + m)$ of the problem may be very large in the topology context. Besides the increase of number of state variables we must now take into account more equilibrium constraints than for the single load case. Moreover, we have to take care of the nasty max-term in the objective function.

Analogously to problem $(\mathrm{P}_{\mathrm{vol}})$ one derives an equivalent minimum volume problem with displacement constraints:

$(\mathrm{P}^{\mathrm{mult}}_{\mathrm{vol}})$ $\qquad$

$$\min_{x\in\mathbb{R}^m,\, u\in\mathbb{R}^{p\cdot n}} \sum_{i=1}^{m} x_i$$

$$\text{s.t.} \quad \boldsymbol{K}(\boldsymbol{x})\boldsymbol{u}_k = \boldsymbol{f}_k \quad \text{for all } k=1,\dots,p,$$
$$\tfrac{1}{2}\boldsymbol{f}^T\boldsymbol{u}_k \leq C \quad \text{for all } k=1,\dots,p,$$
$$\boldsymbol{x} \geq 0$$

We obtain a result parallel to the single load case showing that the solutions of $(\mathrm{P}^{\mathrm{mult}}_{\mathrm{compl}})$ and $(\mathrm{P}^{\mathrm{mult}}_{\mathrm{vol}})$ are the same up to a rescaling.

We now reformulate $(\mathrm{P}^{\mathrm{mult}}_{\mathrm{compl}})$. As for the single load case, we may rewrite compliance as minimum potential energy (cf. Prop. 2.1). Maximization over $\boldsymbol{\lambda}$ does not cause troubles since it can be put in front of minimization over $\boldsymbol{u}$ by means of a minimax-Theorem ([42] Cor. 37.3.2): For each (fixed) $\boldsymbol{x}\in X_V$,

$$\min_{u:\, K(x)u_k=f_k\forall k}\; \max_{\lambda\in\Lambda} \sum_{k=1}^{p} \lambda_k \boldsymbol{f}_k^T\boldsymbol{u}_k = \max_{\lambda\in\Lambda}\; \min_{u:\, K(x)u_k=f_k\forall k} \sum_{k=1}^{p} \lambda_k \boldsymbol{f}_k^T\boldsymbol{u}_k.$$

By this equation and by Prop. 2.1 we get the equivalence of $(\mathrm{P}^{\mathrm{mult}}_{\mathrm{compl}})$ and the following minimax-problem which parallels $(\mathrm{P}_{\mathrm{potEn}})$ (cf. Sec. 2.1):

$(\mathrm{P}^{\mathrm{mult}}_{\mathrm{potEn}})$ $\qquad$ $\displaystyle \min_{x\in X_V} \max_{\lambda\in\Lambda} \max_{u\in\mathbb{R}^{p\cdot n}} \left\{ \sum_{k=1}^{p} \lambda_k \left(\boldsymbol{f}_k^T\boldsymbol{u}_k - \tfrac{1}{2}\boldsymbol{u}_k^T\boldsymbol{K}(\boldsymbol{x})\boldsymbol{u}_k \right) \right\}$

Assume for the moment that m is of moderate size but p is large. Then the numerical solution of $(\mathrm{P}^{\mathrm{mult}}_{\mathrm{potEn}})$ is much easier to perform than the solution of the original problem $(\mathrm{P}^{\mathrm{mult}}_{\mathrm{compl}})$, if $(\mathrm{P}^{\mathrm{mult}}_{\mathrm{potEn}})$ is considered as a nonsmooth problem (compare (2.2) for $p = 1$): We define $C:\mathbb{R}^m \longrightarrow \mathbb{R}\cup\{\infty\}$ by

$$C(\boldsymbol{x}) := \max_{\lambda\in\Lambda} \max_{u\in\mathbb{R}^{p\cdot n}} \left\{ \sum_{k=1}^{p} \lambda_k \left(\boldsymbol{f}_k^T\boldsymbol{u}_k - \tfrac{1}{2}\boldsymbol{u}_k^T\boldsymbol{K}(\boldsymbol{x})\boldsymbol{u}_k \right) \right\} \qquad (3.6)$$

which simply maximizes the negative minimal potential energies (compare (2.2))

$$c_k(\boldsymbol{x}) := \max_{u_k \in \mathbb{R}^n} \{ \boldsymbol{f}_k^T \boldsymbol{u}_k - \tfrac{1}{2} \boldsymbol{u}_k^T \boldsymbol{K}(\boldsymbol{x}) \boldsymbol{u}_k \} \tag{3.7}$$

over all load cases. By this, $(\mathrm{P}_{\mathrm{potEn}}^{\mathrm{mult}})$ can be considered as minimization of C over X_V. Clearly, C cannot be expected to be smooth, as simple examples explicitly show. Thus, for practical treatment, we have to think about an implementable formula for a subgradient (see Sec. 5). Parallel to Prop. 2.2 (and Prop. 2.1) we get the following results:

Proposition 3.1 *(a) C is convex on $\mathbb{R}^m$.*

(b) Let $\boldsymbol{x} \geq 0$. Then $C(\boldsymbol{x}) < \infty$ if and only if there exists $\boldsymbol{u} \in \mathbb{R}^{p \cdot n}$ such that $\boldsymbol{K}(\boldsymbol{x})\boldsymbol{u}_k = \boldsymbol{f}_k$ for all $k = 1, \ldots, p$.

(c) If $\boldsymbol{x} \geq 0$ and $\boldsymbol{K}(\boldsymbol{x})$ is nonsingular then the subdifferential of C at $\boldsymbol{x}$ is

$$\partial C(\boldsymbol{x}) = \mathrm{conv}\{\nabla c_{\bar{k}}(\boldsymbol{x}) \mid \bar{k} \text{ such that } c_{\bar{k}}(\boldsymbol{x}) = \max_{1 \leq k \leq p} c_k(\boldsymbol{x})\}$$

where c_k is given by (3.7) and ∇c_k by Prop. 2.2.

Proof: Since C is the pointwise supremum function (cf. (5.1)) of linear functions in $\boldsymbol{x}$, it is convex. By (3.6), C can be rewritten as

$$C(\boldsymbol{x}) = \max_{1 \leq k \leq m} c_k(\boldsymbol{x})$$

with c_k from (3.7). Thus (b) follows from Prop. 2.1. By Prop. 2.2 and the assumption on $\boldsymbol{K}(\boldsymbol{x})$ we know that each c_k is convex and differentiable at $\boldsymbol{x}$, and thus a standard result (cf. Th. 5.4(a)) applies where ∇c_k is given by Prop. 2.2. $\square$

Summarizing, this proposition shows that $(\mathrm{P}_{\mathrm{potEn}}^{\mathrm{mult}})$ is a convex optimization probem with simple linear constraints. Therefore, a numerical procedure which is able to compute local optima will automatically lead to a *global* optimum of $(\mathrm{P}_{\mathrm{potEn}}^{\mathrm{mult}})$ (see e.g. [15]). Since $(\mathrm{P}_{\mathrm{potEn}}^{\mathrm{mult}})$ is equivalent to $(\mathrm{P}_{\mathrm{compl}}^{\mathrm{mult}})$, we obtain a *global* optimum for $(\mathrm{P}_{\mathrm{compl}}^{\mathrm{mult}})$ though $(\mathrm{P}_{\mathrm{compl}}^{\mathrm{mult}})$ is a non-convex problem. The main trick was to get rid of uncomfortable equilibrium constraints by rewriting them as (maximal negative) potential energies. Note that *infeasibility* of a structure w.r.t. elastic equilibrium results in an infinite function value (see Prop. 3.1(b)). Since we are *minimizing C* in $(\mathrm{P}_{\mathrm{potEn}}^{\mathrm{mult}})$, the procedure will automatically avoid such points $\boldsymbol{x}$. This, however, will only work in practice if C is "continuous", i.e., does not "jump to infinity" whenever $\boldsymbol{x}$ converges to a structure that cannot carry all load cases. This behaviour is excluded by the following

Theorem 3.2 *Let $(\boldsymbol{x}^j)_{j \in \mathbb{N}} \subset \{\boldsymbol{x} \in \mathbb{R}^m \mid \boldsymbol{x} \geq 0\}$ be a sequence of structures converging to some $\bar{\boldsymbol{x}} \geq 0$. Then*

$$\lim_{j \to \infty} C(\boldsymbol{x}^j) \quad \text{exists in } \mathbb{R} \cup \{\infty\},$$

and

$$\lim_{j \to \infty} C(\boldsymbol{x}^j) = C(\bar{\boldsymbol{x}}).$$

For the proof of this theorem it does not suffice to apply standard results for marginal functions like stated, e.g., in [54]. The reason is that for sparse structures $\boldsymbol{x}$ the set of feasible nodal displacements may not be bounded, or, C may not be finite. However, these mathematical difficulties can be circumvented due to the particular situation indicated already in Prop. 2.1: Minimization over $\boldsymbol{u}$ is not a proper optimization... A strictly mathematical proof of Th. 3.2 can be found in [29, 43].

However, formulation $(\mathrm{P}_{\mathrm{potEn}}^{\mathrm{mult}})$ is not appropriate for large-scaled topology problems. But, following the technique of elimination of the design variables described in Sec. 2 (transition from problem $(\mathrm{P}_{\mathrm{potEn}})$ to $(\mathrm{P}_{\mathrm{strEn}})$), we can imagine how to proceed: Consider $(\mathrm{P}_{\mathrm{potEn}}^{\mathrm{mult}})$ in the form

$$(\mathrm{P}_{\mathrm{potEn}}^{\mathrm{mult}}) \qquad \min_{x \in X_V} \max_{\lambda \in \Lambda} \left\{ \sum_{k=1}^{p} \lambda_k \cdot c_k(\boldsymbol{x}) \right\}$$

with c_k from (3.7). Then we easily see that the term enclosed in the brackets $\{\ldots\}$ is convex in $\boldsymbol{x}$ (cf. Prop. 2.2(a), and note $\boldsymbol{\lambda} \geq 0$) and linear in $\boldsymbol{\lambda}$, and thus concave in $\boldsymbol{\lambda}$. Thus, if finiteness was satisfied, the standard minimax-Theorem (cf. [42] Cor. 37.3.2) would allow for the interchange of min and max. A more precise non-trivial consideration shows that this works despite of infinite numbers [29, 43], and we end up with

$$(\mathrm{P}_{\mathrm{potEn}}^{\mathrm{mult}}) \qquad \max_{\lambda \in \Lambda} \min_{x \in X_V} \max_{u \in \mathbb{R}^{p \cdot n}} \left\{ \sum_{k=1}^{p} \lambda_k (\boldsymbol{f}_k^T \boldsymbol{u}_k - \tfrac{1}{2} \boldsymbol{u}_k^T \boldsymbol{K}(\boldsymbol{x}) \boldsymbol{u}_k) \right\}$$

where we used that maximization over $\boldsymbol{u}_k$ can be done independently for each k ("separability of optimization"). Now, for fixed $\boldsymbol{\lambda} \in \Lambda$, the completely analogous technique eliminating design variables $x_1, \ldots, x_m$ can be applied as already outlined in Sec. 2.1 for the single load case (cf. (2.4) to (2.7)). These explicit steps are left as an exercise to the reader. We end up with the following formulation in the displacements only which is a generalization of $(\mathrm{P}_{\mathrm{strEn}})$ (cf. Sec. 2.1):

$$(\mathrm{P}_{\mathrm{strEn}}^{\mathrm{mult}}) \qquad \max_{\lambda \in \Lambda} \max_{u \in \mathbb{R}^{p \cdot n}} \min_{1 \leq i \leq m} \left\{ \sum_{k=1}^{p} \lambda_k (\boldsymbol{f}_k^T \boldsymbol{u}_k - \tfrac{V}{2} \boldsymbol{u}_k^T \boldsymbol{K}_i \boldsymbol{u}_k) \right\}$$

Note that $(\mathrm{P}_{\mathrm{strEn}}^{\mathrm{mult}})$ has only $(p + p \cdot n)$ variables in opposite to $(m + p + p \cdot n)$ in $(\mathrm{P}_{\mathrm{compl}}^{\mathrm{mult}})$ (including $\boldsymbol{\lambda}$). Therefore tackling large scaled topology problems by this formulation is possible if we manage to derive a convex approach. As for the single load case (cf. problem $(\mathrm{P}_{\mathrm{strEn}})$) the price for reduction of variables is nonsmoothness of the objective function. Note that the inside function in $(\boldsymbol{\lambda}, \boldsymbol{u})$ is *not* concave as desired (concavity because of *maximization*-problem). There are two ways to circumvent this problem:

The "dirty trick" of substituting $\boldsymbol{u}_k$ by $\frac{1}{\lambda_k} \tilde{\boldsymbol{u}}_k$ leads to a formulation with a function that is concave jointly in $(\boldsymbol{\lambda}, \tilde{\boldsymbol{u}})$ and that neglects the case $\lambda_k = 0$. From a numerical point of view this is a delicate business which, however, proved to work (cf. [9, 45]).

The disadvantage is that $(p + p \cdot n)$ variables have to be treated in common which is difficult if p is large.

Therefore, we propose a second approach which also illustrates the applicability of nonsmooth formulations. We may treat $(\mathrm{P}_{\mathrm{strEn}}^{\mathrm{mult}})$ by the following *bilevel approach*: We solve problem

$$- \min_{\lambda \in \Lambda}(-\Phi(\boldsymbol{\lambda})) \tag{3.8}$$

where

$$\Phi(\boldsymbol{\lambda}) := \max_{u \in \mathbb{R}^{p \cdot n}} \min_{1 \le i \le m} \Big\{ \sum_{k=1}^{p} \lambda_k \big(\boldsymbol{f}_k^T \boldsymbol{u}_k - \tfrac{V}{2} \boldsymbol{u}_k^T \boldsymbol{K}_i \boldsymbol{u}_k\big)\Big\}. \tag{3.9}$$

By the substitutions in (3.4) and (3.5), evaluation of Φ for fixed $\boldsymbol{\lambda}$ reduces to the solution of a problem in $\boldsymbol{u}$ which is of the form of $(\mathrm{P}_{\mathrm{strEn}})$ (see Sec. 2.1). Numerical treatment can be performed by any of the approaches described in Sec. 2.2 (where a dyadic structure of the $\boldsymbol{K}_i$'s is not needed). Moreover, it can be shown [29, 52] that Φ is concave (i.e., $(-\Phi)$ is convex) on Λ, and also a formula for subgradients of $(-\Phi)$ can be derived. This enables numerical computation of a *global* optimum $\boldsymbol{\lambda}^*$ for problem (3.8). Since a theorem analogous to Th. 2.4 can be shown [29, 52], this leads to a *globally* optimal solution $(\boldsymbol{x}^*, \boldsymbol{u}^*)$ for the original problem $(\mathrm{P}_{\mathrm{compl}}^{\mathrm{mult}})$. Since in problem (3.8) only one variable for each load case is appearing, p may be large (as well as m).

We mention once more that optimization in $\boldsymbol{\lambda}$ contains an interesting aspect from an engineering point of view: The obtained optimal point $\boldsymbol{\lambda}^*$ reflects a qualitative measurement on the influence of each load case on the resulting design (see Ex. 3.4).

3.2   Numerical Examples

In this section we show three numerical examples which illustrate applicability of nonsmooth methods (and the effect of multiple loads). Note that all displayed solutions represent computed *globally* optimal structures.

We have used formulation $(\mathrm{P}_{\mathrm{potEn}}^{\mathrm{mult}})$, i.e., direct minimization of C, as well as approach (3.8). For both approaches the "outside" nonsmooth problems have been solved by the BT("Bundle-Trust")-algorithm (convex version) by Schramm and Zowe [55] (see also Sec. 5). A comparison of computational behaviour and CPU-times of both approaches can be found in [53]. It shows that for topology problems, where m is suitable small compared to $p \cdot n$, formulation $(\mathrm{P}_{\mathrm{potEn}}^{\mathrm{mult}})$ can be recommended. In the case where $p \cdot n$ is small compared to m, formulation (3.8) is preferred. This observation is, of course, directly related to the dimensions of the "inside" and the "outside" problem of each formulation.

Example 3.3 The first example deals with a 3D cantilever which has a triangular cross-sectional area. Its ground structure is shown in Fig. 3(a) and consists of $N = 51$ nodes and $m = 912$ potential bars. Three nodes at one end of the cantilever are fixed while at the opposite three end nodes one force applies at each node, respectively, which (applied at the same time) put torsion on the structure. Figure 3(b) indicates the loads and shows the result for this single load problem (computed via $(\mathrm{P}_{\mathrm{strEn}}^{\mathrm{LP}})$) which is a

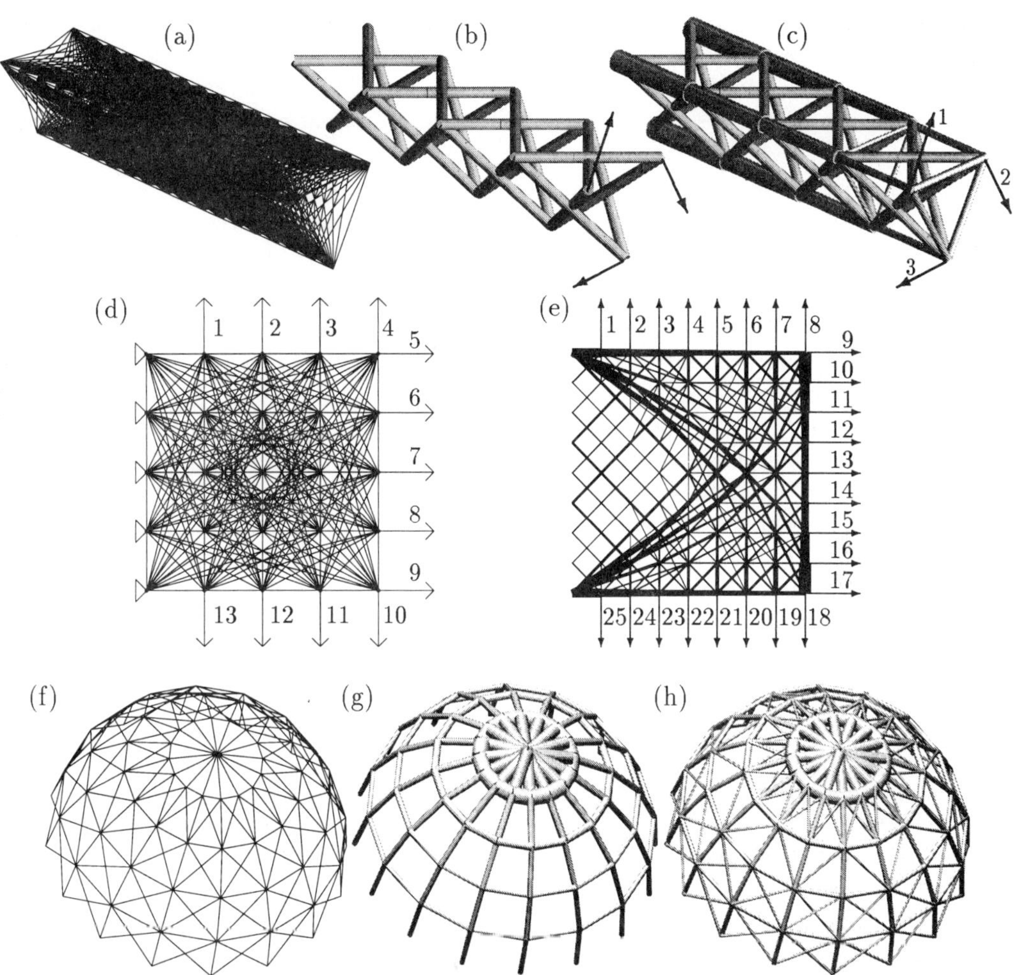

Figure 3 Multiple load test examples (3D Cantilever, Platform, Dome)

structure consisting of $\hat{m} = 24$ bars. Observe that in each side of the structure only those bars are contained which are perpendicular to each other. Figure 3(c) shows the optimal structure where each of the three applied forces is interpreted as a *separate load case*, i.e., applies at a *different* point of time. The optimal design for this multiple load case looks substantially different compared to (b). It consists of $\hat{m} = 48$ bars. For numerics we used formulation (3.8).   ◇

Example 3.4 This example deals with a 2D square domain ("platform") discretized by an $(N_1 \times N_1)$-rectangular nodal grid. The left hand N_1 nodes are fixed while at the remaining boundary nodes load forces apply perpendicularly to the boundary of the domain. As potential bars all connections between all nodes (except overlapping ones) are chosen. Figure 3(d) shows the ground structure for $N_1 = 5$ ($m = 200$ pot. bars) where $p = 13$ loads apply (each arrow symbolizes a different load case).

For $N_1 = 9$ we have $N = 81$ nodes, $m = 2040$ potential bars, and $p = 25$ load cases. We solve problem (3.8), and get the optimal point

$$\boldsymbol{\lambda}^* = (0.0032647, \quad 0.0067743, \quad 0.0141286, \quad 0.0226490, \quad 0.0283064, \quad 0.0365075, \quad 0.0562693,$$
$$\underline{0.3033477}, \quad 0.0178201, \quad 0.0022455, \quad 0.0033889, \quad 0.0034506, \quad 0.0037437, \quad 0.0034304,$$
$$0.0033538, \quad 0.0022536, \quad 0.0178189, \quad \underline{0.3033473}, \quad 0.0562680, \quad 0.0365075, \quad 0.0283062,$$
$$0.0226504, \quad 0.0141344, \quad 0.0067737, \quad 0.0032596)^T \in \mathbb{R}^{25}$$

which is symmetric (within numerical tolerance) as expected. Figure 3(e) shows the corresponding structure consisting of $\hat{m} = 289$ bars (where, due to multiple loads, tension and compression is not illustrated). Note that each function evaluation in problem (3.8) requires the solution of inside nonsmooth problem (3.9) in $p \cdot n = 25 \cdot 144 = 3600$ variables. Observe that the components λ_8^* and λ_{18}^* (underlined) are the (much) largest ones. This indicates that load cases no. 8 and no. 18 are the scenarios most difficult to treat, since they influence the objective most. This fits completely with the geometry (comp. Fig. 3(e)). This illustrates that an optimal point $\boldsymbol{\lambda}^*$ of problem (3.8) contains ranking information on critical load cases. $\diamond$

Example 3.5 In this example we consider a 3D truss-dome that consists of 5 "floors" (containing 16 nodes each) and an additional top node. The ground structure is shown in Fig. 3(f) ($N = 81$ nodes; $m = 272$; $n = 195$). Application of a single vertical load at the top node results in the mechanism shown in Fig. 3(g).

To avoid this unstable situation we simulate small movements of the load vector by multiple loads: We construct $p := 256$ load vectors (applied again at the top node) which point at 256 points, respectively, distributed regularly on a circle that is located on the floor of the dome.

We use formulation ($P_{\text{potEn}}^{\text{mult}}$) computing in the 272 design variables (since m is "small" and p is large). On purpose we choose a bad starting point to see if convexity of C (cf. Th. 3.1) is able to guide the algorithm to an optimal structure.

Figure 3(h) shows the result which in opposite to the single load solution (g) uses almost every potential bar, and thus a mechanism is avoided. Optimal minimax compliance of the multiple load solution (h) is about 17% larger than optimal compliance of the single load structure in (g). $\diamond$

4 Simultaneous Geometry/Topology Optimization

Truss design optimization must deal with unknown bar thicknesses as well as with unknown positions of nodal points. As already mentioned in Sec. 1.2.1, this distinction between topology and geometry optimization can be viewed as purely formal:

By the ground structure approach (cf. Sec. 1.2.1) one can perform a simultaneous optimization of topology and geometry by working with a dense, large scaled ground

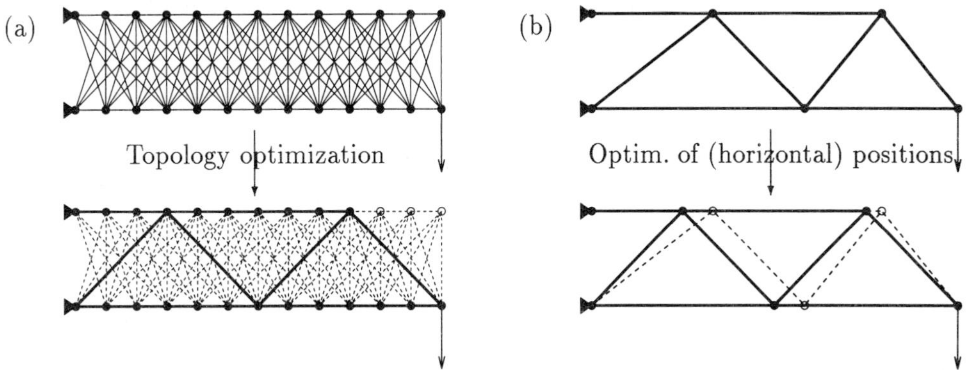

Figure 4 Optimization of Geometry by Ground Structure (a) and Pure Geometry Optimization (b)

structure with many nodal points. By this, the obtained structure automatically defines nodal positions which are close to optimality in a theoretical optimization problem where also nodal positions are varied. Figure 4(a) illustrates this process based on a dense ground structure. The disadvantage of this approach clearly is the large number of design variables which cannot be numerically treated in a complicated realistic setting. Moreover, the nodal positions obtained are only approximating the optimal positions in a problem with free nodes since the choice of nodes is restricted to the grid defined by the ground structure. As an advantage we note the much more simple mathematical structure of the optimization problem. Without design variables modelling nodal positions the large scaled pure topology problem may be solved for simple settings by exploiting the mathematical structure of the problem (cf., e.g., Sec. 2.1).

Opposite to that, we may consider "pure" optimization of geometry. Geometry optimization is historically separated from topology optimization. The reason is that classical methods for geometry optimization require a structure whose number of elements is small since geometrical design variables enter the problem in a nonlinear way. Therefore, sometimes, geometry is optimized *after* a topology optimization has been performed (see, e.g., [11]). In this way, the nodal points can be "corrected" with respect to the topology computed. This is illustrated in Fig. 4(b). The advantage of this approach is that complicated and realistic problem settings can be tackled due to a small number of unknowns. Moreover, classical methods working with direct sensitivity analysis based on elastic equilibrium can be used since the global stiffness matrix is nonsingular. This happens due to the fact that the topology aspect (i.e., zero bar volumes) is not contained in the problem. As a disadvantage we note that the obtained structure generally lacks to be optimal simultaneously w.r.t. topology and geometry (since the topology has been fixed before).

In the following we try to tackle the problem in a simultaneous approach. Though the problem formulation as such is clear, its numerical treatment is difficult. In the following we stress a methodology which uses nonsmooth optimization techniques. Since this approach may be viewed as a general approach to problems with few distinct "groups of unknowns", we illustrate it by the simple single load maximum stiffness problem formulation (cf. problem (P_{compl}) in Sec. 2.1). We follow the development in [56, 57, 58].

4.1 Formulation as Bilevel Problem

Once more we consider problem (P_{compl}) from the beginning of Sec. 2.1 (by Th. 2.8 also each other formulation from Sec. 2.1 could serve). For the N nodal points we introduce additional variables $\boldsymbol{y}_1, \ldots, \boldsymbol{y}_N \in \mathbb{R}^{\dim}$ where dim $= 3$ for spatial and dim $= 2$ for plane trusses. Each $\boldsymbol{y}_j$ denotes the geometrical shift of the position of node j in the ground structure relative to its original location. In a nonlinear way it enters each (geometrical part of the) element stiffness matrix, $\boldsymbol{K}_i = \frac{E_i}{\ell_i^2}\gamma_i\gamma_i^T$, for bars i whose end node is j: Length ℓ_i depends on $\boldsymbol{y}_j$ and also the angle of the bar relative to a global coordinate system may change with $\boldsymbol{y}_j$, i.e., γ_i depends on $\boldsymbol{y}_j$. By collecting all geometry variables $\boldsymbol{y}_j$ in one vector $\boldsymbol{y} \in \mathbb{R}^{N \cdot \dim}$ we may write

$$\boldsymbol{K}_i = \boldsymbol{K}_i(\boldsymbol{y}) = \frac{E_i}{\ell_i(\boldsymbol{y})^2}\gamma_i(\boldsymbol{y})\gamma_i(\boldsymbol{y})^T \tag{4.1}$$

for all $i = 1, \ldots, m$. Elastic equilibrium (1.7) becomes

$$\boldsymbol{K}(\boldsymbol{x},\boldsymbol{y})\boldsymbol{u} = \boldsymbol{f} \quad \text{where } \boldsymbol{K}(\boldsymbol{x},\boldsymbol{y}) := \sum_{i=1}^{m} x_i \boldsymbol{K}_i(\boldsymbol{y})$$

for fixed bar volumes $\boldsymbol{x} \geq 0$ and fixed nodal positions $\boldsymbol{y}$. We mention that the analysis and the numerical approach below does not rely on the dyadic form (4.1) of the element stiffness matrices, and allows also for application to the optimization of continuum mechanical structures where $\boldsymbol{K}_i(\boldsymbol{y})$ are general (positive semi-definite) element stiffness matrices in a discretized problem.

Typically only few nodal positions, say k, are varied, and $\boldsymbol{y}$ is restricted by so-called "geometrical constraints". Therefore we may assume that some neighbourhood $Y \subset \mathbb{R}^k$ of $0 \in \mathbb{R}^k$ is given such that each $\boldsymbol{y} \in Y$ symbolizes a feasible geometry.

With these settings problem (P_{compl}) from Sec. 2.1 becomes

$$(\text{P}_{compl}^{geotopo}) \qquad \min_{y \in \mathbb{R}^k, x \in \mathbb{R}^m, u \in \mathbb{R}^n} \tfrac{1}{2}\boldsymbol{f}^T\boldsymbol{u}$$
$$\text{s.t.} \quad \boldsymbol{K}(\boldsymbol{x},\boldsymbol{y})\boldsymbol{u} = \boldsymbol{f},$$
$$\boldsymbol{x} \in X_V,$$
$$\boldsymbol{y} \in Y.$$

Figure 5 illustrates this problem where topology optimization is indicated by a (sparse) ground structure, and optimization of nodal positions is indicated by "boxes" around the initial nodal points. These boxes form the set Y of feasible nodal positions $\boldsymbol{y}$. Generally, the boxes for each node may intersect. Note that due to variable geometry in this problem m and n (and k) are regarded to be of only moderate size (cf. above).

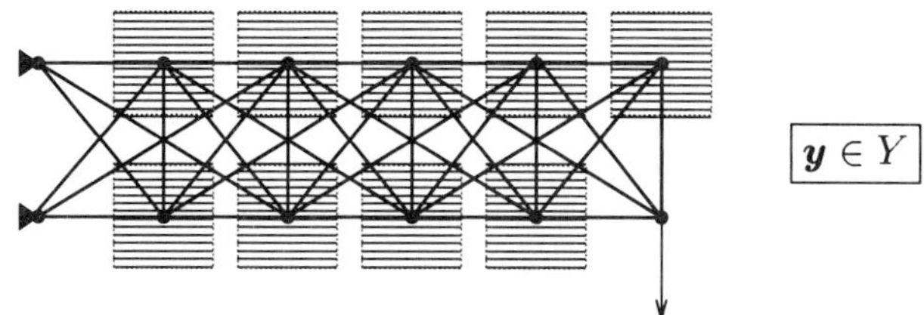

Figure 5 Simultaneous Optimization of Geometry and Topology

However, since $\boldsymbol{y}$ enters $\boldsymbol{K}_i(\boldsymbol{y})$ in a non-linear way, mathematical advantages exploited for (P$_{\mathrm{compl}}$) seem to be lost. Tests show that a direct numerical treatment of this formulation with standard software for nonlinear optimization fails even for moderate problem size.

From Sec. 2.2 we know that the problem in $(\boldsymbol{x}, \boldsymbol{u})$ for fixed $\boldsymbol{y}$ can be expressed and numerically treated in various ways. We denote by $h(\boldsymbol{y})$ the resulting objective value for problem (P$_{\mathrm{compl}}$) if $\boldsymbol{y} \in Y$ is fixed, i.e., minimal (half) compliance (which can be obtained by an optimal structure $\boldsymbol{x}(\boldsymbol{y})$ and a corresponding vector of displacements $\boldsymbol{u}(\boldsymbol{y})$):

$$h(\boldsymbol{y}) := \min_{x,u}\left\{\tfrac{1}{2}\boldsymbol{f}^T\boldsymbol{u} \mid \boldsymbol{x} \in X_V,\ \boldsymbol{K}(\boldsymbol{x},\boldsymbol{y})\boldsymbol{u} = \boldsymbol{f}\right\} \tag{4.2}$$

(Instead of (P$_{\mathrm{compl}}$) any other formulation from Sec. 2.1 could serve, cf. Th. 2.8). By this formal *separation of variables* problem (P$_{\mathrm{compl}}^{\mathrm{geotopo}}$) can be equivalently written as

$$(\mathrm{P}_{\mathrm{bilevel}}^{\mathrm{geotopo}}) \qquad\qquad \min_{y\in Y} h(\boldsymbol{y})$$

which (formally) is a low-dimensional problem in the variable $\boldsymbol{y}$ only. This problem type is sometimes referred to as *bilevel problem* since function h is minimized whose evaluation itself requires the solution of an optimization problem (in $\boldsymbol{x}$ and $\boldsymbol{u}$). Note that problem (P$_{\mathrm{bilevel}}^{\mathrm{geotopo}}$) is substantially different from an approach which alternatingly optimizes in $\boldsymbol{y}$ (with structure $\boldsymbol{x}$ fixed) and in $\boldsymbol{x}$ (with geometry $\boldsymbol{y}$ fixed). This is illustrated by the following simple academic example in two variables:

Example 4.1 Consider the problem of minimizing the function $f(x,y) := x+y$ on the set $X := \{(x,y) \mid x,y \geq 0,\ x + 2y \geq 2\} \subset \mathbb{R} \times \mathbb{R}$. As a starting point we may choose $(10,0)$, and we first optimize in variable x while component $y = \bar{y} = 0$ is fixed. We end up at the point $(\bar{x},\bar{y}) = (2,0)$ at which $\bar{x} + 2\bar{y} = 2$. Now, fixing component $x = \bar{x} = 2$, and optimizing in y we recognize that $(\bar{x},\bar{y})$ is already optimal. However, this point is not optimal for the problem: Point $(x^*,y^*) := (0,1)$ gives $f(x^*,y^*) = 1 < 2 = f(\bar{x},\bar{y})$. Considering the problem as a bilevel problem, we have

$$h(y) := \min_{x:\,(x,y)\in X} f(x,y) = \begin{cases} f(2-2y,y) & = & 2-y, & \text{if } y \in [0,1], \\ f(0,y) & = & y, & \text{if } y > 1. \end{cases} \tag{4.3}$$

Minimizing h over the set $\{y \mid \exists x : (x, y) \in X\} = [0, \infty[$ yields the optimal point $y^* = 1$, and thus leads to the optimal point (x^*, y^*) of the original problem. Analogously, x could play the role of the "master variable". This illustrates that "component-wise search" checks optimality only into the directions of the unit vectors while the bilevel approach takes all directions into account. Note that by def. (4.3), $h(y) \equiv |y - 1| + 1$, and thus is nonsmooth. $\diamond$

From perturbation theory it is known that h generally depends in a nonconvex and nonsmooth way on the variable y (convexity in Ex. 4.1 is due to simplicity of the example). Hence, when treating $(\mathrm{P}^{\mathrm{geotopo}}_{\mathrm{bilevel}})$ in a numerical approach, we can only expect to obtain *local* optima of h. We summarize this situation in the following theorem which makes clear that $(\boldsymbol{x}, \boldsymbol{u})$ is globally optimal w.r.t. fixed $\boldsymbol{y}$ while $\boldsymbol{y}$ may be a local optimum of h (for a proof see [58]):

Theorem 4.2 *The triple* $(\boldsymbol{y}^*, \boldsymbol{x}^*, \boldsymbol{u}^*)$ *is a solution of* $(\mathrm{P}^{\mathrm{geotopo}}_{\mathrm{compl}})$ *which is local with respect to* $\boldsymbol{y}$ *and global with respect to* $(\boldsymbol{x}, \boldsymbol{u})$ *if and only if* $\boldsymbol{y}^* \in Y$ *is a local minimizer of* h *in* $(\mathrm{P}^{\mathrm{geotopo}}_{\mathrm{bilevel}})$, *and* $(\boldsymbol{x}^*, \boldsymbol{u}^*)$ *is a global minimizer of the min-problem in (4.2) for fixed* $\boldsymbol{y} := \boldsymbol{y}^*$.

As already mentioned above, the minimum-value function h is nonsmooth (this can also be verified by easy analytical examples). Hence, when minimizing h in $(\mathrm{P}^{\mathrm{geotopo}}_{\mathrm{bilevel}})$ on the upper level, one has to resort to nonsmooth software. Modern effective nonsmooth optimization methods use subgradients (cf. Sec. 5) instead of gradients. Analogously to routines treating a smooth problem, the user has to provide a subroutine which computes at each iterate $\boldsymbol{y}^k$ the informations needed, i.e.,

$$the\ function\ value\ h(\boldsymbol{y}^k),\ and \tag{4.4}$$

$$one\ (arbitrary)\ subgradient\ \boldsymbol{g}^k \in \mathbb{R}^k. \tag{4.5}$$

Here, step (4.5) replaces the sensitivity analysis.

Since under (standard) regularity assumptions it can be proved [54] that h is locally Lipschitz-continuous (cf. def. (5.7)), the subdifferential $\partial h(\boldsymbol{y}^k)$ is given by (5.8).

Note that (4.4) and (4.5) are the only tasks a user of a nonsmooth method has to take care of. Moreover, (4.4) and (4.5) show the two-level nature of $(\mathrm{P}^{\mathrm{geotopo}}_{\mathrm{bilevel}})$: While h is minimized by a nonsmooth solver, (4.4) requires the solution of problem (4.2) on a lower level for fixed $\boldsymbol{y} = \boldsymbol{y}^k$, and (4.5) provides "gradient" information $\boldsymbol{g}^k$ on h (at the iterate $\boldsymbol{y} = \boldsymbol{y}^k$) for the solver. We mention that (4.4) and (4.5) have to be computed at each iteration point $\boldsymbol{y}^k$, i.e., repeatedly. Therefore a methodology has to be known which can solve (4.4) (i.e., min-problem (4.2)) in reasonable time. Here we refer to Sec. 2.2.

Though (4.5) seems to be the more difficult part, it typically happens that computation of $\boldsymbol{g}^k \in \partial h(\boldsymbol{y}^k)$ is given "for free": The following result holds under standard regularity assumptions (see, e.g., [54, 59]; this result is a generalization of Th. 5.4 for the llc-case (cf. end of Sec. 5)). For simplicity we restrict to the truss case where each $\boldsymbol{K}_i(\boldsymbol{y})$ is dyadic and given by (4.1):

$$\partial h(\boldsymbol{y}^k) = \mathrm{conv}\Big\{ \sum_{i=1}^{m} x_i^k \boldsymbol{\gamma}_i(\boldsymbol{y}^k)^T \boldsymbol{u}^k \nabla \boldsymbol{\gamma}_i(\boldsymbol{y}^k)^T \boldsymbol{u}^k \Big| (\boldsymbol{x}^k, \boldsymbol{u}^k) \text{ solves (4.2) for fixed } \boldsymbol{y}^k \Big\}$$

Thus, whenever a solution $(\boldsymbol{x}^k, \boldsymbol{u}^k)$ of (4.2) for fixed $\boldsymbol{y}^k$ has been computed, a subgradient $\boldsymbol{g}^k$ is immediately given, e.g., by

$$\boldsymbol{g}^k := \sum_{i=1}^{m} x_i^k \boldsymbol{\gamma}_i(\boldsymbol{y}^k)^T \boldsymbol{u}^k \nabla \boldsymbol{\gamma}_i(\boldsymbol{y}^k)^T \boldsymbol{u}^k. \tag{4.6}$$

Note that minimization (4.2) will provide only *one* optimal solution $(\boldsymbol{x}^k, \boldsymbol{u}^k)$, and thus only one subgradient $\boldsymbol{g}^k$ of h at $\boldsymbol{y}^k$ is known. However, as already mentioned above, most nonsmooth codes (as, e.g., Bundle methods; see Sec. 5) are tailored precisely to this situation (cf. (4.5)).

4.2   Numerical Examples

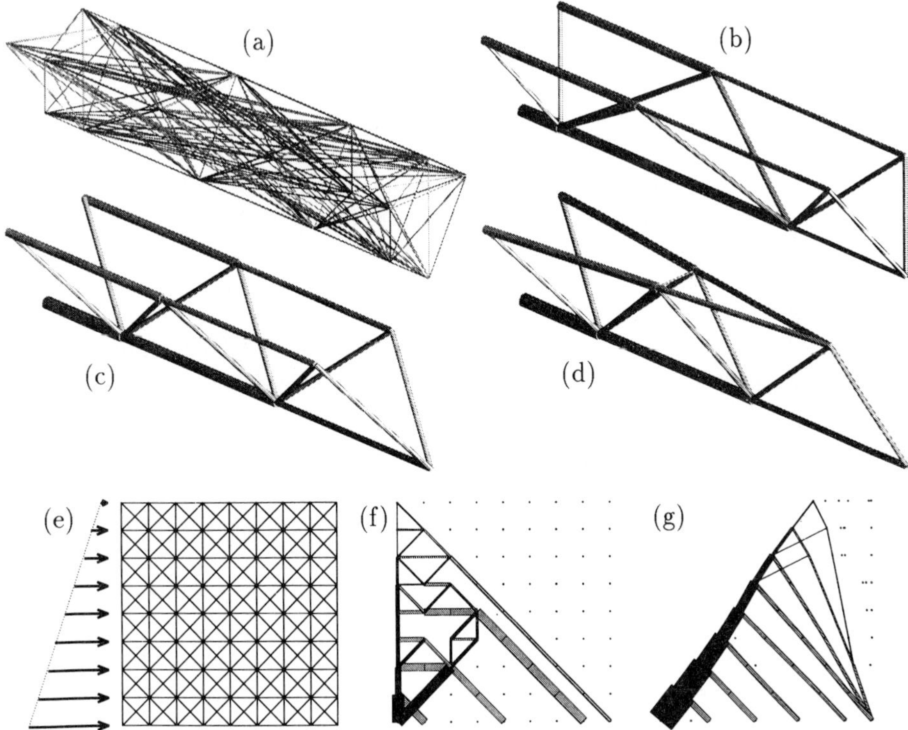

Figure 6 Geo-Topo test examples (3D Cantilever, Dam support)

We present two numerical examples which show the practical applicability of problem $(P_{\text{bilevel}}^{\text{geotopo}})$. Nonsmooth problems $(P_{\text{bilevel}}^{\text{geotopo}})$ have been computed with an implementation of the BT("Bundle-Trust")-algorithm (non-convex version) by Schramm and Zowe [55]. At each iterate, sub-problem (4.4) (i.e., (4.2)) has been solved in the LP-form $(P_{\text{strEn}}^{\text{LP}})$ (see Sec. 2.1) by a standard LP-solver, while sub-problem (4.5) reduces (as outlined above) to computation (4.6). Note, that all solutions shown represent local-global-global("lgg") solutions of $(P_{\text{compl}}^{\text{geotopo}})$ in the sense of Th. 4.2.

Example 4.3 The first example deals with a 3D cantilever arm whose ground structure consists of $N = 18$ nodes and $m = 303$ potential bars. On purpose the (initial) locations of the nodal points along the construction have been chosen non-equidistantly. Figure 6(a) shows the ground structure where the three left hand nodes are fixed support nodes, and thus $n = 45$. At the right hand bottom node a force applies pointing vertically to the ground. Fig. 6(b) displays the (global!) result of the pure topology-problem (P_{compl}), i.e., *without* optimizing nodal positions in y. Figure 6(c) shows a lgg-solution of $(P_{\text{compl}}^{\text{geotopo}})$ (via $(P_{\text{bilevel}}^{\text{geotopo}})$) where Y was chosen in a way that nodal positions were allowed to vary only along the length axis of the construction. In Fig. 6(d) the lgg-result of $(P_{\text{bilevel}}^{\text{geotopo}})$ is shown where nodal positions could vary in each horizontal direction. With suitable choices for V and E, the values of compliances obtained are 6.944 (b), 6.563 (c), and 6.326 (d), respectively. $\diamond$

Example 4.4 This example demonstrates an attractive feature of the bilevel concept: Different from the ground structure approach, also nodes under load can move in the minimization on the upper level.

We consider a structure that can be considered as a skeleton of a dam. The initial mesh of potential nodes and bars, the distribution of the load, and the boundary conditions are depicted in Fig. 6(e). Again, the (global!) solution of (P_{compl}) without optimization of nodal positions is shown in Fig. 6(f). Black bars indicate tension, grey ones are under compression. Figure 6(g) shows the lgg-solution of $(P_{\text{compl}}^{\text{geotopo}})$ where horizontal changes of nodal positions were allowed. Note that the horizontal positions of *all* nodes (except that of the two lower corners) are optimized, i.e., also the locations of the load nodes.

We obtain a structure which in its lower part (due to geometry of fixed support nodes) tries to mimic perpendicular bars under tension and compression. This fits to known theoretical results (cf. [3, 12]). $\diamond$

5   An Excursion to Nonsmooth Optimization

In this section we give a brief introduction to nonsmooth optimization and to the tools usually used by the most effective methods known today. Since this introduction is intended for beginners, we try to focus rather on illustration of concepts than on mathematical technicalities. This is the reason why we mainly restrict ourselves to the convex case.

Introductions to and overviews on nonsmooth optimization (mathematics as well as methodologies) can be found, e.g., in [60, 61]. The latter also includes an extensive and commented bibliography. For mathematical questions we refer to [15, 42] and [59].

5.1 Nonsmooth Optimization and Typical Applications

When speaking about "nonsmooth functions" we mean functions which are continuous but which may have "kinks" at some points, i.e., which are not necessarily differentiable at each point where they are defined.

A simple example of such a function is the *pointwise supremum-function*: Let I be a non-empty index set, and for each $i \in I$ let a function $f_i : \mathbb{R}^n \longrightarrow \mathbb{R} \cup \{\infty\}$ be given. Then $\phi : \mathbb{R}^n \longrightarrow \mathbb{R} \cup \{\infty\}$,

$$\phi(\boldsymbol{x}) := \sup_{i \in I} f_i(\boldsymbol{x}), \tag{5.1}$$

is called *pointwise supremum function (of the f_i's, $i \in I$)*. Figure 7(a) sketches ϕ for a

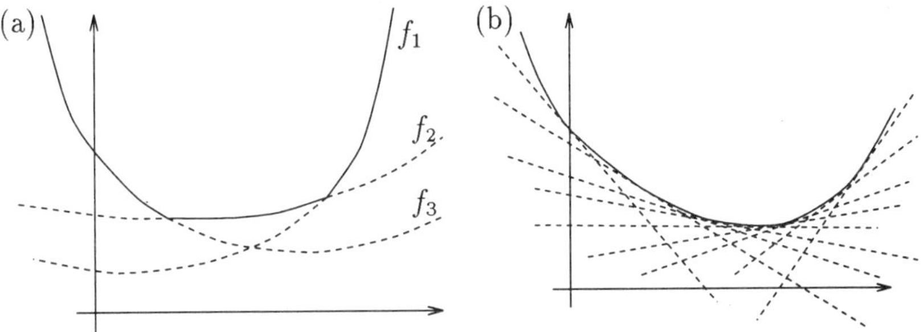

Figure 7 Pointwise supremum function of finitely many (a) and infinitely many (b) functions

discrete set I of $|I| = 3$ functions f_i, while in (b) the situation is indicated where I is infinite. From Fig. 7(a) we see immediately that ϕ is generally nonsmooth.

We will use the pointwise supremum-function in the case of convex functions f_i. Then ϕ is also convex as can easily be shown (see [42] Th. 5.5). We refer once more to Figs. 7(a) and (b) where this situation is mimicked.

A typical application of nonsmooth optimization is the solution of a minimax problem of type

$$\min_{x} \max_{y} f(\boldsymbol{x}, \boldsymbol{y})$$

where f is a function in the two variables $\boldsymbol{x}$ and $\boldsymbol{y}$. This problem can be equivalently rewritten as

$$\min_{x} h(\boldsymbol{x}) \quad \text{where } h(\boldsymbol{x}) := \max_{y} f(\boldsymbol{x}, \boldsymbol{y}).$$

Here h is a pointwise supremum function, and thus, is generally nonsmooth. Examples for such problems described in previous sections are formulations (P_{potEn}), (P_{strEn}), ($P_{\text{potEn}}^{\text{mult}}$), ($P_{\text{strEn}}^{\text{mult}}$). A completely analogous technique can be applied to min-min-problems as outlined in Sec. 4 (cf. problem ($P_{\text{bilevel}}^{\text{geotopo}}$)). Usage of the nonsmooth approach is done in order to exploit convexity of h (if satisfied) or easy-to-solve sub-problems hidden in h for fixed $\boldsymbol{x}$.

Other typical applications of nonsmooth optimization are *Lagrangian relaxation* and *exact penalty methods* [15].

5.2 Failure of Smooth Methods

Since nonsmooth functions usually possess kinks only at isolated points (cf. Fig. 7(a)) or at points on some lower-dimensional "line", one might conclude that numerical solution of a nonsmooth optimization problem can be performed by any algorithm of nonlinear optimization which uses gradient information. However, there exist simple counter-examples [60, 61] which demonstrate the failure of algorithms requiring differentiability when applied to nonsmooth functions. In this situation, such methods usually generate a sequence of points converging to some non-optimal kink.

Besides this disaster the stopping-criterion is a problem (cf. also [60, 61]): Smooth algorithms usually stop if the norm of the gradient at the present iterate is small enough. However, in a nonsmooth optimization problem this does not make sense: Think, e.g., of minimizing $f : (x \mapsto |x|)$ on $\mathbb{R}$. Then at each point $x \neq 0$ the function f is differentiable, and $|f'(x)| = 1$.

Another common idea to circumvent the problem is "to smooth" the nonsmooth function such that the "kinks in the graph become round". A popular scheme among structural engineers is the Kreisselmeier-Steinhauser-Function. However, there is a delicate numerical balance which hardly can be overcome (by *nature* of the problem!): If the smoothed function $\tilde{f}$ is far away from original function f then the result of the optimization will be useless. If, on the other hand, $\tilde{f}$ is close to f then we will observe exactly the same numerical problems as if directly dealing with f, i.e., the gradient "jumps". This fact is even worse since in most applications the (true) minimum (of f) will be at a point where f is nonsmooth. This is obvious by geometry and by the fact that, if nonsmoothness arises in an optimization model, then, typically, it is exactly this nonsmoothness which plays the key-role in modelling of the underlying real life problem.

Summarizing, application of algorithms requiring smoothness is prohibited(!) in the nonsmooth case, and we *must* deal with algorithms that are able to handle kinks, i.e., points at which f lacks to be differentiable.

5.3 Generalization of Gradient: The Subgradient

In the following we briefly introduce the concept of subgradients and (later on in Sec. 5.4) in a particular methodology which is used by modern effective methods of nonsmooth optimization. For simplicity of presentation (and since our basic interest is computation of *global* optima), we concentrate on the convex case. However, most concepts can also be generalized to the non-convex case (see below).

As the basic tool in nonsmooth optimization the *subgradient* is used. It generalizes the gradient of the smooth case.

Definition 5.1 *Let* $f : \mathbb{R}^n \longrightarrow \mathbb{R}$ *be convex, and let* $\boldsymbol{x} \in \mathbb{R}^n$. *Then each vector* $\boldsymbol{g} \in \mathbb{R}^n$ *with*

$$\boldsymbol{g}^T(\boldsymbol{y} - \boldsymbol{x}) \leq f(\boldsymbol{y}) - f(\boldsymbol{x}) \quad \text{for all } \boldsymbol{y} \in \mathbb{R}^n \tag{5.2}$$

is called "subgradient of f at $\boldsymbol{x}$". The set of subgradients at $\boldsymbol{x}$,

$$\partial f(\boldsymbol{x}) := \{\,\boldsymbol{g} \in \mathbb{R}^n \mid \boldsymbol{g}\ \text{is subgradient of } f \text{ at } \boldsymbol{x}\,\},$$

is called "subdifferential of f at $\boldsymbol{x}$".

We may interpret each subgradient $\boldsymbol{g}$ as the first n components of a normal vector $(\boldsymbol{g}^T, -1)^T \in \mathbb{R}^{n+1}$ of a hyperplane $H(\boldsymbol{x})$ which supports the epigraph of f at the point $(\boldsymbol{x}^T, f(\boldsymbol{x}))^T \in \mathbb{R}^n \times \mathbb{R}$. Due to non-differentiability there may be infinitely many hyperplanes which do so, and thus $\partial f(\boldsymbol{x})$ may be a set. Figure 8(a) shows the situation at a point $\boldsymbol{x}$ where f is nonsmooth. In this way, from a theoretical point of view, each

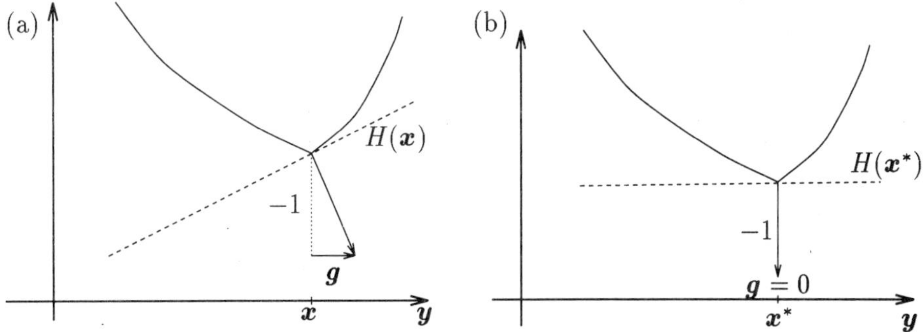

Figure 8 Interpretation of subgradient by supporting hyperplane (a); Interpretation of optimality condition (b)

$\boldsymbol{g} \in \partial f(\boldsymbol{x})$ gives rise to a "weak linearization" of f at $\boldsymbol{x}$ where the linearizing function is $(\boldsymbol{y} \mapsto f(\boldsymbol{x}) + \boldsymbol{g}^T(\boldsymbol{y} - \boldsymbol{x}))$. The analogous situation in the smooth case in terms of gradients is obvious.

The following proposition collects some basic properties of the subdifferential. For proofs see, e.g., [42]. As usual, ∇f denotes the gradient of f.

Proposition 5.2 *Let $f : \mathbb{R}^n \longrightarrow \mathbb{R}$ be convex, and let $\boldsymbol{x} \in \mathbb{R}^n$. Then the following statements hold:*

(a) The subdifferential $\partial f(\boldsymbol{x})$ of f at $\boldsymbol{x}$ is a non-empty closed convex bounded set.

(b) If $\partial f(\boldsymbol{x}) = \{\boldsymbol{g}\}$ then f is differentiable at $\boldsymbol{x}$ and $\nabla f(\boldsymbol{x}) = \boldsymbol{g}$.

(c) If f is differentiable at $\boldsymbol{x}$ then $\partial f(\boldsymbol{x}) = \{\nabla f(\boldsymbol{x})\}$.

Note that by (b) and (c) the concept of subgradients properly generalizes differentiability.

The following theorem shows that subgradients can help to overcome the problem of a stopping criterion (cf. above) if the subdifferential is known. It generalizes the well-known condition "$\nabla f(\boldsymbol{x}^*) = 0$" for the smooth case.

Theorem 5.3 *Let f be convex on $\mathbb{R}^n$. Then x^* is a minimizer of f if and only if $0 \in \partial f(x^*)$.*

Proof: Plugging in $g = 0$ into (5.2) we obtain $0 \leq f(y) - f(x^*)$ for all $y \in \mathbb{R}^n$. $\square$

This optimality condition can easily be interpreted by geometry: Figure 8(a) illustrates (put $x := x^*$) that a subgradient $g = 0$ leads to a normal vector of $H(x^*)$ which is vertically pointing down, i.e., $H(x^*)$ is a horizontal hyperplane. Since the epigraph of f lies above $H(x^*)$ by convexity of f, $f(x^*)$ must be (one of) the "lowest point(s) of the graph", and thus x^* is optimal. This situation is shown in Fig. 8(b).

We mention that for nonsmooth constrained optimization problems there are generalizations of KKT-conditions (cf. [15]) even for the non-convex case [59].

5.4   Methods of Nonsmooth Optimization

During the last 15 years a wide variety of optimization techniques for nonsmooth functions has been developed, especially for the convex case [60, 61]. However, only a few basic concepts led to numerically stable implementations that can be used without further nasty experiments on control parameters. One of these methodologies is the *bundle concept* which is briefly described below.

As already mentioned, effective nonsmooth methods make use of subgradients which mimic the gradient in a kink. Analogously to solvers for smooth problems the user has to provide a subroutine that calculates at an arbitrary (feasible) point x

- the function value $f(x)$ and
- one (arbitrary) subgradient $g \in \partial f(x)$.   (5.3)

Thus for application of nonsmooth methods an implementable formula for computation of an (arbitrary) subgradient is needed. As an illustration we state a result for the pointwise supremum function:

Theorem 5.4 *Consider a pointwise supremum function ϕ (cf. def. (5.1)) where for each $i \in I$ the function $f_i : \mathbb{R}^n \longrightarrow \mathbb{R}$ is convex and continuously differentiable.*

(a) If I is finite then for each $x \in \mathbb{R}^n$, $\partial \phi(x) = \text{conv}\{\nabla f_i(x) \mid i : f_i(x) = \phi(x)\}$.

(b) Let $I \subset \mathbb{R}^k$ be infinite but compact. Moreover, let the mappings $F : ((x, i) \mapsto f_i(x))$ and $\nabla_x F(.\,,.)$ be continuous on $\mathbb{R}^n \times I$. Then the formula in (b) also holds.

Proof: See, e.g., Prop. 2.1 and eq. (1.8) in [61]. $\square$

We mention that knowledge of the *whole* subdifferential is usually *not* given in practical applications (For example, in Th. 5.4(b), usually only *one single* index $i \in I$ with $f_i(x) = \phi(x)$ can be computed in practice). However, requirement (5.3) fits to this difficulty: Only *one* (arbitrary) subgradient has to be computed at an iteration point x.

In the following we present a very simple prototype method which illustrates how a nonsmooth optimization method takes subgradient information into account. Though this algorithm should be considered only as a basic concept, it shows the general methodology of many nonsmooth optimization techniques used today. We refer to it as the *cutting plane algorithm*. In fact, cutting plane methods exist in several versions, sometimes tuned to a particular application [15]. For computing optimal points (in any sense) the common idea of cutting plane methods is to start with a large set containing some optimal points. Then hyperplanes are constructed cutting off parts of the present set without cutting off optimal points. In this way, the considered set "converges" to the set of optimal points.

Since our intention is to illustrate the general methodology of a nonsmooth method, we restrict ourselves to the simple case where we seek the minimum of a nonsmooth function f in a cube.

Algorithm 5.5 *Let $R > 0$ be a given number, and put $X := \{\boldsymbol{x} \in \mathbb{R}^n \mid |x_i| \leq R \, \forall i\}$. Let $f : X' \longrightarrow \mathbb{R}$ be convex where X' is an open convex set with $X' \supset X$. We want to find a point $\boldsymbol{x}^* \in X$ with*

$$f(\boldsymbol{x}^*) = \min_{x \in X} f(\boldsymbol{x}). \tag{5.4}$$

Choose $p \geq 1$ starting point(s) $\boldsymbol{x}^1, \ldots, \boldsymbol{x}^p \in X$. For all $j = 1, \ldots, p-1$ compute $f(\boldsymbol{x}^j)$ and one (arbitrary) subgradient $\boldsymbol{g}^j \in \partial f(\boldsymbol{x}^j)$. Put $k := p$.

1. Compute $f(\boldsymbol{x}^k)$ and one (arbitrary) subgradient $\boldsymbol{g}^k \in \partial f(\boldsymbol{x}^k)$.

2. Solve the problem

$$\min_{x \in \mathbb{R}^n, v \in \mathbb{R}} v \tag{5.5}$$
$$\text{s.t.} \quad v \geq f(\boldsymbol{x}^k) + \boldsymbol{g}^{jT}(\boldsymbol{x} - \boldsymbol{x}^k) \qquad \text{for all } j = 1, \ldots, k,$$
$$-R \leq x_i \leq R \qquad \text{for all } i = 1, \ldots, n,$$

and obtain an optimum denoted by $(\boldsymbol{x}^{k+1}, v^{k+1}) \in \mathbb{R}^n \times \mathbb{R}$.

3. If $f(\boldsymbol{x}^{k+1}) = v^{k+1}$ then stop.
   Otherwise put $k := k + 1$, and go to Step 1.

Algorithm 5.5 is illustrated in Fig. 9: In (a) we see function f to be minimized. We choose $\boldsymbol{x}^1$ and $\boldsymbol{x}^2$, and corresponding subgradients $\boldsymbol{g}^1, \boldsymbol{g}^2$ give us the linearizations of f indicated by dashed lines in (b). Solving (5.5) is minimization of the piece-wise linear function shown in (c). We obtain the point $\boldsymbol{x}^3$. New subgradient $\boldsymbol{g}^3$ at $\boldsymbol{x}^3$ (cf. (d)) improves the linear model from (c) to (e), and solving (5.5) results in $\boldsymbol{x}^4$. And so on...

Theorem 5.6 *Let f be convex. Alg. 5.5 is well-defined since each of the subproblems (5.5) possesses a solution. If Alg. 5.5 terminates at Step 3, then $\boldsymbol{x}^{k+1}$ is optimal for problem (5.4), i.e.,*

$$f(\boldsymbol{x}^{k+1}) = \min_{x \in X} f(\boldsymbol{x}) =: f^*. \tag{5.6}$$

Otherwise, the sequence $(v^k)_{k \in \mathbb{N}}$ is converging monotonically increasing to f^*, and also $(f(\boldsymbol{x}^k))_{k \in \mathbb{N}}$ converges to f^*. Moreover, each limit point $\bar{\boldsymbol{x}}$ of some subsequence $(\boldsymbol{x}^{k_j})_{j \in \mathbb{N}}$ of $(\boldsymbol{x}^k)_{k \in \mathbb{N}}$ satisfies $f(\bar{\boldsymbol{x}}) = f^*$, i.e., is optimal for problem (5.4).

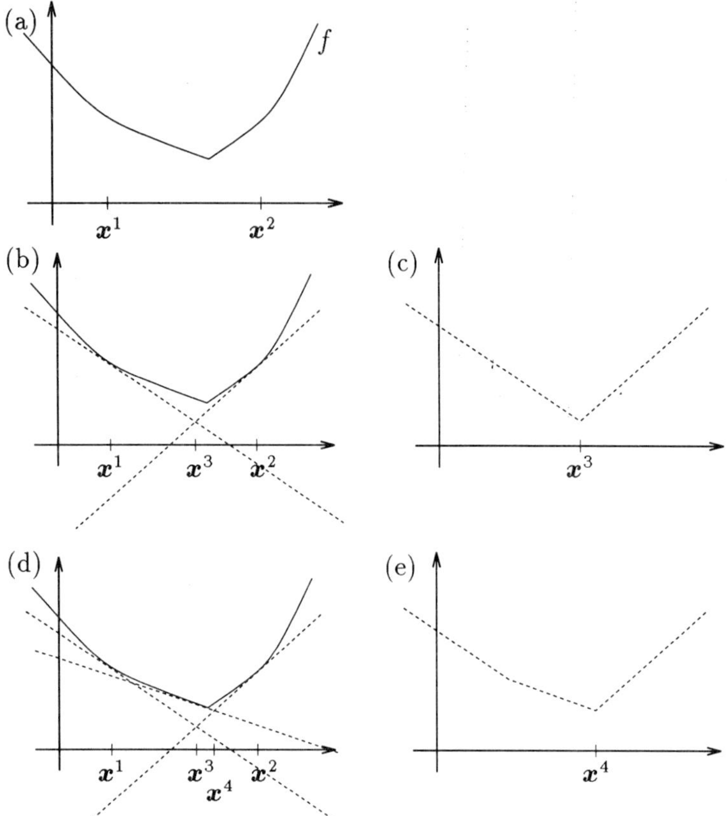

Figure 9 Illustration of cutting-plane-algorithm

The proof of this theorem is left as an exercise. It uses only standard tools from analysis (or, see, [15, 61]).

Note that subproblem (5.5) is a linear programming problem in $(\boldsymbol{x}, v)$ for which effective solvers exist. However, for growing iteration index k, (5.5) possesses more and more constraints. Thus, in practice, an active-set-strategy must be used to circumvent this problem. Another disadvantage is that the numerical condition of the subproblem becomes worse and worse due to "similarity" of subgradients.

However, Alg. 5.5 provides a conceptual basis which is realized in many methods of nonsmooth optimization, like, e.g., so-called *bundle-methods*. In simple words these techniques can be described as follows (cf. [60, 61]): Collect subgradients at several iteration points. This set of subgradients (called "bundle") is used to build up a simple (e.g. piece-wise linear) model of the original function that is minimized. Then this model is treated instead of the original function (either in a primal or in a dual approach). Mostly, an additional strategy assures that only points are considered where the model "is similar" to the original function. This may be, e.g., a trust-region-concept [55] (primal approach) or a strategy which controls the choice of the subgradients [61] (dual approach). At a solution point of this sub-optimization problem again a subgradient is evaluated. This subgradient is then added to the bundle improving the model of the function, and so on.

We mention that there exist also other strategies for minimizing nonsmooth functions like, e.g., so-called *subgradient methods* which mimic smooth steepest-descent- or quasi-Newton methods [61].

For simplicity, in the above we have restricted ourselves to the convex case. We mention that all definitions from above can be generalized to the case where f is locally Lipschitz-continuous: A function $f : \mathbb{R}^n \longrightarrow \mathbb{R}$ is called *locally Lipschitz-continuous ("llc")* if for each closed and bounded set $X \subset \mathbb{R}^n$ there exists some number $L_X > 0$ such that

$$|f(\boldsymbol{x}) - f(\boldsymbol{x}')| \leq L_X \cdot \|\boldsymbol{x} - \boldsymbol{x}'\| \quad \text{for all } \boldsymbol{x}, \boldsymbol{x}' \in X \tag{5.7}$$

where $\| \, . \, \|$ denotes (e.g.) the euclidean norm.

We immediately see that each llc-function is continuous. It can be proved [42, Th. 10.4] that each convex function is llc. For llc-functions the subdifferential is defined by the set

$$\partial f(\boldsymbol{x}) := \text{conv}\{\boldsymbol{g} \in \mathbb{R}^n \mid \lim_{i \to \infty} \boldsymbol{x}^i = \boldsymbol{x}, \; \nabla f(\boldsymbol{x}^i) \text{ exists,} \\ \nabla f(\boldsymbol{x}^i) \text{ converges, } \boldsymbol{g} = \lim_{i \to \infty} \nabla f(\boldsymbol{x}^i)\}. \tag{5.8}$$

It can be shown [59, 61] that for convex functions Def. 5.1 and definition (5.8) coincide (and thus the notation is well-defined). Moreover, Prop. 5.2 also holds for llc f. These facts already indicate that the concept of subgradients helps also in the more general llc case. Algorithmic concepts of minimization algorithms can be carried over from the convex to the general llc situation. However, if convexity does not hold, then it can only be expected that *locally* optimal points can be computed in numerical procedures. As an example for a bundle algorithm which works for llc-functions, and which guarantees to terminate at a local optimum, we refer to [55].

Acknowledgements

All lectures of the author in the CISM course have been coordinated with M. P. Bendsøe, Techn. Univ. of Denmark, and J. Zowe, Univ. of Erlangen-Nuremberg, Germany. The author wants to thank them for providing material and ideas. Moreover, the author is deeply indebted to M. Kočvara, Univ. of Erlangen-Nuremberg, for his kind permission of using some of his material, and for his careful help in erasing logical, typographical, and linguistic errors. Thanks also go to J. Logo, Techn. Univ. of Budapest, for his hints on misleading expressions.

References

[1] Dorn, W., Gomory, R. and Greenberg, M.: Automatic Design of Optimal Structures, J. de Mechanique, 3 (1996), 25–52.

[2] Ringertz, U.: A Branch and Bound Algorithm for Topology Optimization of Truss Structures, Eng. Opt., 10 (1986), 111–124.

[3] Lewiński, T., Zhou, M. and Rozvany, G.I.N.: Extended Exact Solutions for Least-Weight Truss Layouts – Part I and Part II, Int. J. Mech. Sci., 36 (1994), 375–419.

[4] Taylor, J.E.: Truss Topology Design for Elastic/Softening Materials, in: loc. cit. [62], 451–467.

[5] Pedersen, P.: Topology Optimization of Three-dimensional Trusses, in: loc. cit. [62], 19–30.

[6] Topping, B.H.V.: Topology Design of Discrete Structures, in: loc. cit. [62], 517–534.

[7] Zhou, M. and Rozvany, G.I.N.: Iterative COC methods, in: loc. cit. [63], 27–75.

[8] Ben-Tal, A. and Bendsøe, M.P.: A New Method for Optimal Truss Topology Design, SIAM J. Opt., 3 (1993), 322–358.

[9] Bendsøe, M.P.: Methods for Optimization of Structural Topology, Shape and Material, Springer, Berlin, New York 1995.

[10] Kirsch, U.: Optimal Topologies of Truss Structures, Comp. Meth. Appl. Mech. Eng., 72 (1989), 15–28.

[11] Kirsch, U.: Fundamental Properties of Optimal Topologies, in: loc. cit. [62], 3–18.

[12] Rozvany, G.I.N., Bendsøe, M.P. and Kirsch, U.: Layout Optimization of Structures, Appl. Mech. Rev., Vol. 48, No. 2 (1995).

[13] Haug, E.J. and Arora, J.S.: Applied Optimal Design, Wiley & Sons, New York 1979.

[14] Smith, O. da Silva: Generation of 3D-Ground Structures for Truss Topology Optimization, in: loc. cit. [64], 147–152.

[15] Bazaraa, M.S., Sherali, H.D. and Shetty, C.M.: Nonlinear Programming, 2nd Edition, Wiley & Sons, New York 1993.

[16] Svanberg, K.: On Local and Global Minima in Structural Optimization, in: New Directions in Optimum Structural Design (Eds. A. Atrek, R.H. Gallagher, K.M. Ragsdell O.C. Zienkiewicz), Wiley & Sons, New York, 1984, 327–341.

[17] Fleury, C.: Sequential Convex Programming for Structural Optimization Problems, in: loc. cit. [63], 531–553.

[18] Nguyen, V.H., Strodiot, J.J. and Fleury, C.: A Mathematical Convergence Analysis of the Convex Linearization Method for Engineering Design Problems, Eng. Opt., 11 (1987), 195–216.

[19] Fleury, C.: CONLIN: An Efficient Dual Optimizer Based on Convex Approximation Concepts, Struct. Opt., 1 (1989), 81–89.

[20] Svanberg, K.: Method of Moving Asymptotes – A New Method for Structural Optimization, Int. J. Num. Meth. Eng., 24 (1987), 359–373.

[21] Svanberg, K.: A Globally Convergent Version of MMA Without Linesearch, in: loc. cit. [64], 9–16.

[22] Achtziger, W., Bendsøe, M., Ben-Tal, A. and Zowe, J.: Equivalent Displacement Based Formulations for Maximum Strength Truss Topology Design, Impact of Computing in Science and Engineering, 4 (1992), 315–345.

[23] Svanberg, K.: Optimal Truss Sizing Based on Explicit Taylor Series Expansion, Struct. Opt., 2 (1990), 153–162.

[24] Svanberg, K.: Global Convergence of the Stress Ratio Method for Truss Sizing, Struct. Opt., 8 (1994), 60–68.

[25] Taylor, J.E.: Maximum Strength Elastic Structural Design, Proc. ASCE, 95 (1969), 653–663.

[26] Taylor, J.E. and Rossow, M.P.: Optimal Truss Design Based on an Algorithm Using Optimality Criteria, Int. J. Solids Struct., 13 (1977), 913–923.

[27] Cox, H.L.: The Design of Structures for Least Weight, Pergamon Press, Oxford 1965.

[28] Hemp, W.S.: Optimum Structures, Clarendon Press, Oxford, U.K 1973.

[29] Achtziger, W.: Optimierung von einfach und mehrfach belasteten Stabwerken, Bayreuther Mathematische Schriften, 46 (1993), in German.

[30] Oberndorfer, J.M., Achtziger, W. and Hörnlein, H.R.E.M.: Two Approaches for Truss Topology Optimization: A Comparison for Practical Use, Struct. Opt., 11 (1996), 137–144.

[31] Sankaranaryanan, S., Haftka, R. and Kapania, R.K.: Truss Topology Optimization with Stress and Displacement Constraints, in: loc. cit. [62], 71–78.

[32] Cheng, G. and Jiang, Z.: Study on Topology Optimization with Stress Constraints, Eng. Opt., 20 (1992), 129–148.

[33] Ringertz, U.: Newton Methods for Structural Optimization, Technical Report, No. 88-19 (1988), Department of Leightweight Structures, The Royal Institute of Technology, Stockholm, Sweden.

[34] Hörnlein, H.R.E.M.: Ein Algorithmus zur Strukturoptimierung von Fachwerkkonstruktionen, Diplomarbeit, Ludwigs-Maximilian-Universität, München, Germany 1979, in German.

[35] Pedersen, P.: On the Minimum Mass Layout of Trusses, Symposium on Structural Optimization, AGARD-CP-36-70 (1970), AGARD Conf. Proc.

[36] Smith, O. da Silva: Topology Optimization of Trusses with Local Stability Constraints and Multiple Loading Conditions, Technical Report, No. 517 (1996), Danish Center for Applied Mathematics and Mechanics (DCAMM), Technical University of Denmark, Lyngby, Denmark.

[37] Achtziger, W.: Truss Topology Optimization Including Bar Properties Different for Tension and Compression, Struct. Opt., 12 (1996), 63–74.

[38] Petersson, J. and Klarbring, A.: Saddle Point Approach to Stiffness Optimization of Discrete Structures Including Unilateral Contact, Control and Cybernetics, 3 (1994), 461–479.

[39] Klarbring, A., Petersson, J. and Rönnquist, M.: Truss Topology Optimization Involving Unilateral Contact, J. Opt. Theory Appl., 87(1) (1995).

[40] Kočvara, M., Zibulevsky, M. and Zowe, J.: Mechanical Design Problems with Unilateral Contact, Technical Report, No. 190 (1996), Institute of Applied Mathematics, University of Erlangen-Nuremberg, Germany, submitted.

[41] Wilkinson, J.H.: The Algebraic Eigenvalue Problem, Clarendon Press, Oxford 1965.

[42] Rockafellar, R.T.: Convex Analysis, Princeton University Press, Princeton, N.J. 1970.

[43] Achtziger, W.: Multiple Load Truss Topology and Sizing Optimization: Some Properties of Minimax Compliance, Technical Report 1994-41 (1994), Institute of Mathematics, The Technical University of Denmark, submitted.

[44] Jarre, F., Kočvara, M. and Zowe, J.: Interior Point Methods for Mechanical Design Problems, Technical Report, No. 173 (1996), Institute of Applied Mathematics, University of Erlangen-Nuremberg, Germany, submitted.

[45] Zibulevsky, M. and Ben-Tal, A.: On a New Class of Augmented Lagrangian Methods for Large Scale Convex Programming Problems, Technical Report, 2/93 (1993), Opt. Lab., Technion (Israel Inst. of Technology), Haifa, Israel.

[46] Ben-Tal, A. and Nemirovskii, A.: Potential Reduction Polynomial Time Method for Truss Topology Design, SIAM J. Opt., 4 (1994), 596–612.

[47] Ben-Tal, A. and Roth, G.: A Truncated log Barrier Algorithm for Large-Scale Convex Programming and Minmax Problems: Implementation and Computational Results, Optim. Meth. Software, 6 (1996), 283–312.

[48] Murty, K.G.: Linear Programming, Wiley & Sons, New York 1983.

[49] Levy, R.: Fixed Point Theory and Structural Optimization, Eng. Opt., 17 (1991), 251–261.

[50] Maxwell, J.C.: On Reciprocal Figures, Frames and Diagrams of Forces, Scientific Papers, 2 (1890), Cambridge Univ. Press, Cambridge, U.K., 175–177.

[51] Michell, A.G.M.: The Limits of Economy of Material in Frame Structures, Philosophical Magazine, Series 6, Vol. 8 (1904), 589–597.

[52] Achtziger, W.: Minimax Compliance Truss Topology Subject to Multiple Loadings, in: loc. cit. [62], 43–54.

[53] Achtziger, W.: Multiple Load Truss Optimization: Properties of Minimax Compliance and Two Nonsmooth Approaches, in: loc. cit. [64], 123–128.

[54] Gauvin, J. and Dubeau, F.: Differential Properties of the Marginal Function in Mathematical Programming, Math. Prog. Study, 19 (1982), 101–119.

[55] Schramm, H. and Zowe, J.: A Version of the Bundle Idea for Minimizing a Nonsmooth Function: Conceptual Idea, Convergence Analysis, Numerical Results, SIAM J. Opt., Vol. 2 (1992), 121–152.

[56] Ben-Tal, A., Kočvara, M. and Zowe, J.: Two Nonsmooth Methods for Simultaneous Geometry and Topology Design of Trusses, in: loc. cit. [62], 31–42.

[57] Kočvara, M. and Zowe, J.: How To Optimize Mechanical Structures Simultaneously with Respect to Topology and Geometry, in: loc. cit. [64], 135–140.

[58] Kočvara, M. and Zowe, J.: How Mathematics Can Help in Design of Mechanical Structures, in: Numerical Analysis (Eds. D. Griffiths and G. Watson), Longman Scientific and Technical, 1996, 76–93.

[59] Clarke, F.H.: Optimization and Nonsmooth Analysis, Wiley & Sons, New York 1983.

[60] Zowe, J.: Nondifferentiable Optimization, in: Computational Mathematical Programming (Ed. K. Schittkowski), Springer, Berlin, 1985, 323–359.

[61] Lemaréchal, C.: Nondifferentiable Optimization, in: Handbooks in Operations Research and Management Science (Eds. G.L. Nemhauser, A.H.G. Rinnooy Kan and M.J. Todd), Vol. 1, Elsevier Science Publishers, North-Holland, 1989, 529–572.

[62] Bendsøe, M.P. and Mota Soares, C.A. (Eds.): Topology Optimization of Structures, Kluwer Academic Publishers, Dordrecht 1993.

[63] Rozvany, G.I.N. (Ed.): Optimization of Large Structural Systems, NATO ASI Series, Kluwer Academic Publishers, Dordrecht, The Netherlands 1993.

[64] Olhoff, N. and Rozvany, G.I.N. (Eds.): WCSMO-1, First World Congress of Structural and Multidisciplinary Optimization, Pergamon, Elsevier, Oxford, U.K. 1995.

THE HOMOGENIZATION METHOD FOR TOPOLOGY AND SHAPE OPTIMIZATION

G. Allaire

DRN/ DMT/ SERMA, CEA Saclay, Gif sur Yvette, France

and

University of Paris 6, Paris, France

Abstract

This paper is devoted to an elementary introduction to the homogenization theory and its application to topology and shape optimization of elastic structures. It starts with a brief survey of periodic homogenization, H- or G-convergence, and the mathematical modeling of composite materials. Then, these notions are used for minimum compliance and weight design of elastic structures in two or three space dimension. Theoretical, as well as numerical, aspects of the homogenization method are investigated.

1 Introduction.

This series of five lectures is devoted to a presentation of the so-called homogenization method for topology and shape optimization of elastic structures. Since homogenization is a vast field which is probably not so well-known by people interested in structural optimization, we spend quite a lot of time in introducing it, along with its application to the modeling of composite materials, before we get to its use in shape optimization problems. Each lecture corresponds to a given section below.

The two first lectures are devoted to a brief introduction to the mathematical theory of homogenization. Roughly speaking, homogenization is a rigorous version of what is known as averaging. In other words, homogenization extracts homogeneous effective parameters from disordered or heterogeneous media. For a more advanced presentation of homogenization (which is a huge and fascinating field since its beginning in the early seventies), the reader is referred to the classical books [12], [17], [21], [62], and [55].

The third lecture is concerned with the mathematical modeling of composite materials by using the homogenization theory. In short, the effective properties of a composite material are defined as the homogenized coefficients of a fine mixture of given phases. Since very often the precise microgeometry of such mixtures is unknown, except possibly for the volume fractions of each phase, an important problem is to estimate these effective properties by bounding them independently of the underlying microscopic arrangement. It turns out that optimal bounds and optimal composites are of paramount importance for the sequel.

Finally, the fourth and fifth lectures describe the application of homogenization to the shape and topology optimization of elastic structures. The main idea is to replace the difficult "layout" problem of material distribution by a much easier "sizing" problem for the density and effective properties of a perforated composite material obtained by cutting small holes in the original homogeneous material. We focus on both the theoretical aspects (the so-called relaxation process) and the numerical aspects of the homogenization method.

2 Introduction to the homogenization theory: the periodic case.

Homogenization has first been developed for periodic structures. Indeed, in many fields of science and technology one has to solve boundary value problems in periodic media. Quite often the size of the period is small compared to the size of a sample of the medium, and, denoting by ϵ their ratio, an asymptotic analysis, as ϵ goes to zero, is called for. Starting from a microscopic description of a problem, we seek a macroscopic, or effective, description. This process of making an asymptotic analysis and seeking an averaged formulation is called homogenization. This first lecture will focus on the homogenization of periodic structures. The method of two-scale asymptotic expansions is presented, and its mathematical justification will be briefly discussed.

2.1 Setting of the problem.

We consider a model problem of linear elasticity in a periodic medium (for example, a composite material where stiff fibers are included in a matrix of a more compliant phase). To fix ideas, the periodic domain is called Ω (a bounded open set in $\mathbb{R}^N$ with $N \geq 1$), its period ϵ (a positive number which is assumed to be very small in comparison with the size of the domain), and the rescaled unit cell $Y = (0,1)^N$ (see Figure 1). The Hooke's law in Ω is not constant but varies periodically with period ϵ in each direction. It is a fourth order tensor $A(y)$, where $y = x/\epsilon \in Y$ is the fast periodic variable, while $x \in \Omega$ is the slow variable. Equivalently, x is also called the macroscopic variable, and y the microscopic variable. If the Hooke's law A is isotropic, with positive bulk and shear moduli κ and μ, we have

$$A(y) = (\kappa(y) - \frac{2\mu(y)}{N})I_2 \otimes I_2 + 2\mu(y)I_4,$$

but, A can be any fourth order tensor satisfying the usual symmetries of linear elasticity, $A_{ijkl} = A_{klij} = A_{jikl} = A_{ijlk}$ for $1 \leq i,j,k,l \leq N$, and the usual coerciveness assumption, i.e. there exist a positive constant $\alpha > 0$ such that, for any constant symmetric matrix ξ with entries $\xi_{ij} \in \mathbb{R}^N$ and at any point $y \in Y$,

$$\alpha|\xi|^2 \leq \sum_{i,j,k,l=1}^{N} A_{ijkl}(y)\xi_{ij}\xi_{kl}.$$

The tensor $A(y)$ is a periodic function of y, with period Y, and it may be discontinuous in y (to model the discontinuity of the elastic properties from one phase to the other).

Denoting by $f(x)$ the external load (a vector function defined in Ω), and enforcing a Dirichlet boundary condition (for simplicity), our model problem of elasticity reads

$$\begin{cases} -\text{div}\ \left(A\left(\frac{x}{\epsilon}\right)e(u_\epsilon)\right) = f & \text{in } \Omega \\ u_\epsilon = 0 & \text{on } \partial\Omega, \end{cases} \tag{1}$$

where u_ϵ is the unknown displacement, and $e(u_\epsilon)$ is the deformation tensor $1/2(\nabla u_\epsilon +^t \nabla u_\epsilon)$.

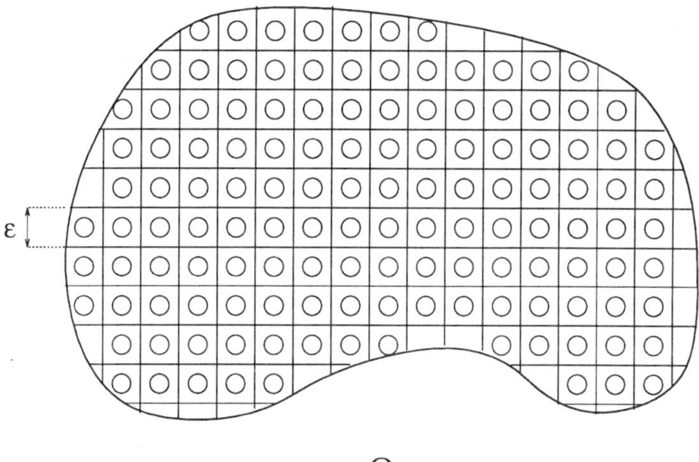

Figure 1: Periodic domain Ω.

Remark 2.1 *From a mathematical point of view, problem (1) is well posed in the sense that, if the force $f(x)$ belongs to the space $L^2(\Omega)^N$ of square integrable vectors on Ω, then the Lax-Milgram lemma implies existence and uniqueness of the solution u_ϵ in the Sobolev space $H_0^1(\Omega)^N$ of vectors functions which belong to $L^2(\Omega)^N$ along with their first derivatives. More smoothness can be obtained for the solution u_ϵ, but it is not necessary in the sequel.*

From a mechanical point of view, the domain Ω with its material properties $A\left(\frac{x}{\epsilon}\right)$ is highly heterogeneous with periodic heterogeneities of lengthsize ϵ. Usually one does not need the full details of the elastic displacement, but rather some global of averaged behavior of the domain Ω submitted to the force f. From a numerical point of view, solving equation (1) by any method will require too much efforts if ϵ is small since the number of elements (or degrees of freedom) for a fixed level of accuracy grows like $1/\epsilon^N$. It is thus preferable to average or homogenize the properties of Ω and compute an approximation of u_ϵ on a coarse mesh. This process of averaging the displacement solution of (1) and finding the effective properties of the domain Ω is precisely what is called homogenization.

In order to find this homogenized or averaged behavior of Ω, we perform an asymptotic analysis of equation (1) as the period ϵ goes to 0. The solution u_ϵ is written as a series in power of ϵ

$$u_\epsilon = \sum_{i=0}^{+\infty} \epsilon^i u_i.$$

The first term u_0 of this series will be identified with the solution of the so-called homogenized equation whose Hooke's law A^* can be exactly computed. Note that A^* is a constant tensor describing a homogeneous medium. Numerical computations on the homogenized equation does not require a fine mesh since the heterogeneities of size ϵ have been averaged out. This homogenized tensor A^* is usually not a usual average (arithmetic or harmonic) of $A(y)$... Various estimates will confirm this asymptotic analysis by telling in which sense u_ϵ is close to u_0 as ϵ is small.

Remark 2.2 *From a mathematical point of view, homogenization can be interpreted as follows. Rather than studying equation (1) for the precise value ϵ which is the physical lengthscale of the heterogeneities, we embed this problem in a sequence of problems indexed by the period ϵ which is now regarded as a small parameter going to zero. Therefore, we now consider the sequence of solutions $(u_\epsilon)_{\epsilon>0}$ as a sequence in the Sobolev space $H_0^1(\Omega)^N$.*

Multiplying equation (1) by u_ϵ, integrating by parts, and using Poincaré inequality (i.e. the L^2 norm of u_ϵ is bounded by that of ∇u_ϵ) yields the following a priori estimate

$$\|u_\epsilon\|_{H_0^1(\Omega)^N} \leq C\|f\|_{L^2(\Omega)^N} \tag{2}$$

where C is a positive constant which does not depend on ϵ. It implies that the sequence u_ϵ is bounded in $H_0^1(\Omega)^N$. Therefore, up to extracting a subsequence, it converges weakly in $H_0^1(\Omega)^N$ to some limit u (and by Rellich' theorem it converges also strongly to u in $L^2(\Omega)^N$). Finally, the homogenization process amounts to find what is the equation satisfied by u. Of course, u will turn out to coincide with u_0, the first term in the series defined above, and it is therefore the solution of the homogenized equation.

2.2   Two-scale asymptotic expansions.

The method of two-scale asymptotic expansions is a formal method which allows to homogenize a great variety of models or equations posed in a periodic domain. There are many good reference books on this topic, see in particular [12], [17], and [55]. As already said, the starting point is to consider the following ansatz, or *two-scale asymptotic expansion*, for the solution u_ϵ of equation (1)

$$u_\epsilon(x) = \sum_{i=0}^{+\infty} \epsilon^i u_i\left(x, \frac{x}{\epsilon}\right), \tag{3}$$

where each term $u_i(x, y)$ is a function of both variables x and y, periodic in y with period $Y = (0, 1)^N$. This series is plugged in the equation, and the following derivation rule is used

$$\nabla\left(u_i\left(x, \frac{x}{\epsilon}\right)\right) = \left(\epsilon^{-1}\nabla_y u_i + \nabla_x u_i\right)\left(x, \frac{x}{\epsilon}\right),$$

where ∇_y and ∇_x denote the partial derivative with respect to the first and second variable of $u_i(x, y)$. For example, one has

$$\nabla u_\epsilon(x) = \epsilon^{-1}\nabla_y u_0\left(x, \frac{x}{\epsilon}\right) + \sum_{i=0}^{+\infty} \epsilon^i \left(\nabla_y u_{i+1} + \nabla_x u_i\right)\left(x, \frac{x}{\epsilon}\right).$$

Equation (1) becomes a series in ϵ

$$-\epsilon^{-2}\left[\mathrm{div}_y\left(Ae_y(u_0)\right)\right]\left(x, \tfrac{x}{\epsilon}\right)$$

$$-\epsilon^{-1}\left[\mathrm{div}_y\left(A(e_x(u_0) + e_y(u_1))\right) + \mathrm{div}_x\left(Ae_y(u_0)\right)\right]\left(x, \tfrac{x}{\epsilon}\right) \tag{4}$$

$$-\sum_{i=0}^{+\infty} \epsilon^i \left[\mathrm{div}_x\left(A(e_x(u_i) + e_y(u_{i+1}))\right) + \mathrm{div}_y\left(A(e_x(u_{i+1}) + e_y(u_{i+2}))\right)\right]\left(x, \tfrac{x}{\epsilon}\right) \quad = f(x).$$

Identifying each coefficient of (4) as an individual equation yields a cascade of equations (a series of the variable ϵ is zero for all value of ϵ if each coefficient is zero). It turns out that the three first equations are enough for our purpose. The ϵ^{-2} equation is

$$-\mathrm{div}_y\ (A(y)e_y(u_0(x,y))) = 0,$$

which is nothing else than an equation in the unit cell Y with periodic boundary condition. In this equation, y is the variable, and x plays the role of a parameter. There exists a unique solution of this equation up to a constant (with respect to y). This implies that u_0 is a function which *does not depend on y*, i.e. there exists a function $u(x)$ which depends only on x such that

$$u_0(x,y) \equiv u(x).$$

Since $e_y(u_0) = 0$, the ϵ^{-1} equation is

$$- \mathrm{div}_y\ (A(y)e_y(u_1(x,y))) = \mathrm{div}_y\ (A(y)e_x(u(x))), \tag{5}$$

which is an equation for the unknown u_1 in the periodic unit cell Y. Again, it is a well-posed problem wich admits a unique solution, up to a constant, as soon as the right hand side is known. Equation (5) allows to compute u_1 in terms of u, and it is easily seen that $u_1(x,y)$ depends linearly on the first derivative $\nabla_x u(x)$.

Finally, the ϵ^0 equation is

$$- \mathrm{div}_y\ (A(y)e_y(u_2(x,y))) = \mathrm{div}_y\ (A(y)e_x(u_1)) + \mathrm{div}_x\ (A(y)(e_y(u_1) + e_x(u))) + f(x), \tag{6}$$

which is an equation for the unknown u_2 in the periodic unit cell Y. Equation (6) admits a solution (unique up to a constant) if a compatibility condition is satisfied (the so-called Fredholm alternative). Indeed, integrating the left hand side of (6) over Y, and using the periodic boundary condition for u_2, we obtain a zero mean value for equation (6). This implies that the right hand side *must have zero average over Y*

$$\int_Y [\mathrm{div}_y\ (A(y)e_x(u_1)) + \mathrm{div}_x\ (A(y)(e_y(u_1) + e_x(u))) + f(x)]\, dy = 0,$$

which simplifies in

$$- \mathrm{div}_x \int_Y A(y)\,(e_y(u_1) + e_x(u))\, dy\ =\ f(x)\ \text{ in } \Omega. \tag{7}$$

Since $u_1(x,y)$ depends linearly on $\nabla_x u(x)$, equation (7) is simply an equation for $u(x)$ involving only the second order derivatives of u. The next section is devoted to an interpretation of this equation as the homogenized equation.

Remark 2.3 *This method of two-scale asymptotic expansions is mathematically formal. In other words, it yields heuristically the homogenized equation, but it is not a rigorous proof of the homogenization process. The reason is that the ansatz (3) is not always correct. Indeed, it is usually false after the two first terms. For example, it does not include possible boundary layers in the vicinity of $\partial\Omega$ (for details, see e.g. [39]).*

2.3   The cell and the homogenized problems.

The method of two-scale asymptotic expansions give rise to a couple of equations (5) (7) that have a mathematical, as well as physical, interpretation. Let us introduce the so-called *cell problems*. We denote by $(e_i)_{1\le i\le N}$ the canonical basis of $\mathbb{R}^N$. Defining

$$e_{ij} = \frac{1}{2}\,(e_i \otimes e_j + e_j \otimes e_i),$$

we obtain a basis of the space of symmetric matrices. For each matrix e_{ij}, consider the following elasticity problem in the periodic unit cell

$$\begin{cases} -\text{div}_y \ (A(y)(e_{ij} + e_y(w_{ij}(y)))) = 0 & \text{in } Y \\ y \to w_{ij}(y) & Y\text{-periodic,} \end{cases} \tag{8}$$

where $w_{ij}(y)$ is the displacement created by a mean deformation equal to e_{ij}.

By linearity, it is not difficult to compute $u_1(x, y)$, solution of (5), in terms of $u(x)$ and $w_{ij}(y)$

$$u_1(x, y) = \sum_{i=1}^{N} \sum_{j=1}^{N} \left(\frac{\partial u_i}{\partial x_j}(x) + \frac{\partial u_j}{\partial x_i}(x) \right) w_{ij}(y). \tag{9}$$

Inserting this expression in equation (7), we obtain the *homogenized equation* for u

$$- \text{div}_x \ (A^* e_x(u(x))) = f(x) \text{ in } \Omega, \tag{10}$$

that we supplement with Dirichlet boundary condition on $\partial\Omega$. The homogenized Hooke's law A^* is defined by its entries

$$A^*_{ijkl} = \int_Y A(y)(e_{ij} + e(w_{ij})) \cdot (e_{kl} + e(w_{kl})) \, dy, \tag{11}$$

or equivalently, by a simple integration by parts in Y,

$$A^*_{ijkl} = \int_Y (A(y)e_y(w_{ij})_{kl} + A_{ijkl}(y)) \, dy.$$

The constant tensor A^* describes the effective or homogenized properties of the heterogeneous material $A\left(\frac{x}{\epsilon}\right)$. Remark that it does not depend on the choice of domain Ω, force f, or boundary condition on $\partial\Omega$.

Remark 2.4 *From a mathematical point of view, it is possible to rigorously justify the fact that equation (10) is the homogenized equation associated to the original equation (1). For example, in [12], [17], and [55], it is proved that the sequence $u_\epsilon(x)$ of solutions of (1) converges weakly in $H_0^1(\Omega)^N$ to a limit $u(x)$ which is the unique solution of the homogenized problem 10) . This justifies the first term in the ansatz of u_ϵ, but one can justify the two first terms and improve the weak convergence of u_ϵ. More precisely, $(u_\epsilon(x) - u(x) - \epsilon u_1 (x, \frac{x}{\epsilon}))$ converges strongly to 0 in $H_0^1(\Omega)^N$ (one can even estimate the speed of convergence in terms of ϵ).*

Technically, the most general and powerful method for proving such an homogenization theorem is the so-called energy method of Tartar [59]. Another method (which is simpler but restricted to the periodic setting) is the two-scale convergence introduced by Nguetseng [47] and Allaire [2].

2.4 A variational characterization of the homogenized coefficients.

The homogenized Hook's law A^* is defined in terms of the solutions of the cell problems by equation (11). It is convenient to give another definition of A^* involving standard variational principles. Recall that both the original and the homogenized Hooke's law $A(y)$ and A^* are symmetric fourth order tensors. Therefore, A^* is completely determined by the knowledge of the quadratic form $A^*\xi \cdot \xi$ where ξ is any symmetric matrix. From definition (11) it is not difficult to check that

$$A^*\xi \cdot \xi = \int_Y A(y)(\xi + e(w_\xi)) \cdot (\xi + e(w_\xi)) \, dy, \tag{12}$$

where w_ξ is the solution of

$$\begin{cases} -\text{div}_y \left(A(y) \left(\xi + e_y(w_\xi(y)) \right) \right) = 0 & \text{in } Y \\ y \to w_\xi(y) & Y\text{-periodic.} \end{cases} \qquad (13)$$

It is well-known that equation (13) is the Euler-Lagrange equation of the following variational principle : find w which minimizes

$$\int_Y A(y) \left(\xi + e(w) \right) \cdot \left(\xi + e(w) \right) dy$$

over all periodic displacement fields w. In other words, $A^* \xi \cdot \xi$ is given by the minimization of the elastic energy (primal formulation)

$$A^* \xi \cdot \xi = \min_{w(y) \in H^1_\#(Y)^N} \int_Y A(y) \left(\xi + e(w) \right) \cdot \left(\xi + e(w) \right) dy$$

where $H^1_\#(Y)^N$ is the Sobolev space of periodic displacements in Y.

There is also a dual formulation for the elastic energy, in terms of stresses rather than deformations. For any symmetric matrix σ, this yields a definition of A^{*-1}

$$A^{*-1} \sigma \cdot \sigma = \min_{\substack{\tau(y) \in L^2_\#(Y)^{N^2} \\ \text{div } \tau = 0 \text{ and } \int_Y \tau dy = 0}} \int_Y A^{-1}(y) \left(\sigma + \tau(y) \right) \cdot \left(\sigma + \tau(y) \right) dy$$

where $L^2_\#(Y)^{N^2}$ is the space of periodic square integrable admissible stresses in Y.

3 Introduction to the homogenization theory: the general case.

This second lecture focus on the general setting of homogenization when no geometric assumptions are available (like periodicity, or ergodicity in a probabilistic framework) . It turns out that homogenization can be applied to any kind of disordered media, and is definitely not restricted to the periodic case (although the nice "explicit" formulae of the periodic setting for the homogenized Hooke's law have no analogue). We introduce the notion of G- or H-convergence which is due to DeGiorgi and Spagnolo [25], [56], [57], and has been further generalized by Murat and Tartar [45], [46], [59] (see also the textbooks [48], [62]). It allows to consider any possible geometrical situation without any specific assumptions like periodicity or randomness. The G- or H-convergence turns out to be the adequate notion of convergence for effective properties that will be the key tool in the study of optimal shape design problems.

Eventually, let us mention that there is also a stochastic theory of homogenization (see [38], [22], [50]) and a variational theory of homogenization (the Γ-convergence of De Giorgi, [23], [24], see also the book [21]) that will not be described below.

3.1 Definition of G-, or H-convergence.

The G-convergence is a notion of convergence associated to sequences of symmetric operators (typically, these operators are applications giving the solution of a partial differential equation in terms of the right hand side). The G means Green since this type of convergence corresponds roughly to the convergence of the associated Green functions. The H-convergence is a generalization of the G-convergence to the case of non-symmetric operators (it provides also an easier mathematical framework, but we shall not dwell on that). The H stands for Homogenization since it is an important tool of that theory. Since the elasticity equations have symmetric coefficients, the underlying operators are symmetric too. Therefore in the sequel, we use only the notation G-convergence for simplicity.

The main result of the G-convergence is a compactness theorem in the homogenization theory which states that, for any bounded and uniformly coercive sequence of coefficients of a symmetric second order elliptic equation, there exist a subsequence and a G-limit (i.e. homogenized coefficients) such that, for any source term, the corresponding subsequence of solutions converges to the solution of the homogenized equation. In practical terms, it means that the mechanical properties of an heterogeneous medium (like its conductivity, or elastic moduli) can be well approximated by the properties of a homogeneous or homogenized medium if the size of the heterogeneities are small compared to the overall size of the medium.

The G-convergence can be seen as a mathematically rigorous version of the so-called *representative volume element* method for computing effective or averaged parameters of heterogeneous media.

We introduce the notion of G-convergence for the specific case of the elasticity system with a Dirichlet boundary condition, but all the results hold for a larger class of second order elliptic operators and boundary conditions. Let Ω be a bounded open set in $\mathbb{R}^N$, and let α, β be two positive constants such that $0 < \alpha \leq \beta$. We introduce the set $\mathcal{M}(\alpha, \beta, \Omega)$ of all possible symmetric fourth order tensors defined on Ω with uniform coercivity constant α and $L^\infty(\Omega)$-bound β. Furthermore, $A \in \mathcal{M}(\alpha, \beta, \Omega)$ satisfies the usual symmetries of a Hooke's law, $A_{ijkl} = A_{klij} = A_{jikl} = A_{ijlk}$ for $1 \leq i, j, k, l \leq N$. The coerciveness and boundedness of $A(x)$ is equivalent to

$$\alpha|\xi|^2 \leq \sum_{i,j,k,l=1}^{N} A_{ijkl}(x)\xi_{ij}\xi_{kl} \leq \beta|\xi|^2.$$

We consider a sequence $A_\epsilon(x)$ of Hooke's law in $\mathcal{M}(\alpha, \beta, \Omega)$, indexed by a sequence of positive numbers ϵ going to 0. Here, ϵ is not associated to any specific lengthscale or statistical property of the elastic medium. In other words, no special assumptions (like periodicity or stationarity) are placed on the sequence A_ϵ.

For a given force $f(x) \in L^2(\Omega)^N$, there exists a unique displacement u_ϵ, solution in the Sobolev space $H_0^1(\Omega)^N$ of the following elasticity equation

$$\begin{cases} -\text{div } (A_\epsilon(x)e(u_\epsilon)) = f(x) & \text{in } \Omega \\ u_\epsilon = 0 & \text{on } \partial\Omega, \end{cases} \tag{14}$$

where $e(u_\epsilon)$ is the deformation tensor $1/2(\nabla u_\epsilon +^t \nabla u_\epsilon)$. The G-convergence of the sequence A_ϵ is defined below as the convergence of the corresponding solutions u_ϵ.

Definition 3.1 *The sequence of tensors $A_\epsilon(x)$ is said to G-converge to a limit $A^*(x)$, as ϵ goes to 0, if, for any force $f \in L^2(\Omega)^N$ in (14), the sequence of solutions u_ϵ converges weakly in $H_0^1(\Omega)$ to a limit u which is the unique solution of the homogenized equation associated to A^*:*

$$\begin{cases} -\text{div } (A^*(x)e(u)) = f(x) & in \ \Omega \\ u = 0 & on \ \partial\Omega. \end{cases} \tag{15}$$

Remark that, by definition, the homogenized tensor A^* is independent of the applied force f. We shall see that it is also independent of the boundary condition and of the domain.

This definition makes sense because of the compactness of the set $\mathcal{M}(\alpha, \beta, \Omega)$ with respect to the G-convergence, as stated in the following theorem.

Theorem 3.2 *For any sequence A_ϵ in $\mathcal{M}(\alpha, \beta, \Omega)$, there exist a subsequence (still denoted by ϵ) and a homogenized limit A^*, belonging to $\mathcal{M}(\alpha, \beta, \Omega)$, such that A_ϵ G-converges to A^*.*

The G-convergence of a general sequence A_ϵ is always stated *up to a subsequence* since A_ϵ can be the union of two sequences converging to two different limits. The G-convergence of A_ϵ is not equivalent to any other "classical" convergence. For example, if A_ϵ converges strongly in $L^\infty(\Omega)$ to a limit A (i.e. the convergence is pointwise), then its G-limit A^* coincides with A. But the converse is not true ! On the same token, the G-convergence has nothing to do with the usual weak convergence. Indeed, the G-limit A^* of a sequence A_ϵ is usually different of its weak-* $L^\infty(\Omega)$-limit. For example, a straightforward computation in one dimension ($N = 1$) shows that the G-limit of a sequence A_ϵ is given as the inverse of the weak-* $L^\infty(\Omega)$-limit of A_ϵ^{-1} (the so-called harmonic limit). However, this last result holds true only in 1-D, and no such explicit formula is available in higher dimensions.

The G-convergence enjoys a few useful properties as enumerated in the following proposition.

Proposition 3.3 *Properties of G-convergence.*

1. *If a sequence A_ϵ G-converges, its G-limit is unique.*

2. *Let A_ϵ and B_ϵ be two sequences which G-converge to A^* and B^* respectively. Let $\omega \subset \Omega$ be a subset strictly included in Ω such that $A_\epsilon = B_\epsilon$ in ω. Then $A^* = B^*$ in ω (this property is called the locality of G-convergence).*

3. *The G-limit of a sequence A_ϵ is independent of the source term f and of the boundary condition on $\partial\Omega$.*

4. *Let A_ϵ be a sequence which G-converges to A^*. Then, the associated density of energy $A_\epsilon e(u_\epsilon) \cdot e(u_\epsilon)$ also converges to the homogenized density of energy $A^* e(u) \cdot e(u)$ in the sense of distributions in Ω.*

5. *If a sequence A_ϵ G-converges to a limit A^*, then the sequence of stresses $A_\epsilon e(u_\epsilon)$ converges weakly in $L^2(\Omega)^{N^2}$ to the homogenized stress $A^* e(u)$.*

These properties of the G-convergence implies that the homogenized medium A^* approximates the heterogeneous medium A_ϵ in many different ways. First of all, by definition of G-convergence, the displacements and the deformation tensors are closed (this is the sense of the convergence of u_ϵ to u in the Sobolev space $H_0^1(\Omega)^N$). Then, by application of the above proposition, the stresses and the energy densities are also closed.

Remark also that, by locality of the G-convergence, the homogenized tensor is defined at each point of the domain Ω independently of what may happen in other regions of Ω.

Of course, a particular example of G-convergent sequences A_ϵ is given by periodic media of the type $A\left(\frac{x}{\epsilon}\right)$ as in the previous section.

3.2   The G-closure problem.

A crucial problem in the theory of G-convergence is the following. Assume that we constrain the sequence of Hooke's law $A_\epsilon(x)$ to take only two values A and B, corresponding to two given elastic materials, at each point $x \in \Omega$. Then, what is the possible set of G-limits $A^*(x)$ that can be attained by such sequences ? In other words, what kind of composite materials can be achieved by mixing A and B in any possible manner ? This problem is known as *the G-closure problem*. A variant of this problem is to fix the overall volume fraction of A and B in Ω to be θ and $1 - \theta$, with $0 \leq \theta \leq 1$. This problem is called the G_θ-closure problem.

Surprisingly enough, these two questions have not yet been solved in the context of elasticity (even if A and B are assumed to be isotropic), although the answer is known in the context of conductivity (see [46], [60] , [40]). However, there are a few useful results in this direction that indicates that there is no loss of generality in considering only periodic composite materials (i.e. homogenized tensors A^* obtained by periodic homogenization). The following result is due to Dal maso and Kohn.

Theorem 3.4 *Let $A^*(x)$ be any homogenized Hooke's law (or composite material) obtained by mixing the two constituents A and B in proportions $\theta(x)$ and $(1 - \theta(x))$ in a domain Ω (in other words $A^*(x)$ is any element of the so-called G-closure). Then, at each point $x \in \Omega$, the proportion $\theta(x)$ can take any value between 0 and 1, while $A^*(x)$ can take any value in the set $G_{\theta(x)}$, defined as the set of all periodic homogenized Hooke's law (i.e., obtained by homogenization of a periodic mixture of A and B in proportions $\theta(x)$ and $(1 - \theta(x)))$.*

This theorem states that any homogenized material A^* (obtained with any possible mixing procedure) can locally be attained by a periodic mixture in the same proportions. This feature is, of course, not so surprising, since we already claimed that the G-convergence is local.

Although we do not have an explicit characterization of the set G_θ for $0 \leq \theta \leq 1$, we know part of the boundary of this set. In other words we know some "optimal composites" which leave on ∂G_θ and are therefore extremal. For example, if A^* is isotropic, the celebrated Hashin-Shtrikman bounds hold for its bulk and shear moduli [32]. These bounds are known to be attained by optimal composites which are sequential laminates (see [27] and the next lecture for details).

4   Homogenization and composite materials.

This lecture is devoted to a brief introduction to the theory of composite materials in the context of homogenization. For more details, the reader is referred to [20], [32] for a mechanical point of view, and to [27], [36], [43] for a mathematical point of view.

We begin by introducing a particular family of composite materials, known as finite-rank sequential laminates, that constitute a class of optimal composites having effective properties explicitly given by an algebraic formula. Then, we recall the well-known Hashin-Shtrikman variational principle that permits to obtain a lower bound on the complementary elastic energy of a composite material. It turns out that such a bound is also optimal, i.e. it is realizable, or attained, by a finite-rank sequential laminate. In other words, we exhibit such a laminate which is the strongest possible composite material under a specific load. The derivation of the corresponding explicit formulae is the focus of the last section.

4.1 The lamination formula.

As already said, the G_θ-closure is unknown, but its subset of so-called "finite rank sequential laminates", denoted by L_θ, is both, easy to compute, and rich enough for many applications. Let us first define what are finite rank sequential laminates.

There exists a nice explicit formula for sequential laminates which is due to Francfort and Murat [27] (inspired by some earlier work of Tartar [60]). This formula is purely algebraic, which is in sharp contrast with the periodic formula for which a p.d.e. has to be solved in the unit cell. Throughout this lecture, we assume that both materials A and B are isotropic ; for any symmetric matrix ξ,

$$A\xi = 2\mu_A\xi + \lambda_A(tr\xi)I_2, \quad B\xi = 2\mu_B\xi + \lambda_B(tr\xi)I_2,$$

where I_2 is the identity matrix, and $(\mu_{A,B}, \lambda_{A,B})$ are the Lamé coefficients of the materials. The shear modulus $\mu_{A,B}$ is always positive as well as the bulk modulus $\kappa_{A,B} = \lambda_{A,B} + 2\mu_{A,B}/N$. We assume that B is weaker than A

$$\mu_B < \mu_A, \quad \kappa_B < \kappa_A.$$

We now give the lamination formula in terms of A^{-1} and B^{-1} (since we shall use it for stresses), but a similar formula holds in terms of A and B (if one wants to use it with deformations).

Proposition 4.1 *Let A^* be a rank-p sequential laminate of material A around a core of material B, in proportion θ and $(1-\theta)$ respectively, with lamination directions $(e_i)_{1\leq i\leq p}$ and lamination parameters $(m_i)_{1\leq i\leq p}$ satisfying $0 \leq m_i \leq 1$ and $\sum_{i=1}^p m_i = 1$ (these parameters are related to the volume fractions of material A at each step of the lamination process). Then*

$$(1-\theta)\left(A^{*-1} - A^{-1}\right)^{-1} = \left(B^{-1} - A^{-1}\right)^{-1} + \theta\sum_{i=1}^p m_i f_A^c(e_i) \tag{16}$$

where $f_A^c(e_i)$ is a fourth order tensor defined, for any symmetric matrix ξ, by the quadratic form

$$f_A^c(e_i)\xi \cdot \xi = A\xi \cdot \xi - \frac{1}{\mu_A}|A\xi e_i|^2 + \frac{\mu_A + \lambda_A}{\mu_A(2\mu_A + \lambda_A)}((A\xi)e_i \cdot e_i)^2,$$

where (μ_A, λ_A) are the Lamé coefficients of A.

A proof of the lamination formula (16) would parallel that of Proposition 4.2 in [27] using complementary energy instead of primal energy (see also [36], [43]). Note that $f_A^c(e_i)$ is a degenerate Hooke's law in the sense that it is a non negative, semi definite, fourth order tensor.

Let us briefly indicate the main steps of the proof of Proposition 4.1. It starts with an explicit computation of the homogenized Hooke's law of a special periodic composite obtained by a single lamination of A and B. Recall from the first lecture that A^{*-1} is given by

$$A^{*-1}\sigma \cdot \sigma = \min_{\substack{\tau(y)\in L^2_\#(Y)^{N^2} \\ \text{div } \tau(y)=0, \ \int_Y \tau(y)dy=0}} \int_Y (\chi(y)A + (1-\chi(y))B)^{-1}(\sigma + \tau(y)) \cdot (\sigma + \tau(y))\, dy \tag{17}$$

where $L^2_\#(Y)^{N^2}$ is the space of periodic square integrable admissible stresses in Y, and $\chi(y)$ is the characteristic function of phase A. In the case where $\chi(y)$ corresponds to a layer of A and B, the computation of (17) amounts to a simple algebraic calculation since the deformation and stress tensors turn out to be constant in each phase. The result is (see Theorem 4.1 in [27] for details)

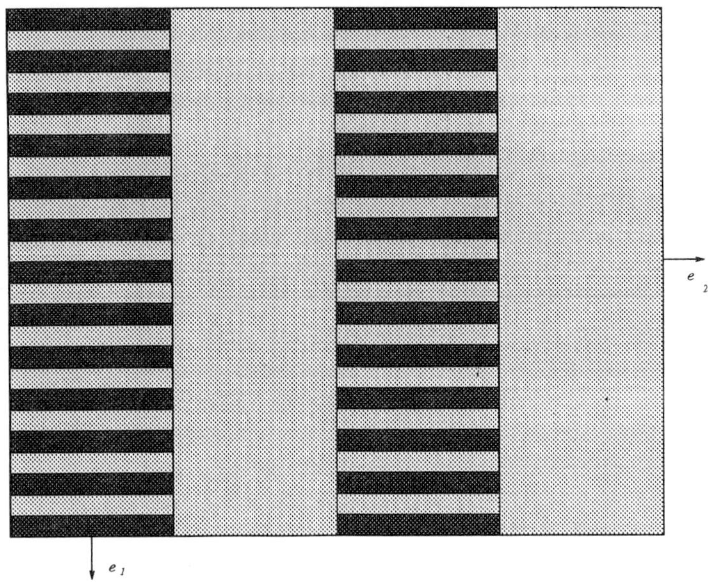

Figure 2: A rank-2 sequential laminate.

Lemma 4.2 *Let A^* be a single lamination of material A and B, in proportion θ and $(1 - \theta)$ respectively, in the direction e. Then*

$$(1 - \theta)\left(A^{*-1} - A^{-1}\right)^{-1} = \left(B^{-1} - A^{-1}\right)^{-1} + \theta f_A^c(e) \tag{18}$$

where $f_A^c(e)$ is defined as in Proposition 4.1.

Formula (18) is written in a strange manner (why not a formula giving directly A^{*-1} ?), but it allows to reiterate the lamination process. Indeed, it is possible to laminate this first composite materials, that we denote by A_1^*, with A again to obtain a new composite denoted by A_2^*. By induction, we obtain A_p^* by lamination of A and A_{p-1}^*, in proportions θ_p $(1 - \theta_p)$, and in the direction e_p

$$(1 - \theta_p)\left(A_p^{*-1} - A^{-1}\right)^{-1} = \left(A_{p-1}^{*-1} - A^{-1}\right)^{-1} + f_A^c(e_p).$$

Replacing $\left(A_{p-1}^{*-1} - A^{-1}\right)^{-1}$ in the above formula by its formula, and so on up to $A_0^* \equiv B$, yields a formula equivalent to (16). The overall volume fraction of material B is

$$1 - \theta = \prod_{i=1}^{p}(1 - \theta_i),$$

which is a definition of θ, and the parameters m_i and θ_i are easily seen to be related by a one-to one map. Remark that we always laminate an intermediate composite with material A. This implies that A plays the role of a matrix phase and B that of a core phase. Of course, A and B can be interchanged, but this yields a new sequential laminate. Globally, A_p^* can be seen as a mixture of A and B in different layers having a large separation of scales (see Figure 2).

4.2 The Hashin-Shtrikman variational principle.

The Hashin-Shtrikman variational principle [32] is a method for bounding effective properties of composite materials. It starts from the usual definition of A^{*-1}

$$A^{*-1}\sigma \cdot \sigma = \min_{\mathrm{div}\,\tau(y)=0,\ \int_Y \tau(y)dy=0} \int_Y (\chi(y)A + (1-\chi(y))B)^{-1} (\sigma + \tau(y)) \cdot (\sigma + \tau(y))\, dy$$

where $\chi(y)$ is the characteristic function of phase A. Since this definition involves periodic functions on the unit cell, it is tempting to use Fourier analysis to evaluate it. This is indeed the main idea behind the Hashin-Shtrikman variational principle. We use this method to obtain a so-called *lower bound on the effective complementary energy.*

Proposition 4.3 *Let A^* be a composite material obtained by any possible mixing of A and B in proportions θ and $(1-\theta)$. Assume that B is a weaker material than A. Then, for any stress σ,*

$$A^{*-1}\sigma \cdot \sigma \geq HS(\sigma) = A^{-1}\sigma \cdot \sigma + \tag{19}$$

$$(1-\theta)\max_{\eta} \left(2\sigma \cdot \eta - (B^{-1} - A^{-1})^{-1} \eta \cdot \eta - \theta g(\eta) \right),$$

where η is any symmetric matrix, and $g(\eta)$ is a non-local term defined by

$$g(\eta) = \max_{|k|=1} f_A^c(k)\eta \cdot \eta,$$

where k is any unit vector, and f_A^c is defined as in Proposition 4.1.

Formula (19) seems at first a little mysterious since it involves two maximization process over constant matrices η and vectors k. However it is simpler than the exact definition of A^* since there is no partial differential equation to solve. Indeed, $HS(\sigma)$ can even be evaluated explicitly in some cases (see the last section). Anyway, the precise expression of the lower bound (19) is irrelevant since the only information about it that we shall use in the sequel concerns its attainability by finite-rank sequential laminates (see the next section).

Let us briefly indicate how Proposition 4.3 can be established. By adding and subtracting a reference energy $A^{-1}\sigma \cdot \sigma$, we obtain

$$A^{*-1}\sigma \cdot \sigma = \min_{\substack{\mathrm{div}\,\tau(y)=0 \\ \int_Y \tau(y)dy=0}} \left\{ \int_Y (1-\chi(y)) (B^{-1} - A^{-1}) (\sigma + \tau(y)) \cdot (\sigma + \tau(y))\, dy \right.$$

$$\left. + \int_Y A^{-1} (\sigma + \tau(y)) \cdot (\sigma + \tau(y))\, dy \right\}. \tag{20}$$

Using the positivity of $B^{-1} - A^{-1}$ and convex duality, the first integral in the right hand side is rewritten

$$\sup_{\eta(y)} \int_Y (1-\chi(y)) \left(2\eta(y) \cdot (\sigma + \tau(y)) - (B^{-1} - A^{-1})^{-1} \eta(y) \cdot \eta(y) \right) dy.$$

One can get a lower bound by specializing to constant tensors η

$$\sup_{\eta(y)} \geq 2(1-\theta)\eta \cdot \sigma - (1-\theta)(B^{-1} - A^{-1})^{-1} \eta \cdot \eta + \int_Y 2(1-\chi(y))\eta \cdot \tau(y)\, dy. \tag{21}$$

Substitution in (20) yields after some simplification

$$A^{*-1}\sigma \cdot \sigma \geq A^{-1}\sigma \cdot \sigma + 2(1-\theta)\sigma \cdot \eta - (1-\theta)\left(B^{-1} - A^{-1}\right)^{-1}\eta \cdot \eta \tag{22}$$

$$+ \inf_{\tau(y)} \int_Y \left(A^{-1}\left(\sigma + \tau(y)\right) \cdot \left(\sigma + \tau(y)\right) + 2(1-\chi(y))\eta \cdot \tau(y)\right)\, dy.$$

The above infimum in τ is easily computed by Fourier analysis (see e.g. Proposition 2.1 in [8]). Denoting by $\hat{\chi}(k)$ the Fourier component at frequency k of the characteristic function $\chi(y)$, it is exactly equal to

$$- \sum_{k \neq 0} \mid \hat{\chi}(k) \mid^2 f_A^c\left(\frac{k}{|k|}\right)\eta \cdot \eta. \tag{23}$$

Remarking that

$$\sum_{k \neq 0} \mid \hat{\chi}(k) \mid^2 = \theta(1-\theta),$$

the quantity (23) is bounded from below by $-\theta(1-\theta)g(\eta)$. Varying η among all constant symmetric matrices gives the desired bound.

4.3   Energy bounds.

The Hashin-Shtrikman variational principle provides upper and lower bounds for quadratic elastic energies involving A^*. It turns out that these bounds are optimal, i.e. they are realizable for a specific microstructure depending on the matrix σ. Such microstructures, or special choices of A^*, which attain the value of the bound, are called "optimal". As we shall see, optimality can always be achieved in the class of sequential laminates. Formula (19) is therefore called an optimal lower bound on the effective complementary energy. In other words, it is the elastic energy at the stress σ, of the most rigid possible composite of density θ, which turns out to be chosen as a sequential laminate of rank N.

For the special case of elastic energies as in (19), it is proved in [10] that the minimum over G_θ is actually attained within the subset L_θ of finite rank sequential laminates.

Theorem 4.4 *Whenever B is a weaker material than A, the optimal bound (19) is given by*

$$HS(\sigma) = \min_{A^* \in G_\theta} A^{*-1}\sigma \cdot \sigma = \min_{A^* \in L_\theta} A^{*-1}\sigma \cdot \sigma, \tag{24}$$

where L_θ is the set of all effective Hooke's law of finite rank sequential laminates defined through (16). Furthermore, optimality in the right hand side of (24) is achieved by a rank-N sequential laminate (in space dimension N) with lamination directions coinciding with the eigendirections of the symmetric matrix σ.

As already mentioned, the first part of Theorem 4.4 may be found [10], while the second part is to be found in [8]. For details, the reader is referred to e.g. formulae (6.11), (6.18), and (7.6) and Remark 3.7 in [8].

We can compute the energy bounds when $B = 0$ (i.e. the weak material is degenerate and becomes a hole). For simplicity, from now on we denote by (μ, λ) the Lamé coefficients of material A. As is well known, most materials have a non negative Poisson ratio, i.e., $\lambda \geq 0$. Since this last hypothesis greatly simplifies the computations (at least in the three dimensional setting), we shall assume henceforth that

$$\mu > 0, \lambda \geq 0.$$

The computation of the optimal lower bound amounts to a simple optimization of the lamination parameters m_i of a rank-N sequential laminate, while the lamination directions e_i are kept fixed and equal to the eigendirections of σ. However, the lamination formula (16) yields A^{*-1} at the price of a non trivial inversion of a sum of degenerate Hooke's law. Inverting this sum in full generality is a difficult task. In any case we need only to address the class of so-called *orthogonal* rank-N sequential laminates which, by definition, admit an orthonormal basis of $\mathbb{R}^N$ as lamination directions $(e_i)_{1 \leq i \leq N}$. The following result was proved in [4].

Lemma 4.5 *The inverse Hooke's law A^{*-1} of an orthogonal rank-N sequential laminate is given by the following quadratic form*

$$A^{*-1}\sigma \cdot \sigma = A^{-1}\sigma \cdot \sigma + \frac{1-\theta}{2\mu\theta}G(\alpha_i, \sigma) \tag{25}$$

with

$$G(\alpha_i, \sigma) = \sum_{i,j=1, i \neq j}^{N} \frac{\sigma_{ij}^2}{(1-m_i-m_j)} + \sum_{i=1}^{N} \alpha_i \sigma_{ii}^2$$
$$- \frac{\lambda}{2\mu+N\lambda}\left(\sum_{i=1}^{N} \sigma_{ii}\right)^2 + \frac{\lambda}{2\mu+N\lambda}\frac{\left(\sum_{i=1}^{N}(\alpha_i-1)\sigma_{ii}\right)^2}{(1-\frac{\lambda}{2\mu+N\lambda}\sum_{i=1}^{N}\alpha_i)},$$

where σ_{ij} denotes the entries of a symmetric matrix σ in the orthonormal basis of lamination directions, and the parameters $(\alpha_i)_{1 \leq i \leq N}$ are defined by

$$\alpha_i = \left(1 - \frac{2\mu m_i}{2\mu+\lambda}\right)^{-1}.$$

Remark 4.6 *The quadratic form (25) defines a coercive Hooke's law A^* in dimension $N \geq 3$ as soon as none of the parameters m_i are zero, that is whenever all lamination directions have been put to use. (Indeed, $m_i > 0$ for $1 \leq i \leq N$ implies that $1-m_i-m_j > 0$ for $1 \leq i,j \leq N$ and $i \neq j$.) Thus, in three dimensions, an orthogonal rank-3 laminate is a realistic composite material. On the contrary, in two dimensions, we always have $1 - m_i - m_j = 0$! Thus, formula (25) only holds for stresses σ which are diagonal in the basis of lamination directions (i.e., such that $\sigma_{ij} = 0$). In other words, in 2-D, an orthogonal rank-2 laminate cannot support a stress whose eigendirections are not aligned with the lamination directions.*

Thanks to the above lemma a simple minimization in m_i yields (see [9] for the 2-D case, and [4], [29] for the 3-D case) :

$$A^{*-1}\sigma \cdot \sigma = A^{-1}\sigma \cdot \sigma + \frac{1-\theta}{\theta}g^*(\sigma),$$

where

$$g^*(\sigma) = \frac{1}{2\mu}\min_{m_i} G(\alpha_i, \sigma).$$

Proposition 4.7 *In two dimensions,*

$$g^*(\sigma) = \frac{\lambda}{4\mu(\mu+\lambda)}\left(|\sigma_1| + |\sigma_2|\right)^2 \tag{26}$$

where σ_1 and σ_2 are the eigenvalues of the stress σ (a two by two symmetric matrix in 2-D). Furthermore, the associated optimal rank-2 sequential laminate is characterized by its parameters

$$m_1 = \frac{|\sigma_2|}{|\sigma_1| + |\sigma_2|}, m_2 = \frac{|\sigma_1|}{|\sigma_1| + |\sigma_2|}. \tag{27}$$

Proposition 4.8 *In three dimensions, if $\sigma_1 \leq \sigma_2 \leq \sigma_3$ are the eigenvalues of σ, then*

1. in the case where $0 \leq \sigma_1 \leq \sigma_2 \leq \sigma_3$

$$g^*(\sigma) = \frac{2\mu + \lambda}{4\mu(2\mu + 3\lambda)} (\sigma_1 + \sigma_2 + \sigma_3)^2 \ \ if \ \sigma_3 \leq \sigma_1 + \sigma_2 \tag{28}$$

$$g^*(\sigma) = \frac{1}{2\mu} \left((\sigma_1 + \sigma_2)^2 + \sigma_3^2\right) - \frac{\lambda}{2\mu(2\mu + 3\lambda)} (\sigma_1 + \sigma_2 + \sigma_3)^2 \ \ if \ \sigma_3 \geq \sigma_1 + \sigma_2 \tag{29}$$

2. in the case where $\sigma_1 \leq 0 \leq \sigma_2 \leq \sigma_3$

$$g^*(\sigma) = \frac{2\mu + \lambda}{4\mu(2\mu + 3\lambda)} \left(\sigma_3 + \sigma_2 - \frac{\mu + 2\lambda}{\mu + \lambda}\sigma_1\right)^2 \ \ if \ \begin{cases} \sigma_3 + \sigma_2 \geq -\frac{\mu}{\mu+\lambda}\sigma_1 \\ \sigma_3 - \sigma_2 \leq -\frac{\mu}{\mu+\lambda}\sigma_1 \end{cases} \tag{30}$$

$$g^*(\sigma) = \frac{1}{2\mu} \left((\sigma_3 + \sigma_2)^2 + \sigma_1^2\right) - \frac{\lambda}{2\mu(2\mu + 3\lambda)} (\sigma_1 + \sigma_2 + \sigma_3)^2 \ \ if \ \sigma_3 + \sigma_2 \leq -\frac{\mu}{\mu+\lambda}\sigma_1 \tag{31}$$

$$g^*(\sigma) = \frac{1}{2\mu} \left(\sigma_1^2 + \sigma_2^2 + \sigma_3^2\right) - \frac{2\mu}{2\mu(\mu + \lambda)}\sigma_1\sigma_2 - \frac{\lambda}{2\mu(2\mu + 3\lambda)} (\sigma_1 + \sigma_2 + \sigma_3)^2 \ \ if \ \sigma_3 - \sigma_2 \geq -\frac{\mu}{\mu+\lambda}\sigma_1 \tag{32}$$

3. the remaining cases are obtained from (1) and (2) by symmetry, changing σ into $-\sigma$.

Furthermore, optimality in the regime (28) is achieved by a rank-3 sequential laminate with parameters

$$m_1 = \frac{\sigma_3 + \sigma_2 - \sigma_1}{\sigma_1 + \sigma_2 + \sigma_3}, m_2 = \frac{\sigma_1 - \sigma_2 + \sigma_3}{\sigma_1 + \sigma_2 + \sigma_3}, m_3 = \frac{\sigma_1 + \sigma_2 - \sigma_3}{\sigma_1 + \sigma_2 + \sigma_3}; \tag{33}$$

in the regime (29) it is achieved by a rank-2 sequential laminate with parameters

$$m_1 = \frac{\sigma_2}{\sigma_1 + \sigma_2}, m_2 = \frac{\sigma_1}{\sigma_1 + \sigma_2}, m_3 = 0; \tag{34}$$

in the regime (30) it is achieved by a rank-3 sequential laminate with parameters

$$m_1 = \frac{\sigma_3 + \sigma_2 + \frac{\mu}{\mu+\lambda}\sigma_1}{\sigma_3 + \sigma_2 - \frac{\mu+2\lambda}{\mu+\lambda}\sigma_1}, m_2 = \frac{\mu + \lambda}{\mu} \frac{\sigma_3 - \sigma_2 - \frac{\mu}{\mu+\lambda}\sigma_1}{\sigma_3 + \sigma_2 - \frac{\mu+2\lambda}{\mu+\lambda}\sigma_1}, \tag{35}$$

$$m_3 = -\frac{\mu + \lambda}{\mu} \frac{\sigma_3 - \sigma_2 + \frac{\mu}{\mu+\lambda}\sigma_1}{\sigma_3 + \sigma_2 - \frac{\mu+2\lambda}{\mu+\lambda}\sigma_1}; \tag{36}$$

in the regime (31) it is achieved by a rank-2 sequential laminate with parameters

$$m_1 = 0, m_2 = \frac{\sigma_3}{\sigma_2 + \sigma_3}, m_3 = \frac{\sigma_2}{\sigma_2 + \sigma_3}, \tag{37}$$

in the regime (32) it is achieved by a rank-2 sequential laminate with parameters

$$m_1 = \frac{\sigma_2}{\sigma_2 - \sigma_1}, m_2 = \frac{-\sigma_1}{\sigma_2 - \sigma_1}, m_3 = 0. \tag{38}$$

5 Application of homogenization to shape optimization: theoretical aspects.

The typical problem of structural optimization is to find the "best" structure which is, at the same time, of minimal weight and of maximum strength. Of course there is some subjectivity in the definition of what is "best". It depends on many different considerations : what is the underlying mechanical model (linear or non-linear elasticity, plasticity, etc.) ? Are there any constraints on admissible shapes (industrial feasibility, smoothness of the boundary, etc.) ? What kind of stiffness criterion is used (maximum stress, compliance, etc.) ?

Since the focus of this lecture is to discuss the homogenization method for structural optimization, some assumptions are required for the definition of a suitable model problem. First of all, we deliberately forget about any feasibility or smoothness constraints on the shape's boundary. Indeed, the process of homogenization (or relaxation) is intimately linked to the possibility of boundary oscillations (small ribs or holes), which are usually prevented by adding the above type of constraints. Then, to complete as far as possible our analysis, we work in the context of linear elasticity, and we choose the compliance (i.e. the work done by the load) as a global measure of rigidity. Finally, for simplicity we consider a single loading configuration in two or three space dimensions.

5.1 A model problem in shape optimization.

We consider a bounded reference domain $\Omega \in \mathbb{R}^N$ ($N = 2, 3$ is the spatial dimension), occupied by a linearly elastic material with isotropic Hooke's law A (with bulk and shear moduli κ and μ) defined by

$$A = (\kappa - \frac{2\mu}{N})I_2 \otimes I_2 + 2\mu I_4, \quad 0 < \kappa, \mu < +\infty, \tag{39}$$

The domain Ω is subject to surface loadings f on its boundary $\partial\Omega$, and equilibrium of the domain is assumed, i.e.

$$\int_{\partial\Omega} f \, ds = 0.$$

An admissible design ω is a subset of the reference domain Ω obtained by removing one or more holes (the new boundaries created this way are traction-free). The equations of elasticity for the resulting structure are

$$\begin{cases} \sigma = Ae(u) & e(u) = \frac{1}{2}(\nabla u + \nabla^t u) \\ \text{div } \sigma = 0 & \text{in } \omega \\ \sigma \cdot n = f & \text{on } \partial\Omega \\ \sigma \cdot n = 0 & \text{on } \partial\omega \setminus \partial\Omega. \end{cases} \tag{40}$$

The compliance of the design ω is

$$c(\omega) = \int_{\partial\Omega} f \cdot u = \int_{\omega} < Ae(u), e(u) > = \int_{\omega} A^{-1}\sigma \cdot \sigma. \tag{41}$$

Introducing a positive Lagrange multiplier ℓ, the goal is to minimize, over all subsets $\omega \subset \Omega$, the weighted sum $E(\omega)$ of the compliance and the weight (proportional to the volume $|\omega|$), namely to compute

$$\inf_{\omega \subset \Omega} \left(E(\omega) = c(\omega) + \ell|\omega| \right). \tag{42}$$

The Lagrange multiplier ℓ has the effect of balancing the two contradictory objectives of rigidity and lightness of the optimal structure (increasing its value decreases the weight). There exists a different formulation of the same structural optimization problem which will be very helpful in the sequel. It is based on the principle of complementary energy which gives the value of the compliance

$$c(\omega) = \int_{\partial\Omega} f \cdot u = \min_{\substack{\mathrm{div}\ \tau=0\,\mathrm{in}\ \omega \\ \tau\cdot n=f\,\mathrm{on}\ \partial\Omega \\ \tau\cdot n=0\,\mathrm{on}\ \partial\omega\backslash\partial\Omega}} \int_{\omega} A^{-1}\tau \cdot \tau.$$

Extending the admissible stress τ by 0 inside the holes, the compliance is also defined by

$$c(\omega) = \min_{\substack{\mathrm{div}\ \tau=0\ \mathrm{in}\ \Omega \\ \tau\cdot n=f\ \mathrm{on}\ \partial\Omega}} \int_{\Omega} (\chi_\omega(x)A)^{-1}\tau \cdot \tau, \tag{43}$$

where χ_ω is the characteristic function of the design ω. The infimum over designs and the minimum over statically admissible stresses can be switched. Then, for a fixed stress, the inside minimization over $\chi_\omega = 0, 1$ is easy. It yields that (42) is equivalent to

$$\inf_{\substack{div\tau=0\ \mathrm{in}\ \Omega \\ \tau\cdot n=f\ \mathrm{on}\ \partial\Omega}} \left(F(\tau) = \int_\Omega \left\{ \begin{array}{ll} A^{-1}\tau \cdot \tau + \ell & \mathrm{if}\ \tau \neq 0 \\ 0 & \mathrm{if}\ \tau = 0 \end{array} \right) \right. \tag{44}$$

in the sense that minimizers of (42) and (44) (if any) are related by

$$\chi_\omega(x) = 0 \Leftrightarrow \sigma(x) = 0, \quad \chi_\omega(x) = 1 \Leftrightarrow \sigma(x) \neq 0. \tag{45}$$

As is well known in the mathematical community, in absence of any supplementary constraints on the admissible designs ω, the objective function $E(\omega)$ may have no minimizer, i.e. there is no optimal shape (for striking counter-examples on similar, but simpler, problems, we refer to [44], [46], and to [19] for numerical evidence). This can also be guessed from the other formulation (44) where the objective function $F(\tau)$ is obviously not convex and, as we shall see, not even lower semi-continuous (the correct mathematical notion for proving existence theorems). The physical reason for this non-existence is that it is often advantageous to cut infinitely many small holes (rather than just a few big ones) in a given design in order to decrease $E(\omega)$. Thus, achieving the minimum may require a limiting procedure leading to a "generalized" design consisting of composite materials made by microperforation.

To take into account this physical behavior of nearly optimal shapes, we have to enlarge the space of admissible designs by permitting perforated composites from the start : this process is called *homogenization* (or relaxation). Such a composite structure is determined by two functions $\theta(x)$, its local volume fraction of material taking values between 0 and 1, and $A(x)$, its effective Hooke's law corresponding to its microstructure. Of course, we need to find an adequate definition of the *homogenized* (or relaxed) objective function $\tilde{E}(\theta, A)$ which generalizes $E(\omega)$. This is done in the next section by using the theories of homogenization and optimal bounds on the effective properties of composite materials. The ultimate goal is twofold : prove an existence theorem for the relaxed formulation of the above structural optimization problem, and find a new numerical algorithm for computing optimal shapes.

For more details on the mathematical theory of relaxation by homogenization in the context of optimal design, we refer to the pioneering works [37], [41], [46]). In the specific framework of computational structural optimization, we refer e.g. to [3], [5], [7], [6], [9], [13], [14], [15], [16], [33], [34], [58].

5.2 Homogenized formulation.

In this section we describe the homogenization or relaxation process of the structural optimization problem (42) following the articles [9] and [6].

Let $(\omega_\epsilon)_{\epsilon \to 0}$ be a minimizing sequence of nearly optimal shapes for the objective function (42), and denote by χ_{ω_ϵ} their characteristic functions. In the reference domain Ω, we regard it as a fine mixture of the original material A and void (holes). Then, as a result of the homogenization theory (see the previous lectures), there exists an effective behavior of this fine mixture, i.e. a composite material of density $\theta(x)$, taking any value in the interval $[0, 1]$, and a Hooke's law $A(x)$ such that

$$\chi_{\omega_\epsilon}(x) \rightharpoonup \theta(x) \text{ weakly in } L^\infty(\Omega),$$

and

$$\chi_{\omega_\epsilon}(x)A \overset{G}{\to} A^*(x)$$

in the sense of H or $G-$convergence. In truth, the homogenization theory works only for composite materials made of two *non-degenerate* phases. Therefore, to be mathematically rigorous the holes are first filled with a weak material, then homogenization takes place, and, in the end, we have to justify the passing to the degenerate limit. For simplicity we skip these details here.

The above homogenization result implies in particular the convergence of the compliance

$$c(\omega_\epsilon) \to \tilde{c}(\theta, A^*) = \int_\Omega A^*(x)^{-1}\sigma \cdot \sigma \tag{46}$$

where the stress σ is now solution of the following homogenized equation

$$\begin{cases} \sigma = A^*(x)e(u) & e(u) = \frac{1}{2}(\nabla u + \nabla^t u) \\ \text{div } \sigma = 0 & \text{in } \Omega \\ \sigma \cdot n = f & \text{on } \partial\Omega. \end{cases} \tag{47}$$

For a same value θ of the density, there are many different possible effective Hooke's law A^* corresponding to different microstructures (or geometric patterns of the holes), i.e.

$$A^* \in G_\theta$$

where G_θ (the so-called G-closure set at volume fraction θ) is the set of all possible effective Hooke's law with material density θ.

Applying these results, we pass to the limit in the objective function and obtain the homogenized or relaxed functional

$$\lim_{\epsilon \to 0} E(\omega_\epsilon) = \inf_{\omega \subset \Omega} E(\omega) = \min_{\substack{A^* \in G_\theta \\ 0 \leq \theta \leq 1}} \tilde{E}(\theta, A^*),$$

where

$$\tilde{E}(\theta, A^*) = \tilde{c}(\theta, A^*) + \lambda \int_\Omega \theta(x). \tag{48}$$

This relaxed formulation is not entirely explicit since the precise definition of the G-closure set G_θ is unknown ! However, by using again the principle of complementary energy, we can use our knowledge of so-called *optimal bounds* on G_θ (see the previous lecture) that will simplify the relaxed formulation by optimizing the microstructure and eliminating the dependence on A^*. We rewrite the compliance as

$$\tilde{c}(\theta, A^*) = \min_{\substack{\text{div } \tau = 0 \text{ in } \Omega \\ \tau \cdot n = f \text{ on } \partial\Omega}} \int_\Omega A^*(x)^{-1}\tau \cdot \tau. \tag{49}$$

Then, switching the two minimizations and optimizing pointwise the microstructure and density, the relaxed formulation becomes

$$\min_{\substack{div\,\tau=0\,\text{in }\Omega \\ \tau\cdot n=f\,\text{on }\partial\Omega}} \left\{ QF(\tau) = \int_\Omega \min_{\substack{A\in G_\theta \\ 0\le\theta\le 1}} \left(A^{*-1}\tau\cdot\tau + \lambda\theta\right) \right\}. \tag{50}$$

For a fixed stress τ, the minimization of $A^{*-1}\tau\cdot\tau$ on G_θ is a classical problem in the theory of optimal bounds on effective properties of composite materials. It has been solved in 2-D in [8], and in 3-D in [4]. In two dimensions, the result is

$$\min_{A^*\in G_\theta} A^{*-1}\tau\cdot\tau = A^{-1}\tau\cdot\tau + \frac{(\kappa+\mu)(1-\theta)}{4\kappa\mu\theta}\left(|\tau_1|+|\tau_2|\right)^2 \tag{51}$$

where τ_1 and τ_2 are the eigenvalues of the 2 by 2 symmetric matrix τ. Furthermore, optimality in (51) is achieved for a so-called rank-2 sequential laminate aligned with the eigendirections of τ (see the previous lecture for details).

In three dimensions, the result is messier, and we give it in the special case of Poisson's ratio equal to zero, i.e. $3\kappa = 2\mu$ (the general case is not much different in essence, see [4])

$$\min_{A^*\in G_\theta} A^{*-1}\tau\cdot\tau = A^{-1}\tau\cdot\tau +$$

$$+ \begin{cases} \frac{(1-\theta)}{4\mu\theta}\left(|\tau_1|+|\tau_2|+|\tau_3|\right)^2 & \text{if } |\tau_3| \le |\tau_1|+|\tau_2| \\[2ex] \frac{(1-\theta)}{2\mu\theta}\left((|\tau_1|+|\tau_2|)^2+|\tau_3|^2\right) & \text{if } |\tau_3| \ge |\tau_1|+|\tau_2| \end{cases} \tag{52}$$

where the eigenvalues of τ are labeled in such a way that

$$|\tau_1| \le |\tau_2| \le |\tau_3|.$$

Furthermore, optimality in the first regime of (52) is achieved by a rank-3 sequential laminate aligned with the eigendirections of τ, while in the second regime it is achieved by a rank-2 sequential laminate aligned with the two first eigendirections of τ.

After this crucial step, the minimization over θ can easily be done by hand, which completes the explicit calculation of the relaxed formulation. The final result is the following

Theorem 5.1 *Problem (48), or equivalently (50), is the homogenized or relaxed formulation of the original problem (42), or (44), in the sense that there exists at least one "generalized" or homogenized optimal design (θ, A^*), which is the limit of a minimizing sequence of "classical" shapes ω_ϵ, and the minimal values of the original or homogenized energies coincide*

$$\inf_{\omega\subset\Omega} E(\omega) = \min_{\substack{A^*\in G_\theta \\ 0\le\theta\le 1}} \tilde{E}(\theta, A^*).$$

Its proof can be found in [9] for the 2-D case, and in [6] for the 3-D case. Remark that the homogenization process does not change the physics of the problem. Indeed, an homogenized optimal design is just a limit of nearly optimal classical designs, and the homogenized energy is precisely the average of the original energy when the nearly optimal classical designs oscillate (i.e., have many holes or ribs). In particular, any possible solution of the original problem is also a solution of the homogenized problem.

5.3 Comments and generalizations.

There is a wild non-uniqueness for the optimal homogenized designs. In the first place, the optimal microstructure is not always unique. For example, the optimal sequential laminate is not uniquely defined in the case of an hydrostatic stress τ proportional to the identity I_2 (any orthonormal basis of $\mathbb{R}^N$ is a set of eigenvectors of τ and thus a set of lamination directions). It can be checked that different directions lead to different homogenized Hooke's law. But there is another type of non-uniqueness of the microstructure : sequential laminates are not the only known class of optimal microstructures (although probably the easiest to work with). The so-called concentric spheres construction [31] (generalized in [60] to confocal ellipsoids), or the periodic arrangement of adequately shaped inclusions in [61] (see also [30]) are also optimal in specific situations.

Another possible non-uniqueness is that of the homogenized density. For example, in the case of a constant hydrostatic boundary condition $f = pI_2$, the homogenized stress is exactly equal to pI_2 and the average compliance is $\frac{Np^2}{\kappa^*}$ where κ^* is the so-called Hashin-Shtrikman upper bound on the bulk modulus [32]. A generalized optimal design is gievn as a composite material of constant density θ determined by the values of p and ℓ. But there are also an infinite number of classical optimal shapes obtained by the well-known concentric spheres constructions (see e.g. [20]). It amounts to cover the domain Ω by a dense packing of non-overlapping spheres of all sizes. Then, in each sphere, a concentric spherical hole is cut, and its radius is determined in a manner such that the volume fraction of material is precisely θ. This yields a perforated domain Ω with infinitely many disjoint spherical holes of all sizes. It is a classical result that, for such a perforated domain ω under the hydrostatic boundary condition $f = pI_2$, the exact compliance is equal to the homogenized compliance. Therefore, $E(\omega)$ being equal to the minimal relaxed energy, ω is also an optimal shape. Remark that this type of classical optimal shapes would be very difficult to compute numerically, partly because they are not homogenized or averaged. Indeed its boundary is very complex since it involves an infinite number of connected components on various length scales. Therefore, even in this case, the relaxed problem is more practical from a numerical standpoint.

An important feature of the optimal sequential laminates is that they are "smart" materials. The optimal microstructure (namely the rank-N laminate) adapts itself to the stress that it should sustain, by aligning its lamination directions with the principal directions of the stress and adopting in each layer a volume fraction which is controlled by the values of the principal stresses. This correlation between microstructure and stress is a rigorous consequence of the homogenization theory and not a postulate. In particular in 2-D we recover the well-known principle of material economy in frame-structures due to Michell [42].

In two dimensions, when the Lagrange multiplier ℓ goes to infinity, it is easily seen (see [9]) that the relaxed problem is asymptotically equivalent to the so-called *Michell trusses* problem

$$\min_{\substack{div\,\tau=0\,\text{in } \Omega \\ \tau\cdot n=f\,\text{on } \partial\Omega}} \int_\Omega (|\tau_1| + |\tau_2|)\, dx,$$

where τ_1, τ_2 are the eigenvalues of the stress τ. There is a rich literature on this problem (see e.g. [1], [35], [52], [54]). Note that this limiting case of the relaxed formulation may explain the success of our computations, and more precisely the fact that many of our optimal structures look like a network of trusses, or bars, in 2-D.

Remark also that, by virtue of formulae (51) and (52), the pointwise optimization of the density yields $\theta = 0$ if and only if $\tau = 0$, which means that holes are created only where the stress vanish.

In 2-D only one type of optimal laminates, namely rank-2 laminates, are used (although they can degenerate to rank-1 when one of the eigenvalues vanishes). On the contrary, in 3-D there are two distinct regimes of optimal laminates: rank-3 or rank-2 (which in turn can degenerate to rank-1). This can be easily explained as follows. The conditions defining regimes in (52) imply that, when a rank-3 laminate is optimal, the three principal stresses are of the same order of magnitude. This means that the material can be optimally layered in the three principal directions, creating a microstructure made of plate-like holes in a matrix of material. On the other hand, when a rank-2 laminate is optimal, one of the principal stresses is large compared to the other two. In this case, it is more economical not to layer in the direction of the largest principal stress, and simply to translate, in this direction, a planar optimal microstructure which allows the available material to sustain the largest stress in the direction of translation. The corresponding microstructure looks like an array of tubes or channels of holes aligned in the direction of the largest principal stress.

It would be tempting to assume that the 3-D result (52) degenerates into the 2-D result (51) in a plane stress situation, *i.e.*, when one of the principal stresses is equal to zero. But, it is not the case ; rather, if τ is a plane stress with eigenvalues $\tau_1 = 0$, $\tau_2 \neq 0$, and $\tau_3 \neq 0$, then

$$\min_{A^* \in G_\theta} A^{*-1}\tau \cdot \tau \;\; = \;\; \frac{1}{\theta} A^{-1}\tau \cdot \tau,$$

and optimality is attained for a rank-1 laminate in the direction of the first eigenvector. From a practical standpoint, it has the consequence that, if we can use 3-D microstructures for solving a 2-D problem, then it is preferable to use a "varying thickness plate" approach (corresponding to the optimal rank-1 laminate) than a "plane Michell trusses" approach (corresponding the rank-2 laminates, optimal only in 2-D). Hence the qualitative differences that will be evidenced between 2-D and 3-D pictures : in 2-D the optimal microstructures look like a network of trusses or bars, while in 3-D they will appeal to either trusses or plates.

An other interesting limit case in 3-D is that of a uni-axial stress. If τ is a uni-axial stress with eigenvalues $\tau_1 = 0$, $\tau_2 = 0$, and $\tau_3 \neq 0$, then, again,

$$\min_{A^* \in G_\theta} A^{*-1}\tau \cdot \tau \;\; = \;\; \frac{1}{\theta} A^{-1}\tau \cdot \tau,$$

but optimality is now attained for any rank-2 laminate with directions in the plane defined by the two first eigenvectors. This optimal microstructure looks like an array of fibers aligned with the stress and any type of cross-sectional arrangement is admissible.

So far, we concentrated only on the so-called "single load" problem. This means that the elastic structure is optimized for a single configuration of loading forces and may well be totally inadequate for other loads. In practice it is an undesirable feature and it is quite often more realistic to investigate the so-called "multiple loads" problem which amounts to an optimization of the structure for several configurations. Specifically, various surface loadings $f_1, \cdots, f_p$ are given and we consider the minimization problem

$$\inf_{\omega \subset \Omega} \left\{ E_p(\omega) = \sum_{i=1}^{p} c_i(\omega) + \ell|\omega| \right\}$$

where $c_i(\omega)$ is the compliance defined by (41) for the boundary condition f_i. Most of the obtained theoretical results hold true for the multiple loads problem. The homogenized problem is

$$\min_{\substack{A^* \in G_\theta \\ 0 \leq \theta \leq 1}} \left\{ \tilde{E}_P(\theta, A^*) = \sum_{i=1}^{p} \tilde{c}_i(\theta, A^*) + \ell \int_\Omega \theta(x)dx \right\}$$

where $\tilde{c}_i(\theta, A^*)$ is the homogenized compliance defined by (46) for the boundary condition f_i. However, the optimization over the microstructure A^* cannot be done analytically. In other words an explicit formula for the optimal microstructure is not available. We simply know that optimality is attained in the class of sequential laminates, but the direction of laminations and the proportions are not specified. For any number p of loading configurations, the number of laminations is never larger than 3 in 2-D [11], and 6 in 3-D [28]. Therefore, the optimal microstructure has to be determined numerically rather than through an explicit formula.

Other possible generalizations concern the optimization of eigenfrequencies, von Mises stress, maximal displacement, etc. In these cases, a mathematically rigorous treatment of the homogenization method is not available, but numerically there is no obstacle in implementing a homogenized formulation by simply choosing a priori a class of "optimal" composites (for example laminates of any rank).

6   Application of homogenization to shape optimization: computational aspects.

Up to now, using homogenization theory and introducing a relaxed formulation might appear to be just a trick for proving existence theorems. In fact its importance goes much further, and it is at the root of new numerical algorithms for computing optimal shapes. Indeed, it permits to separate the minimization process in two different tasks : first, optimize locally the microstructure (that is the effective Hooke's law A^*) with explicit formula, second, minimize globally on the density $\theta(x)$. This has the effect of transforming a difficult "free-boundary" or layout problem into a much easier "sizing" optimization problem in a fixed domain. It has many advantages : on the one hand, the cost of a computation is very low compared to traditional algorithms since the mesh is the same for any shape in the iterative process of optimization; on the other hand, it behaves as a topology optimizer and the final optimal shape may have a topology completely different from that of the initial guess. As such the homogenization method is usually applied as a pre-processor for classical shape optimization algorithms (see, e.g., [51]) : first, an optimal topology is found by homogenization, then the resulting shape is optimized by a sensitivity analysis of its boundary (for numerical examples, see [49]). Note that classical shape optimization algorithms work with a fixed topology, namely that of the initial guess, and are therefore unable to optimize it (with the noticeable exception of the so-called bubble method [26]).

The key features of homogenization-based algorithms have been first recognized by M. Bendsoe and N. Kikuchi in their pioneering work [16]. Many generalizations have appeared since then. Here, we shall follow our approach advocated in [9], [7], [6].

6.1   A numerical algorithm for 2 and 3-dimensional shape optimization.

The proposed numerical algorithm for shape optimization, is based on the homogenization method as described in the previous lecture. The key idea is to compute "generalized" optimal shapes for the relaxed formulation, rather than "classical" shapes which are merely approximately optimal for the original formulation. Recall that the relaxed or homogenized objective function is

$$\min_{\substack{div\,\tau=0\,\text{in}\ \Omega \\ \tau\cdot n=f\,\text{on}\ \partial\Omega}} \left\{ QF(\tau) = \int_\Omega \min_{\substack{A\in G_\theta \\ 0\le\theta\le 1}} \left(A^{*-1}\tau\cdot\tau + \lambda\theta\right) \right\}. \tag{53}$$

Furthermore, we have explicitly computed the minimizer A^* in the right hand side of (53), in terms of the stress τ.

The relaxed formulation (53) evokes a problem of nonlinear elasticity. The optimal density (a "generalized" shape) is recovered by the optimality condition on θ. A first simple approach (as implemented in [9] for the 2-D case) amounts to solve, in a first step, this nonlinear minimization problem in the stress τ, by using, e.g., a conjugate gradient method. In a second step an optimal density θ is recovered through the optimality condition. Such an approach is not completely satisfactory since the highly non-trivial energy QF is not smooth at $\tau = 0$ which calls for special care in the gradient method. Convergence to the minimum is usually fairly slow. Furthermore, in 3-D the expression of QF is fairly intricate.

Therefore, we prefer another algorithm, the so-called "alternate directions method", that we now describe. It is based on two key ideas. The first one is to consider the above relaxed problem as a minimization problem not only for the stress, but also for the structural parameters, the density θ, and the microstructure A^*. The second key idea is not to try to minimize directly in the triplet of variables (τ, θ, A^*), but rather to adopt an iterative approach and minimize separately and successively in the design variables (θ, A^*) and in the field variable τ. The minimization in τ for fixed design variables amounts to the resolution of a linear elasticity problem for the structure defined by the previous design variables. The minimization in (θ, A^*) for a fixed stress field is explicit in view of the formulae obtained in the previous section. Consequently, the algorithm is structured as follows:

1. Initialization of the design parameters (θ_0, A_0^*) (for example, taking $\theta_0 = 1$ and $A_0^* = A$ everywhere in the domain).

2. Iteration until convergence:

   (a) Computation of τ_n through a linear elasticity problem with $(\theta_{n-1}, A_{n-1}^*)$ as design variables.

   (b) Updating of the design variables (θ_n, A_n^*) by using the stress τ_n in the explicit optimality formulae.

The alternate direction algorithm is apparented to the two previously known methods: that of [16], [58], and that of [9]. It is a version of the well-known optimality criteria algorithm (see e.g. [54]). However, if the minimization over statically admissible stresses is transformed into a maximization over displacements, it yields a min-max problem. Since the integrand does not satisfy any concave-convex type condition, existence of a saddle point is not guaranteed.

On the contrary, here, convergence is always achieved since the above iterative process always decreases the value of the objective function at each iteration. In practice, convergence of this iterative algorithm is detected when the objective function becomes stationary, or when the change in the design variables becomes smaller than some preset threshold.

Our experience shows that this algorithm works very well and converges smoothly in a relatively small number of iterations (between 10 and 100, depending on the desired accuracy). Furthermore, it seems to be insensitive to the choice of initial guess and convergent under mesh refinement, suggesting uniqueness of a global minimum (at least numerically). However, as expected, it usually produces "generalized" optimal designs that include large region of composite materials with intermediate density. For the cantilever problem (see Figure 3), we present the resulting optimal density of material in Figure 4.

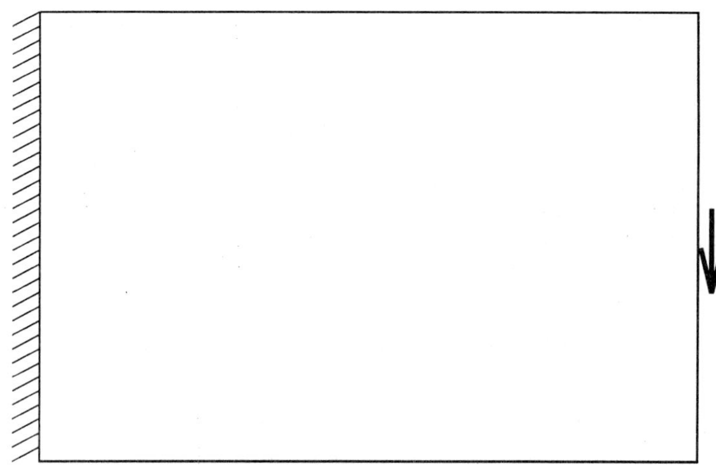

Figure 3: Loading configuration for a cantilever.

6.2   Some technical issues

The generalized Hooke's laws computed at each iteration turn out to be singular, an undesired feature when solving linear elasticity problems. This singular behavior has several sources.

First, we note that the effective tensor is equal to zero when the density vanishes. Implicitly, the corresponding stress field should vanish simultaneously. This problem, which occurs in 2 and 3-D, is easily circumvented by imposing a positive threshold on the density. In practice, the smallest admissible value of θ is fixed around 10^{-3}. Numerical experiments suggest that the choice of 10^{-3} is not important.

We also remark that rank-1 and rank-2 laminates produce degenerate Hooke's laws. In 3-D, the proportions m_i are forced to be greater than zero. Consequently, the algorithm only uses rank-3 laminates, which are non-singular.

In 2-D, rank-1 laminates are eliminated like in the 3-D case. However, the algorithm uses rank-2 laminates as optimal microstructures. The singularity is avoided by adding a small correction term to the composite Hooke's law.

When using $P1$ or $Q1$ finite elements in a displacement formulation, our algorithm is subject to checkerboard instabilities for the density θ similar to those reported in [15], [33], [34]. Such a phenomenon does not occur if a stress-based or complementary formulation is used. Also, these instabilities do not appear if the displacements are computed using higher order elements ($Q2$ for example), while the lamination parameters are computed with only piecewise constant stresses.

The numerical onset of checkerboard patterns is still mysterious. In practice, such instabilities only appear after a large number of iterations, when the convergence criterion is very tight.

In 2-D calculations, we eliminate these instabilities with a method used to filter the pressure in a Stokes flow computation [18]. Once the piecewise constant optimal densities θ_i^n are determined, we project them on super-elements, which are clusters of 4 adjacent elements, so as to eliminate the

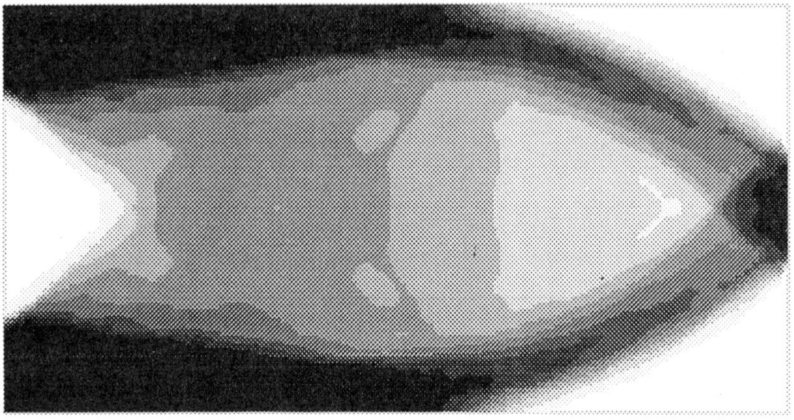

Figure 4: Cantilever: composite design.

checkerboard mode and preserve the overall density. We have not experienced any checkerboard patterns in 3-D calculations.

6.3　Penalization of intermediate densities.

The success of this method is due to the fact that the relaxed design is characterized not only by a density θ but also by a microstructure A which is hidden at the sub-mesh level. The penalization has the effect of reproducing this microstructure at the mesh level (see Figures 5 and 6 for the cantilever problem in 2-D and 3-D respectively). Of course it is strongly mesh-dependent in the sense that the finer the mesh the more complicated the resulting "almost optimal" structure.

As already explained the numerical computations deliver relaxed, or generalized, optimal shapes – a density of material – rather than classical optimal shapes for the original formulation – a characteristic function of the material domain. In other words, our method produces a layout of material, which, as expected, includes large region of composite materials with intermediate density. From a practical standpoint, this is an undesirable feature since the primary goal is to find a real shape – a density taking only the values 0 or 1! This drawback is avoided through a post-processing technique that *penalizes* composite regions. The goal is to deduce, from the optimal densities, a quasi-optimal shape. In loose terms, the solution of the relaxed problem is projected onto the set of classical solutions of the original problem, in the hope that the value of the objective functional will not increase too much in the process.

Figure 5: Cantilever: penalized design.

The strategy is as follows. Upon convergence to an optimal density, we run a few more iterations of our algorithm where we *force* the density to take values close to 0 or 1. This changes the optimal density and produces a quasi-optimal shape. Of course, the procedure is purely numerical and mesh dependent. The finer the mesh, the more detailed the resulting structure will appear at the outset of the penalization process. The method works well, because the relaxed design is characterized not only by a density θ but also by a microstructure A^*, which is hidden at the sub-mesh level. The penalization tends to reproduce the microstructure at the mesh level.

Two penalization techniques for the intermediate composite densities have been used. Both amount to a modification of the explicit optimality formula that expresses the optimal density in terms of the stress. Specifically, instead of updating the density with the true optimal density θ_{opt}, a value θ_{pen} is used. Our first choice for θ_{pen} is

$$\theta_{pen} = \frac{1 - \cos(\pi\theta_{opt})}{2}.$$

The choice of a cosine function for the penalized density is arbitrary. If θ_{pen} is too close to θ_{opt}, the scheme is insensitive to the proposed penalization, while if θ_{pen} is forced too close to 0 or 1, the fine patterns of the shape are destroyed.

In the context of plate thickness optimization, another technique has been proposed [13],[63].

It consists in setting

$$\theta_{pen} = (\theta_{opt}^2/p)^{1/(1+p)} \text{ for some } p < 1.$$

This alternate choice also gives good results. It corresponds to the optimal value of θ for a modified minimization problem, namely

$$\min_{0 \le \theta \le 1} \left\{ F(\tau, \theta) + \ell\theta^p \right\} ,$$

which is supposed to take into account "manufacturing costs" of perforated materials (the "cost" of intermediate densities increases as p decreases from 1 to 0).

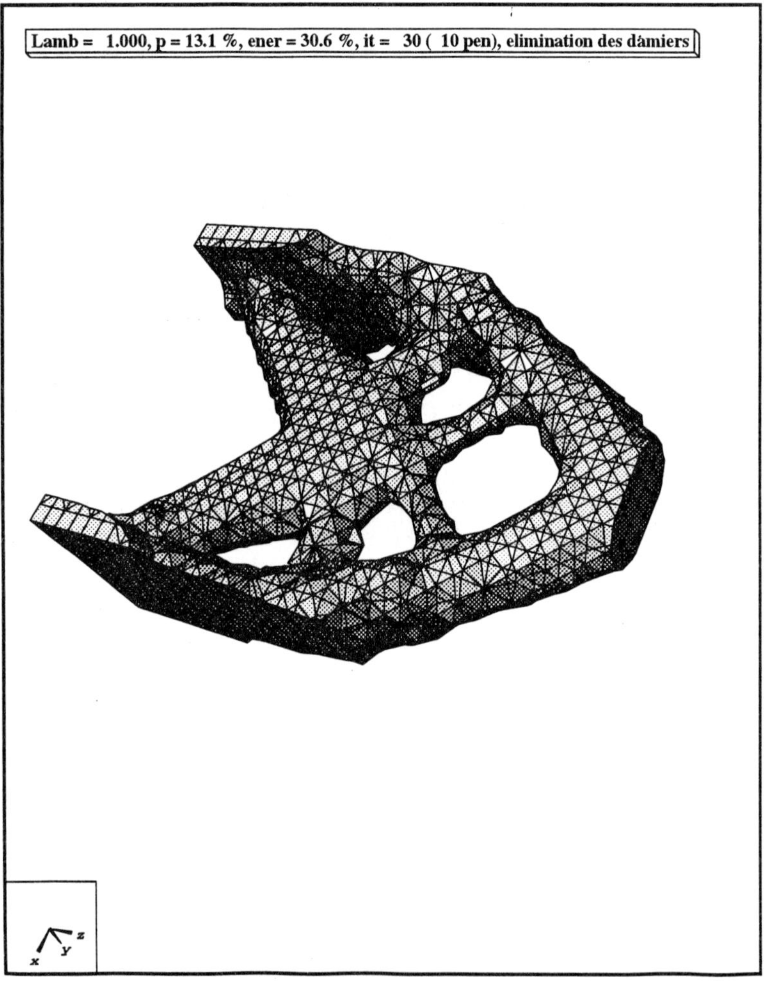

Figure 6: 3-D cantilever: penalized design.

6.4 Homogenization versus "fictitious material".

The preceding numerical algorithm for computing optimal shapes may seem a little bizarre : we spent a long time to introduce homogenization theory, a proper optimal microstructure, and complicated formulas for updating the design variables, and in the end, one could think that we are simply throwing away everything since we penalize and get rid of all the composite materials ! Some authors (see e.g. [54]) have thus been led to propose a simpler approach that is sometimes called "fictitious material" method and coincides with the use of the *convexification* of the original problem. Their argument is the following : the idea of working with a density instead of a real shape is a good one, but, since in the end all intermediate densities are eliminated by penalization of the final result, why not suppress the delicate concept of a real microstructure and rather work simply with the same material A with a varying density (or thickness in the language of plate theory) ? Of course, such an approach has the real advantage of being straightforward to implement. However, as we shall see, its results are not as good as the ones of the homogenization method.

The "fictitious material" approach amounts to consider the following state equation

$$\left\{ \begin{array}{ll} \sigma = \theta(x)Ae(u) & e(u) = \frac{1}{2}(\nabla u + \nabla^t u) \\ \text{div } \sigma = 0 & \text{in } \Omega \\ \sigma \cdot n = f & \text{on } \partial\Omega \end{array} \right. \tag{54}$$

where $\theta(x)$ is a density function taking its values between 0 and 1. The goal is still to minimize, over all possible density, the weighted sum of the compliance and the weight, namely to compute

$$\min_{0 \le \theta(x) \le 1} \left(c(\theta) + \lambda \int_\Omega \theta(x) \right). \tag{55}$$

where the compliance is defined by

$$c(\theta) = \int_{\partial\Omega} f \cdot u = \int_\Omega (\theta(x)A)^{-1} \sigma \cdot \sigma. \tag{56}$$

By using the principle of complementary energy and switching the two minimizations, it is easily seen to have the following equivalent formulation which is nothing but the convexification of the stress based formulation.

$$\inf_{\substack{\text{div } \tau = 0 \text{ in } \Omega \\ \tau \cdot n = f \text{ on } \partial\Omega}} \left(CF(\tau) = \int_\Omega \left\{ \begin{array}{ll} A^{-1}\tau \cdot \tau + \lambda & \text{if } A^{-1}\tau \cdot \tau \ge \lambda \\ 2\sqrt{\lambda A^{-1}\tau \cdot \tau} & \text{if } A^{-1}\tau \cdot \tau \le \lambda \end{array} \right. \right) \tag{57}$$

Since it is a convex minimization problem, existence of optimal solutions is straightforward. By definition, the different energies of the stress τ satisfies $F(\tau) \ge QF(\tau) \ge CF(\tau)$, where the inequalities are strict for most choices of τ.

We have implemented numerically this convex formulation by using the same "alternate directions" strategy as before : for a given density θ, we compute the stress σ solution of the linear elasticity state equation, then we update the design variable θ by the following optimality relationship

$$\theta(x) = \left\{ \begin{array}{ll} 1 & \text{if } A^{-1}\sigma \cdot \sigma \ge \lambda \\ \sqrt{\lambda^{-1}A^{-1}\sigma \cdot \sigma} & \text{if } A^{-1}\sigma \cdot \sigma \le \lambda \end{array} \right. \tag{58}$$

This algorithm converges quickly and smoothly, and we supplement it with the same penalization procedure as before. In general, the fictitious penalized design fails to have the same degree of complexity and detailed patterns as the homogenized penalized design (its energy $E(\omega)$ is indeed larger).

This sensibly worse behavior of the fictitious material approach takes its roots in the fact that there are no implicit microstructure hidden at the sub-mesh level like for the homogenization method. Thus, penalization does not reveal any structure which was waiting to appear at the grid level. In other words, a solution of the convex formulation lies far away from any quasi-minimizer of the original formulation.

Acknowledgments

The 3-D picture has been obtained by the code developed mainly by F. Jouve, at Ecole Polytechnique de Paris, and used by the authors of [6] in the context of shape optimization.

References

[1] W. Achtziger, M. Bendsoe, A. Ben-Tal, J. Zowe, *Equivalent displacement based formulations for maximum strength truss topology design,* IMPACT of Computing in Science and Engineering, 4, pp.315-345 (1992).

[2] G. Allaire, *Homogenization and two-scale convergence,* SIAM J. Math. Anal. **23.6**, pp.1482-1518 (1992).

[3] G. Allaire, *Structural optimization using optimal microstructures,* In "MECAMAT 93 International Seminar on Micromechanics of Materials", Collection de la Direction des Etudes et Recherches d'Electricité de France, Eyrolles, Paris (1993).

[4] G. Allaire, *Explicit lamination parameters for three-dimensional shape optimization,* Control and Cybernetics 23, pp.309-326 (1994).

[5] G. Allaire, *Relaxation of structural optimization problems by homogenization,* "Trends in Applications of Mathematics to Mechanics", M.M.Marques and J.F.Rodrigues Eds., Pitman monographs and surveys in pure and applied mathematics 77, pp.237-251, Longman, Harlow (1995).

[6] G. Allaire, E. Bonnetier, G. Francfort, F. Jouve, *Shape optimization by the homogenization method,* to appear in Numerische Mathematik.

[7] G. Allaire, G. Francfort, *A numerical algorithm for topology and shape optimization,* In "Topology design of structures" Nato ASI Series E, M. Bendsoe et al. eds., 239-248, Kluwer, Dordrecht (1993).

[8] G. Allaire, R.V. Kohn, *Optimal bounds on the effective behavior of a mixture of two well-ordered elastic materials,* Quat. Appl. Math. 51, 643-674 (1993).

[9] G. Allaire, R.V. Kohn, *Optimal design for minimum weight and compliance in plane stress using extremal microstructures,* Europ. J. Mech. A/Solids 12, 6, 839-878 (1993).

[10] M. Avellaneda, *Optimal bounds and microgeometries for elastic two-phase composites,* SIAM J. Appl. Math. 47, 6, 1216-1228 (1987).

[11] M. Avellaneda, G. Milton, *Bounds on the effective elasticity tensor of composites based on two point correlations,* in Proceedings of the ASME Energy Technology Conference and Exposition, Houston, 1989, D. Hui et al. eds., ASME Press, New York (1989).

[12] N. Bakhvalov, G. Panasenko, *Homogenization: Averaging Processes in Periodic Media*, Kluwer, Dordrecht (1989).

[13] M. Bendsoe, *Optimal shape design as a material distribution problem*, Struct. Optim. 1, 193-202 (1989).

[14] M. Bendsoe, *Methods for optimization of structural topology, shape and material*, Springer Verlag (1995).

[15] M. Bendsoe, A. Diaz, N. Kikuchi, *Topology and generalized layout optimization of structures*, In "Topology Optimization of Structures" Nato ASI Series E, M. Bendsoe et al. eds., pp.159-205, Kluwer, Dordrecht (1993).

[16] M. Bendsoe, N. Kikuchi, *Generating Optimal Topologies in Structural Design Using a Homogenization Method*, Comp. Meth. Appl. Mech. Eng. 71, 197-224 (1988).

[17] A. Bensoussan, J.L. Lions, G. Papanicolaou, *Asymptotic Analysis for Periodic Structures*, North-Holland, Amsterdam (1978).

[18] F. Brezzi, M.Fortin, *Mixed and hybrid Finite Element Methods*, Springer, Berlin, 1991.

[19] G. Cheng, N. Olhoff, *An investigation concerning optimal design of solid elastic plates*, Int. J. Solids Struct. 16, pp.305-323 (1981).

[20] R. Christensen, *Mechanics of Composite Materials*, Wiley-Interscience, New York (1979).

[21] G. Dal Maso, *An Introduction to Γ-Convergence*, Birkhäuser, Boston (1993).

[22] G. Dal Maso, L. Modica, *Nonlinear stochastic homogenization and ergodic theory*, Journal für die reine und angewandte Mathematik **368**, pp.28-42 (1986).

[23] E. De Giorgi, *Sulla convergenza di alcune successioni di integrali del tipo dell'area*, Rendi Conti di Mat. **8**, pp.277-294 ,(1975).

[24] E. De Giorgi, *G-operators and Γ-convergence*, Proceedings of the international congress of mathematicians (Warsazwa, August 1983), PWN Polish Scientific Publishers and North Holland, pp.1175-1191 (1984).

[25] E. De Giorgi, S. Spagnolo, *Sulla convergenza degli integrali dell'energia per operatori ellittici del secondo ordine*, Boll. Un. Mat. It. **8**, pp.391-411 (1973).

[26] H. Eschenauer, V. Kobelev, A. Schumacher, *Bubble method of topology and shape optimization of structures*, Struct. Optim. 8, pp.42-51 (1994).

[27] G. Francfort, F. Murat, *Homogenization and Optimal Bounds in Linear Elasticity*, Arch. Rat. Mech. Anal. 94, 307-334 (1986).

[28] G. Francfort, F. Murat, L. Tartar, *Fourth Order Moments of a Non-Negative Measure on S^2 and Application*, to appear in Arch. Rat. Mech. Anal. (1995).

[29] L. Gibianski, A. Cherkaev, *Design of composite plates of extremal rigidity*, Ioffe Physicotechnical Institute preprint (1984).

[30] Y. Grabovsky, R. Kohn, *Microstructures minimizing the energy of a two-phase elastic composite in two space dimensions II: the Vigdergauz microstructure*, to appear.

[31] Z. Hashin, *The elastic moduli of heterogeneous materials*, J. Appl. Mech. 29, 143-150 (1963).

[32] Z. Hashin, S. Shtrikman, *A variational approach to the theory of the elastic behavior of multiphase materials*, J. Mech. Phys. Solids 11, 127-140 (1963).

[33] C. Jog, R. Haber, M. Bendsoe, *A displacement-based topology design method with self-adaptive layered materials*, Topology design of structures, Nato ASI Series E, M. Bendsoe et al. eds., 219-238 , Kluwer, Dordrecht (1993).

[34] C. Jog, R. Haber, M. Bendsoe, *Topology design with optimized, self-adaptative materials*, Int. Journal for Numerical Methods in Engineering 37, 1323-1350 (1994).

[35] U. Kirsch, *Optimal topologies of truss structures*, Comp. Meth. Engrg. 72, pp.15-28 (1989).

[36] R. Kohn, *Recent progress in the mathematical modeling of composite materials*, Composite Materials Response: Constitutive Relations and Damage Mechanisms, G. Sih et al. eds., pp.155-177, Elsevier, New York (1988).

[37] R. Kohn, G. Strang, *Optimal Design and Relaxation of Variational Problems I-II-III*, Comm. Pure Appl. Math. 39, 113-182,353-377 (1986).

[38] S. Kozlov, *Averaging of random operators* Math. USSR Sbornik **37**, pp.167-180 (1980).

[39] J.L. Lions, *Some methods in the mathematical analysis of systems and their control*, Science Press, Beijing, Gordon and Breach, New York (1981).

[40] K. Lurie, A. Cherkaev, *Exact estimates of the conductivity of a binary mixture of isotropic materials*, Proc. Royal Soc. Edinburgh **104A**, pp.21-38 (1986).

[41] K. Lurie, A. Cherkaev, A. Fedorov, *Regularization of Optimal Design Problems for Bars and Plates I,II*, J. Optim. Th. Appl. 37, pp.499-521, 523-543 (1982).

[42] A. Michell, *The limits of economy of material in frame-structures*, Phil. Mag. 8, 589-597 (1904).

[43] G. Milton, *Modeling the properties of composites by laminates*, Homogenization and effective moduli of materials and media, J. Ericksen et al. eds., pp. 150-174, Springer Verlag, New York (1986).

[44] F. Murat, *Contre-exemples pour divers probèmes où le contrôle intervient dans les coefficients*, Ann. Mat. Pura Appl. 112, 49-68 (1977).

[45] F. Murat, L. Tartar, *H-convergence*, Séminaire d'Analyse Fonctionnelle et Numérique de l'Université d'Alger (1977).

[46] F. Murat, L. Tartar, *Calcul des variations et homogénéisation*, in Les méthodes de l'homogénéisation: théorie et applications en physique, Coll. de la Dir. des Etudes et Recherches EDF, pp.319-370, Eyrolles, Paris (1985).

[47] G. Nguetseng, *A general convergence result for a functional related to the theory of homogenization* SIAM J. Math. Anal. **20**, pp.608-629 (1989).

[48] O. Oleinik, A. Shamaev, G. Yosifian, *Mathematical Problems in Elasticity and Homogenization*, Studies in Mathematics an Its Application **26**, Elsevier, Amsterdam (1992).

[49] N. Olhoff, M. Bendsoe, J. Rasmussen, *On CAD-integrated structural topology and design optimization*, Comp. Meth. Appl. Mechs. Engng. 89, pp.259-279 (1992).

[50] G. Papanicolaou, S. Varadhan, *Boundary value problems with rapidly oscillating random coefficients*, Colloquia Mathematica Societatis János Bolyai, North-Holland, Amsterdam, pp.835-873 (1982).

[51] O. Pironneau, *Optimal shape design for elliptic systems*, Springer Verlag, Berlin (1984).

[52] W. Prager, G. Rozvany, *Optimal layout of grillages*, J. Struct. Mech. 5, 1-18 (1977).

[53] G. Rozvany, M. Bendsoe, U. Kirsch, *Layout optimization of structures*, Applied Mechanical reviews 48, 2, pp.41-118 (1995).

[54] G. Rozvany, M. Zhou, T. Birker, O. Sigmund, *Topology optimization using iterative continuum-type optimality criteria (COC) methods for discretized systems*, Topology design of structures, Nato ASI Series E, M. Bendsoe et al. eds., 273-286 , Kluwer, Dordrecht (1993).

[55] E. Sanchez-Palencia, *Non-Homogeneous Media and Vibration Theory*, Springer Lecture Notes in Physics **129** (1980).

[56] S. Spagnolo, *Sulla convergenza delle soluzioni di equazioni paraboliche ed ellittiche*, Ann. Sc. Norm. Sup. Pisa Cl. Sci. (3), **22**, pp.571-597 (1968).

[57] S. Spagnolo, *Convergence in energy for elliptic operators*, Proc. Third Symp. Numer. Solut. Partial Diff. Equat. (College Park 1975), B. Hubbard ed., Academic Press, San Diego, pp.469-498 (1976).

[58] K. Suzuki, N. Kikuchi, *A homogenization method for shape and topology optimization*, Comp. Meth. Appl. Mech. Eng. 93, 291-318 (1991).

[59] L. Tartar, *Quelques remarques sur l'homogénéisation*, Proc. of the Japan-France Seminar 1976 "Functional Analysis and Numerical Analysis", Japan Society for the Promotion of Sciences, pp.469-482 (1978).

[60] L. Tartar, *Estimations Fines des Coéfficients Homogénéisés*, Ennio de Giorgi colloquium, P. Krée ed., Pitman Research Notes in Math. 125, 168-187 (1985).

[61] S. Vigdergauz, *Effective elastic parameters of a plate with a regular system of equal-strength holes*, Mech. Solids 21, 162-166 (1986).

[62] V. Zhikov, S. Kozlov, O. Oleinik, *Homogenization of Differential Operators*, Springer, Berlin, (1995).

[63] M. Zhou, G. Rozvany, *The COC algorithm, Part II: Topological, geometrical and generalized shape optimization*, Comp. Meth. App. Mech. Engrg. 89, 309-336 (1991).

TOPOLOGY AND SHAPE OPTIMIZATION PROCEDURES USING HOLE POSITIONING CRITERIA
THEORY AND APPLICATIONS

H.A. Eschenauer and A. Schumacher
University of Siegen, Siegen, Germany

Abstract

The topology of any constructions, i.e., the position and arrangement of structural elements in a given design space, has strong influence on its structural behaviour. Currently, the topology is still chosen intuitively or by referring to existing constructions („Current Design World State"), or it is selected from a number of different variants. The topology optimization aims at the use of mathematical-mechanical strategies in a design process.

The present paper addresses a simultaneous method of topology and shape optimization, called *Bubble-Method*. Its basic idea is the iterative positioning of new holes into a given design domain. The essential task of this method is a problem-dependent finding of an optimal position vector for a hole that is to be inserted into a body. The criteria required for this purpose are derived from a general optimization problem. The positioning criteria are determined from so-called *characteristic functions* described by stresses, strains, and displacements. The characteristic functions depend on special optimization functionals and the shape of the hole.

In the final sub-chapter a number of application examples are shown, among others a panel truss structure for a radio telescope, a casing of a handsaw grip, and a wing rib of an airplane.

1   Introduction

1.1   Motivation

Owing to the increasing demands on the efficiency, reliability and shortened development cycle of a product, it has become inevitable to solve problems by computer-based procedures. Substantial progress has been achieved in the structural computation of components, especially in view of the versatile FE-method. In many applications, an algorithm-based optimization of the component dimensions (e.g. of the wall-thicknesses of a container) has already become general use, however, applicable methods and strategies are still to be developed for generating best-possible initial layouts for components.

Topology optimization aims at determining constructive solutions for component structures, proceeding from very few prescribed specifications like load cases, boundary conditions, and admissible design spaces (topology domain). In recent years, substantial efforts have been made in the development of topology optimization procedures, and there are several different strategies the use of which is in most cases highly problem-dependent. Topology strategies are to determine an optimal topology according to the defined optimization problem independently of the designer. They shall support the interactive work in the design process, since an isolated optimization calculation often does not yield an optimal result. Thus, it is important to include the designer's creativity especially in those cases where essential demands cannot be modelled sufficiently in the optimization process. Creativity should not be underestimated particularly in complex design processes, and it is also important in topology optimization [34, 45].

1.2   Current research activities in topology optimization

MICHELL (1904) [35] developed a design theory for the topology of bar structures that are optimal with regard to weight. The bars in these structures are all perpendicular to each other and therefore form an optimal arrangement in terms of maximum tensile and compressive stresses. Nowadays, research into topology optimization for more complex structures is yet in its early stages. Important initial steps in this direction were made by W. PRAGER (1969) [38, 39] who solved topology optimization problems by analytical procedures. U. KIRSCH [29] and G. ROZVANY (1989) [42] developed optimality criteria for bar structures. These criteria determine the optimal structure from a defined basic structure containing all feasible bar elements. Truss topology optimization with stress and displacement constraints is dealt with in [43].

Proceeding from continua, further research has been carried out over the past years. ATREK (1989) [2] has developed a procedure which divides the topology domain into many smaller sub-areas, the thicknesses of which are defined as design variables and are then varied by means of a simple optimality criterion. He defined a decision limit for the element thickness according to which all design variables under this limit are set "zero" (hole), and those beyond this limit are set "one" (material). This is called a 0-1-check.

Based on works on the homoginization of porous materials BOURGAT [8], BENDSØE et al. [5,6,7] have developed the *homogenization method* for topology optimization. In this context, they have referred to research activities carried out by ALLAIRE and KOHN (1993) [1]. The homogenization method works without a 0-1-check, since it considers microcells which form the small structural areas of the topology domain. These microcells consist of massive material, and of a no-material area (void) so that a porous material behaviour can be simulated. For the computation, this porous material is then subjected to a homogenization, and thus it is made indistinct. The basic idea of this type of optimization is a variation of the volumetric efficiency (ratio of volume of massive material to the total volume of the microcells in the structural areas). Here, one tries to approach the volumetric efficiency "0" (no material) and "1" (full material). This method represents a further step in the direction of finding topological layouts of mechanical structures.

The topology optimization method addressed in this paper uses an iterative positioning and hierarchically structured shape optimization of new holes, so-called *bubbles*. This means that the boundaries of the structure are taken as parameters, and that the shape optimization of the new bubbles and of the other variable boundaries of the component is carried out as a parameter optimization (ESCHENAUER *et al.* (1994) [14, 15, 16, 19], ROSEN, D.W. (1992) [41]).

2   Fundamentals of a topology optimization problem

2.1   Terms of topology within a design process

The *set-oriented topology* [9, 27, 28] describes those properties of geometrical shapes that remain unchanged even if the figure is subjected to deformations large enough to eliminate all metric and projective properties. The *topological properties* are the most general qualities of a domain.

Shapes that belong into one *topology class* are called *topologically equivalent* (Fig. 2.1.a). A second topology class is defined by the degree to which the areas are connected (Fig.2.1.b). Finally, a third topology class is termed n-fold connected, if (n-1) cuts from one boundary to another are required to transform a given, multiply connected domain into a simply connected domain (Fig. 2.1.c).

In the classical shape optimization of structural components, interrelations between the elements that constitute a domain are maintained, and the isomorphous mapping laws are valid [28]. Topology optimization, i.e., an improving transformation into other topology classes, modifies these interrelations. In view of topology, neither the position of a hole alone nor the shape of this hole alone play the important role. As these properties are decisive for the mechanical behaviour of a component, a structural improvement can only be achieved by a simultaneous treatment of topology and shape optimization.

a) Topologically equivalent domains

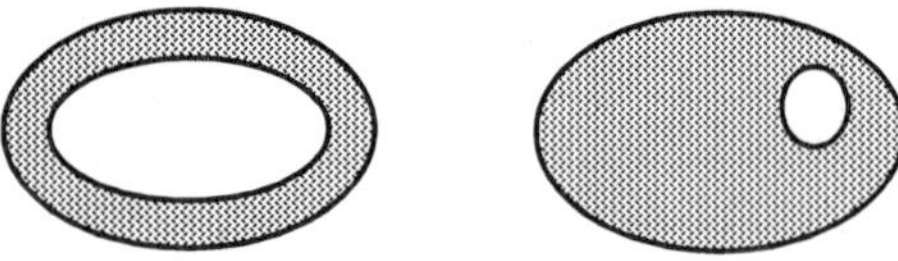

b) Simply, two-fold, and three-fold connected domains

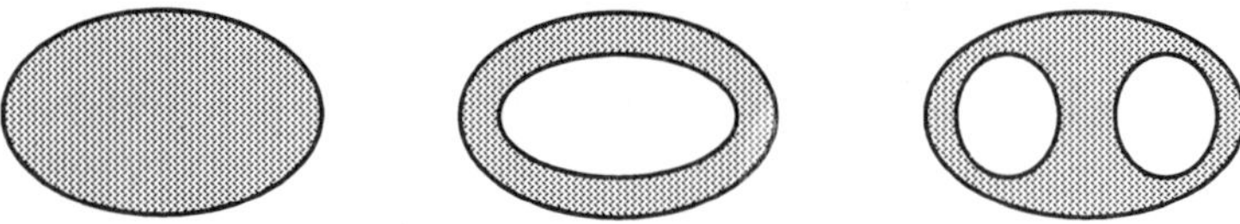

c) Reduction of a three-fold connected domain

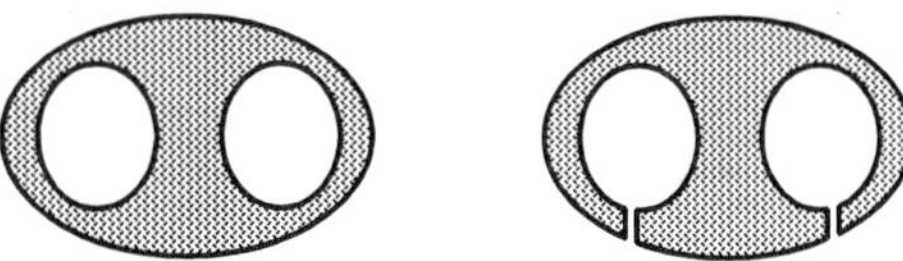

Fig. 2.1: Topological properties of domains

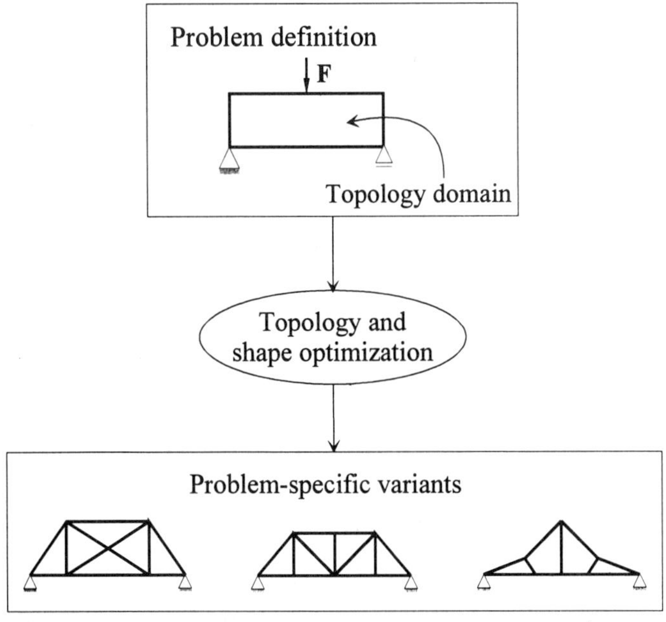

Fig. 2.2: Use of topology and shape optimization in the design process

The effective application of topology and shape optimization requires a practice-relevant definition of the respective optimization problem, where information on the general conditions (size and shape of the topology domain, used material) are necessary in addition to the description of the demands put on a structure (Fig. 2.2). In topology optimization one tries to provide an initial design by using a relatively small number of specifications.

2.2 Mathematical formulation of an optimization problem

In a general optimization problem the vector of the design variables $\mathbf{x}$ is to be chosen in such a way that the objective functional $F(\mathbf{x})$ attains an extreme value, while at the same time equality constraint operators $\mathbf{H}(\mathbf{x})$ and inequality constraint operators $\mathbf{G}(\mathbf{x})$ have to be fulfilled. This task can mathematically be expressed as

$$F^*[x^*] = \operatorname*{Min}_{\mathbf{x}}\left\{F[x] \mid x \in X\right\} \tag{2.1}$$

with

$$X = \left\{\mathbf{x} \in \mathfrak{R}^n \mid \mathbf{H}[\mathbf{x}] = \mathbf{H}^0 , \mathbf{G}[\mathbf{x}] \geq \mathbf{G}^0\right\} ,$$

where the following notations are used:

F	objective functional,
F^*	optimal value of the objective functional,
$\mathbf{x}$	vector of the n design variables,
$\mathbf{x}^*$	optimum configuration of the design variables,
X	admissible design space,
$\mathfrak{R}^n$	n-dimensional set of real numbers,
$\mathbf{H}, \mathbf{G}$	equality and inequality operators,
$\mathbf{H}^0 , \mathbf{G}^0$	bounds of the equality and inequality constraint operators.

The constrained optimization problem can be transformed into an unconstrained optimization problem using the LAGRANGE function

$$L(\mathbf{x}, \lambda) = F(\mathbf{x}) - \sum_{\mu=1}^{m_H} \lambda_\mu H_\mu(\mathbf{x}) - \sum_{\nu=1}^{m_G} \lambda_\nu \left(G_\nu(\mathbf{x}) - \mu_\nu^2\right) , \tag{2.2}$$

where the LAGRANGE-multipliers λ_μ and λ_ν describe the influence of the respective constraints on the minimum of the LAGRANGE-function. In the case of non-active constraints, it holds that $\lambda_\nu = 0$. By introducing slack variables μ_ν^2 describing the distance of the inequality constraints to the corresponding bounds, the inequality constraints are treated as equality constraints. The derivatives of the LAGRANGE-function then yield the conditions for the existence of an optimum:

$$\frac{\partial L(\mathbf{x}, \lambda)}{\partial x_i} = \frac{\partial F(\mathbf{x})}{\partial x_i} - \sum_{\mu=1}^{m_H} \lambda_\mu \frac{\partial H_\mu(\mathbf{x})}{\partial x_i} - \sum_{v=1}^{m_G} \lambda_v \frac{\partial G_v(\mathbf{x})}{\partial x_i} = 0 , \qquad i = 1, n ;$$

$$\frac{\partial L(\mathbf{x}, \lambda)}{\partial \mu_v} = 2\lambda_v \mu_v = 0 , \qquad n = 1, m_H ;$$

$$\frac{\partial L(\mathbf{x}, \lambda)}{\partial \lambda_v} = -G_v(\mathbf{x}) + \mu_v^2 = 0 , \qquad n = 1, m_G ; \qquad\qquad (2.3)$$

$$\frac{\partial L(\mathbf{x}, \lambda)}{\partial \lambda_\mu} = -H_\mu(\mathbf{x}) = 0 , \qquad n = 1, m_H$$

for n design variables.

These KUHN-TUCKER-criteria are necessary conditions for the existence of local minima. For further details on the fundamentals of optimization refer among others to [21, 22, 25].

Problems of this type are in most cases highly nonlinear and non-convex, and thus several local minima can occur in addition to the absolute minimum. In order to reliably determine a global optimum by means of mathematical programming algorithms, a convex problem must exist. However, as objective and constraint functions cannot be treated explicitly in most cases, convexity of the problem cannot be guaranteed, and hence it appears sensible to use different starting points for the optimization.

2.3 Characteristics of topology optimization methods

The methods for topology optimization presented in the references can be evaluated by the following characteristics (Fig. 2.3):

- *Definition of the topology domain*:
  - Methods for the optimization of *discrete structures* (generally bar structures) use as basic structures a set of spatial points connected by as many bars as possible in as many variants as possible. The optimal bars are chosen from this basic structure.
  - Methods for the optimization of *continuous structures* do not need such basic structures and merely require a definition of the available space which partly shows complex boundaries.

- *Type of objective and constraint functions*:
  Most topology optimization methods use the compliance and the weight of a component as objective functions. Some methods are limited to these or related functions (e.g. ei-

genfrequences), because they require analytical sensitivities for optimization, and these sensitivities are sometimes difficult to calculate for all objective and constraint functions.

- *Definition of design variables*:
Except for methods with parameterized boundary description, topology optimization is synonymous with the existence of many design variables. As a rule, the topology space consists of many smaller structural domains that are described by at least one design variable each.

- *Applied solution algorithms*:
Very different optimization algorithms are employed to solve the optimization problem. Methods of Mathematical Programming (MP) and the Optimality Criteria (OC) solve topology optimization problems without particular requirements concerning the interrelations of design variables as it would be necessary in rule-based methods.

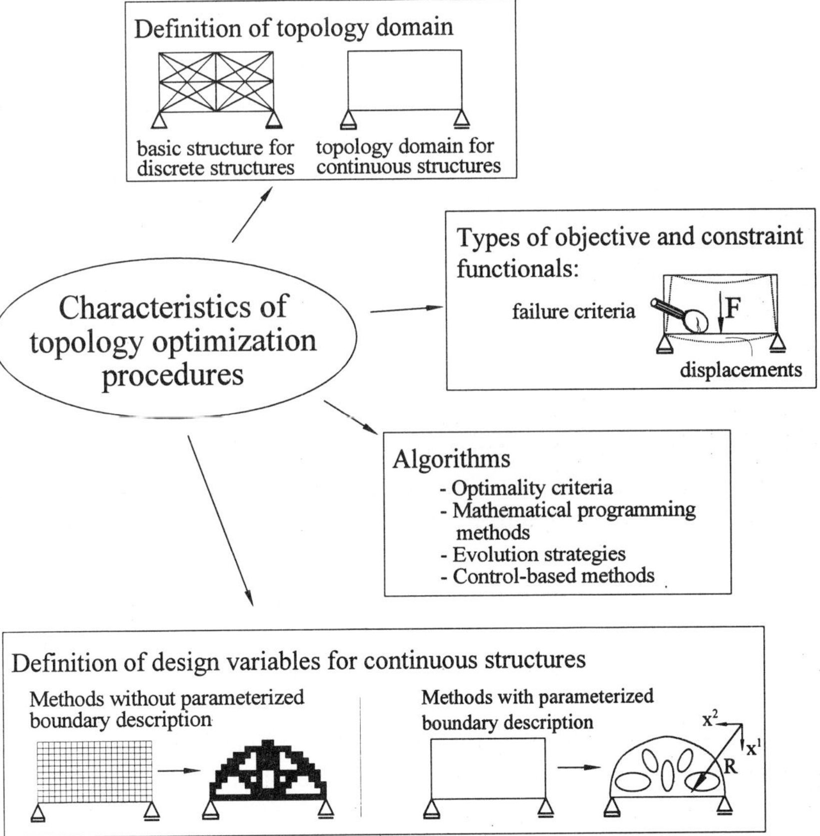

Fig. 2.3: Characteristics of the topology optimization method

3   *Bubble-Method* - topology optimization with hole positioning

3.1   Global and local domain variation

The *bubble method* combines global with local domain variation. The global domain variation varies the boundary of the domain, complying with the homeomorphical mapping rules. The interrelations between the elements constituting the domain are maintained, i.e., the topology class of the domain is unchanged (shape optimization). By means of the local domain variation the topology class is modified by inserting small holes (topology optimization).

a) Global domain variation

The solutions of the global domain variation (Fig. 3.1) are based on the theory of the *variation with variable domain* [9], where a domain functional is considered expressed by an arbitrary function f_Γ :

$$J_\Gamma = \int_\Omega f_\Gamma \, d\Omega \ . \tag{3.1}$$

Using GREEN's rule, the variation of this domain functional can be written as an integral of the product of a function f_Γ and the variation of the boundary δs (distance between the old and the new boundary considered in the direction of the normal) along the boundary of the domain Γ:

$$\delta J_\Gamma = \int_\Gamma f_\Gamma \, \delta s \, d\Gamma = 0 \ . \tag{3.2}$$

This problem formulation corresponds to the first variation of the functional J_Γ (3.1). The majority of numerical methods for domain variation is based on this first variation. In order to explicitly determine the function f_Γ, one can employ existing methods based on the formulation of the problem by means of the LAGRANGE-function (2.2), the variation of multidimensional problems, and on the fundamental equations of structural mechanics.

b) Local domain variation

The aim of a local domain variation is to determine the optimal position of a new hole (bubble) within the structure. The bubble is then inserted, and by that the topology class is increased. Thus, the confinement to the existing topology class is no longer given.

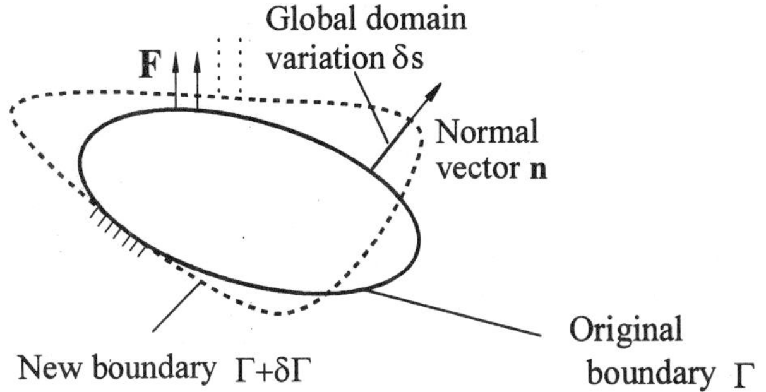

Fig. 3.1: Global domain variation

The insertion of a hole changes the states of stress and deformation, respectively, in the elastic body. If an infinitesimally small hole is inserted, it can be treated as a singular disturbance. Two elastic bodies described by the domains Ω^p and Ω^{p+1} (Fig. 3.2), respectively, possess identical characteristics, that means, the domain Ω^{p+1} has an infinitesimal hole with the coordinate vector $\mathbf{r}$ and the radius r_B in the case of a circular hole.

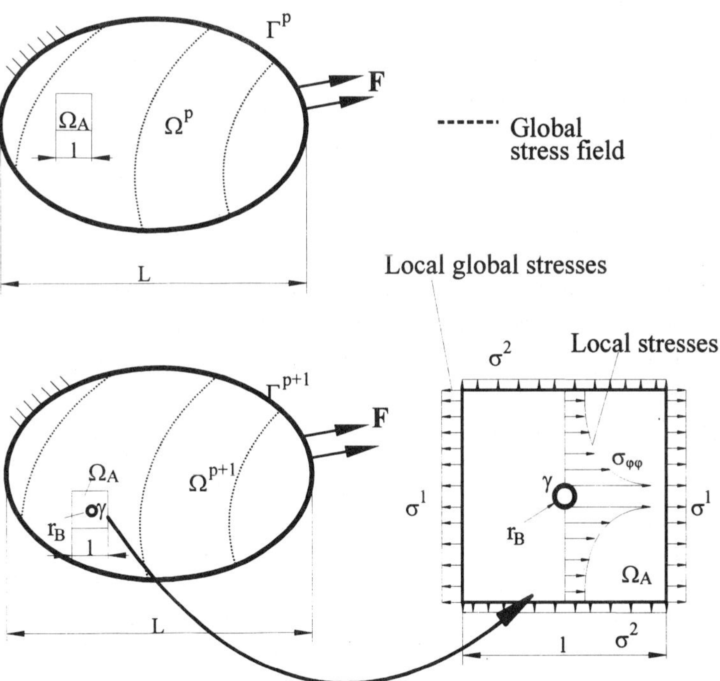

Fig. 3.2: Local variation of the domain by inserting a hole (bubble)

The difference of an optimization functional G for the two bodies then reads as follows:

$$\Delta G_v = G_v(\Omega^{p+1}) - G_v(\Omega^p) , \qquad (3.3)$$

where it is assumed that ΔG_v vanishes if $r_B \to 0$. This assumption is to be checked for each ΔG_v by means of a convergence test. If ΔG_v does not vanish for $r_B \to 0$, a minor disturbance gains strong influence on the domain functional. In this case (which will not be treated in this paper), we are dealing with an ill-posed problem that has to be solved by special methods. For a well-posed problem (i.e., ΔG_v vanishes for $r_B \to 0$), the radius of the circular bubble r_B around $r_B = 0$ can be expanded in a TAYLOR-series:

$$\Delta G_v = \left(\frac{\partial \Delta G_v}{\partial r_B}\right)_{r_B=0} r_B + \left(\frac{\partial^2 \Delta G_v}{\partial r_B^2}\right)_{r_B=0} \frac{r_B^2}{2!} + \cdots + \left(\frac{\partial^n \Delta G_v}{\partial r_B^n}\right)_{r_B=0} \frac{r_B^n}{n!} . \qquad (3.4)$$

The first term (not equal zero) describes the sensitivity of the functional with respect to the positioning of the infinitesimal hole. The solution is simplified by introducing an evaluation domain Ω_A (Fig. 3.2), where the radius of the bubble r_B shall be substantially smaller than the dimensions of the evaluation domain, and these dimensions shall be much smaller than the component dimensions ($r_B \ll l \ll L$). The new hole causes a stress increase in the vicinity of the boundary of the hole, but it decays rapidly according to the Principle of DE SAINT VENANT.

Based on the above assumptions we define that

- the global stress field of the component remains practically unchanged by the infinitesimal bubble (the global stress field depends on the outer shape of the component and on the external loads),

- a local stress concentration is generated in the vicinity of the bubble so that the local stress field in the small evaluation domain Ω_A depends on the shape of the bubble and on the mean value of the global stress in the domain.

Thus, it is sufficient to evaluate Ω_A in order to determine the optimal position. By means of GREEN's rule, the variation of the domain integral is to be reduced to a boundary integral, and the local variational problem can be written as an integral of a function f_γ over the boundary of the inserted, infinitesimal bubble:

$$\delta J_\gamma = \int_\gamma f_\gamma \, \delta s_B \, d\gamma = 0, \qquad (3.5)$$

where the function f_γ depends on the considered objective and constraint functions of the optimization problem, and on the mechanical conditions. It is required in order to determine the optimal position of the hole.

c) Variation of the total problem

The variation of the total problem can be written as the sum of the global domain variation (3.2) and the local domain variation (3.5):

$$\delta J = \int_{\Gamma} f_{\Gamma}\,\delta s\,d\Gamma + \int_{\gamma} f_{\gamma}\,\delta s_B\,d\gamma = 0 \; . \qquad (3.6)$$

In order to solve the simultaneous shape and topology optimization problem, this task is processed by special algorithms.

3.2 Global domain variation by direct shape strategies

Shape optimization problems can be solved by means of indirect and direct strategies. In *indirect* strategies, the necessary conditions for the optimal shape are derived using variational principles, and the resulting differential equations are then solved, in general by means of approximation methods because of their nonlinearity.

The *direct* solution strategy is easier to apply in many applications. Here, the shape optimization problem is transformed into a parameter optimization problem using approach functions.

Based on the general description of a shape optimization problem, a domain variation problem can be formulated as follows [13, 17, 50]:

$$F^{*}\!\left[\Gamma_{var}^{*}\!\left(\xi^{\alpha}\right)\right] = \operatorname*{Min}_{\Gamma_{var}}\!\left\{ F\!\left[\Gamma_{var}\!\left(\xi^{\alpha}\right)\right] \;\middle|\; \Gamma_{var}\!\left(\xi^{\alpha}\right) \in X \right\} \qquad (3.7)$$

with

$$X = \left\{ \Gamma_{var}\!\left(\xi^{\alpha}\right) \in \Re^{3} \;\middle|\; \mathbf{H}\left[\Gamma_{var}\!\left(\xi^{\alpha}\right)\right] = \mathbf{H}^{0} \; , \; \mathbf{G}\left[\Gamma_{var}\!\left(\xi^{\alpha}\right)\right] \geq \mathbf{G}^{0} \right\} ,$$

and the following notations

$F,\ F^{*}$	objective functionals (see (2.1)),
$\Gamma_{var}\!\left(\xi^{\alpha}\right)$	variable boundary of the structure,
$\Gamma_{var}^{*}\!\left(\xi^{\alpha}\right)$	optimal boundary configuration of the structure,
ξ^{α}	GAUSSIAN surface parameters $\alpha = 1,2$,
$\mathbf{H},\ \mathbf{G}$	equality and inequality operators,
$\mathbf{H}^{0},\ \mathbf{G}^{0}$	bounds of the equality and inequality operators,
X	admissible design space,
$\Re^{3}$	three-dimensional topology domain.

This problem formulation assumes that the boundaries of a three-dimensional space are represented by surfaces, and that the two GAUSSIAN surface parameters ξ^α $(\alpha = 1,2)$ uniquely describe each point on the boundary. For the curve representation of a two-dimensional domain, one merely requires one describing parameter ξ.

The boundary of a body $\Gamma_{var}\left(\xi^\alpha\right)$ is described by the approach functions $\mathbf{R}(\xi^\alpha, \mathbf{x})$. The shape optimization problem (3.7) can thus be written as a parameter optimization problem:

$$F^*\left[\mathbf{R}^*(\xi^\alpha, \mathbf{x}^*)\right] = \underset{\mathbf{R}(\xi^\alpha, \mathbf{x})}{\text{Min}} \left\{ F\left[\mathbf{R}(\xi^\alpha, \mathbf{x})\right] \;\middle|\; \mathbf{R}(\xi^\alpha, \mathbf{x}) \in X \right\} \qquad (3.8)$$

with

$$X = \left\{ \mathbf{R}(\xi^\alpha, \mathbf{x}) \in \Re^3 \;\middle|\; \mathbf{H}\left[\mathbf{R}(\xi^\alpha, \mathbf{x})\right] = \mathbf{H}^0 \,,\, \mathbf{G}\left[\mathbf{R}(\xi^\alpha, \mathbf{x})\right] \geq \mathbf{G}^0 \right\},$$

and with the additional notations

$\mathbf{R}(\xi^\alpha, \mathbf{x})$ approach functions for the description of the component boundaries,

$\mathbf{R}^*(\xi^\alpha, \mathbf{x}^*)$ optimal configuration of the approach functions,

$\mathbf{x}$ vector of the design variables.

The use of general approach functions $\mathbf{R}(\mathbf{x})$ like NURBS (**N**on-**U**niform-**R**ational-**B**-Splines) [20] allows to vary the structural boundaries by the vector of the design variables $\mathbf{x}$ only. The vector consists of the parameters of the approach functions, as, e.g., the coordinates of the control points of splines.

a) Flexible approach functions for shape description

In order to describe so-called free-shape surfaces and -lines, approach functions are required that provide high flexibility with only a small number of free parameters. Flexibility is an important feature of these functions as they allow a sufficient variety of shapes in the optimization. Besides that, one also has to guarantee stability against oscillations.

The boundary of three-dimensional components is generally described by parameterized approach functions $\mathbf{R}(\xi^\alpha, \mathbf{x})$ depending on the GAUSSIAN surface parameters ξ^α $(\alpha = 1,2)$ and on the design variable vector $\mathbf{x}$. The parameter representation of a straight line between two points as an example reads in dependence on the GAUSSIAN line parameter ξ:

$$\mathbf{R}(\xi, \mathbf{P}_0, \mathbf{P}_1) = \mathbf{P}_0 + (\mathbf{P}_1 - \mathbf{P}_0) \cdot \xi. \qquad (3.9)$$

The line is completely described by $\xi \in \left[0,1\right]$.

The non-parameterized description of a line is carried out implicitly by the interrelations of the space coordinates $x^i = (i = 1,2,3)$. In contrast to the non-parameterized description, a parameterized formulation can also describe lines and surfaces parallel to a space coordinate axis.

There are a large number of different approach functions in the field of Computer Aided Geometrical Design (CAGD) among which the B-Spline are the most familiar ones [10].

In practical applications, free-shape curves and surfaces are increasingly described by NURBS, since they are a good compromise between the conflicting demands for flexibility and stability. Free-shape geometries as well as ruled geometries can equally be approximated by NURBS.

The recursive formula for NURBS has been developed in the scope of fundamental investigations into B-Spline curves:

$$R(\xi, \mathbf{x}) = \sum_{i=0}^{n} \frac{N_{i,j}(\xi) \cdot w_i}{\sum_{l=0}^{n} N_{l,j}(\xi) \cdot w_l} \mathbf{P}_i \qquad (3.10)$$

with

$$N_{i,j}(\xi) = \frac{(\xi - t_i) N_{i,j-1}(\xi)}{t_{i+j-1} - t_i} + \frac{(t_{i+j} - \xi) N_{i+1,j-1}(\xi)}{t_{i+j} - t_{i+1}} ,$$

$$N_{i,1}(\xi) = 1 , \qquad \text{for } \xi \in \left[t_i , t_{i+1} \right] ,$$

$$N_{i,1}(\xi) = 0 , \qquad \text{for } \xi \notin \left[t_i , t_{i+1} \right]$$

with the following notations

$\mathbf{P}_i$ vector of the space coordinates of a control point i,

$\xi \triangleq \xi^1$ GAUSSIAN line parameter,

$N_{i,j}(\xi)$ base functions,

j degree of the base functions,

w_i weighting factor of the i-th control point,

t_i i-th component of the knot vector $\mathbf{t}$,

$n + 1$ number of control points.

This approach implies a variation of the space coordinate vector, of the weighting factors, and of the components of the knot vector $\mathbf{t}$. The latter quantities described in the following are used as design variables in a parameter optimization problem.

The *knot vector* $\mathbf{t}$ consists of the j-fold defined starting and end values, and of the intermediate values. For the special case $j = 3$, the intermediate values describe the point where the curve touches the control polygon. Fig. 3.3 illustrates the variation of the knot vector, where the coordinates and the weightings of the control points are kept constant. By variation of $t_{A4} = 1.0$ and $t_{A5} = 2.0$ to $t_{B4} = 1.4$ and $t_{B5} = 1.6$, respectively, the length of the curve section between the two touching points is reduced. For the purpose of illustration, the components of the knot vector could be interpreted as fixing points of a strain-rigid rope.

By increasing the *weighting factor*, the curve is moved towards the respective control point, and vice versa. If all weighting factors possess the same value, they vanish. In contrast to the knot vector, the curve can here be interpreted as a flexible rubber band which is drawn in the direction of the weighted control point. Large-scale differences in the single weighting factors lead to several undesired strong changes of curvature in the whole curve area. The strong changes of curvature can be avoided if weighting factors in the interval $w_i \in [0.2 , 5]$ are used. While the variation of the weighting factors and of the knot vector only changes the curve within the polygon, the coordinates of the control points modify the polygon itself. A geometrical interpretation of this variation can be obtained easily.

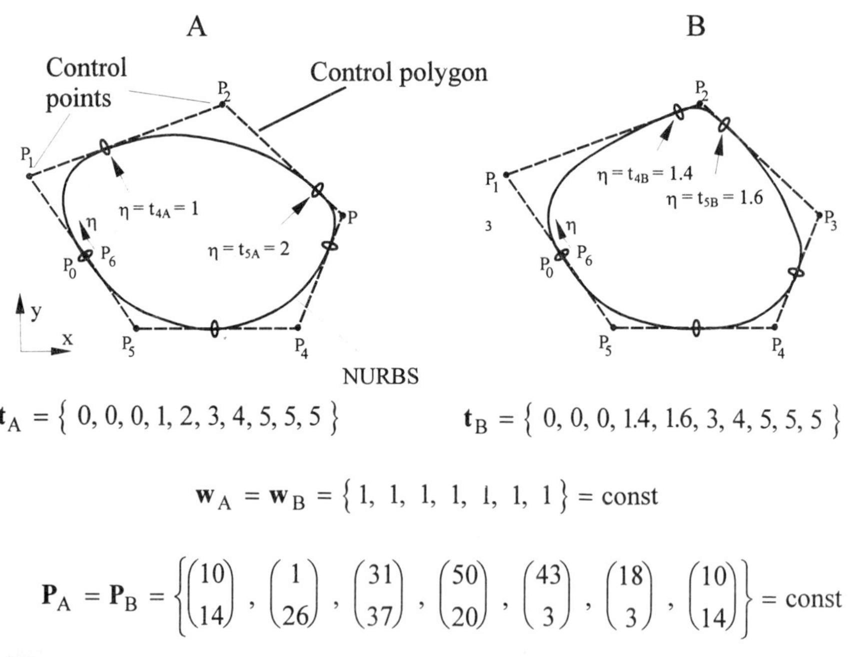

$$\mathbf{t}_A = \{\, 0, 0, 0, 1, 2, 3, 4, 5, 5, 5\, \} \qquad \mathbf{t}_B = \{\, 0, 0, 0, 1.4, 1.6, 3, 4, 5, 5, 5\, \}$$

$$\mathbf{w}_A = \mathbf{w}_B = \{\, 1,\ 1,\ 1,\ 1,\ 1,\ 1,\ 1\, \} = \text{const}$$

$$\mathbf{P}_A = \mathbf{P}_B = \left\{ \begin{pmatrix} 10 \\ 14 \end{pmatrix} , \begin{pmatrix} 1 \\ 26 \end{pmatrix} , \begin{pmatrix} 31 \\ 37 \end{pmatrix} , \begin{pmatrix} 50 \\ 20 \end{pmatrix} , \begin{pmatrix} 43 \\ 3 \end{pmatrix} , \begin{pmatrix} 18 \\ 3 \end{pmatrix} , \begin{pmatrix} 10 \\ 14 \end{pmatrix} \right\} = \text{const}$$

Fig. 3.3: Variation of the knot vector $\mathbf{t}$ of the NURBS-curve

b) Implementation of NURBS into commercial geometry modeling systems

The application of flexible NURBS approach functions for boundary description is still in its initial state [46, 47, 52]. The NURBS are very often approximated piece-wise by implemented spline functions (generally cubic splines) [20]. For this purpose, problem-dependent definitions are to be set up with regard to the mapping exactness. If very sharp edges are demanded, a large number of cubic splines is required for a satisfactory representation. This approximation must be seen as an interim solution until the commercial geometry modeling systems also provide NURBS-curves.

c) Standard-defined approach functions

In the case of highly complex contours, a description with only one NURBS-curve is not sufficient, and hence the boundary has to be approximated by several NURBS. By applying third-degree NURBS-curves with five control points, e.g., the boundaries can be assembled in a C^0 - and C^1 - steady manner.

In order to keep the number of design variables low at the beginning of the optimization process it is sensible to carry out standard definitions for the boundaries of the bubbles, which are then refined successively in the course of the optimization. By using very flexible NURBS, this refinement can be achieved easily in the scope of a program-system. The following distinction can be drawn with regard to the positioning of holes:

- *Positioning within a body*:
  Since many optimized truss structures exclusively possess triangular holes, a triangle is recommended as standard shape.

- *Positioning at the boundary of a body*:
  If a hole is positioned at a boundary, one can deduce that the corresponding boundary description has to be refined, and that a shape optimization in the same topology class can be repeated. If a hole is to be positioned at a non-variable boundary (e.g., at a clamped edge), it is advised to implement a notch. In case the symmetry of a component is employed in the calculation, and if the positioning criterion determines a position on the symmetry line, a notch can be used, which then has to fulfill C^0 - and C^1 - steadiness at the symmetry plane.

d) Successive refinement of the standard definitions of the NURBS approach functions

Whenever a hole is positioned at a variable component boundary, the corresponding approach function has to be refined successively in accordance with the following hierarchy:

1. Definition of the coordinates of the control points calculated in the standard shapes from the coordinates of other control points;
2. Variation of the knot vectors;
3. Variation of the weighting factors;

4. Insertion of further control points;
5. Introduction of an additional NURBS standard curve.

The approach functions are refined owing to the positioning of a hole at the boundary of a component. There are further possibilities of adapting the geometry model during an optimization process.

e) Sensitivity analysis

The next step is the sensitivity analysis, which presents a very important task in the optimization loop [4, 24]. When using commercial structural analysis programs, one requires a sensitivity analysis that is independent of the structural analysis program. The shape optimization in the *bubble method* is carried out using the finite differences method [15, 16] and the variational sensitivity analysis [11, 30, 31]. The first method requires a high degree of computing time and is relatively inaccurate, but it can be applied to arbitrary optimization functionals. The second method is suitable for a certain type of optimization functionals, because it does not have to carry out a new structural analysis for each design variable.

3.3 Explicit determination of hole positioning

An elastic body is subjected to different loads, and can be described by a mixed boundary-value problem with given stresses

$$t^j_{(\Gamma_\tau)} = (\tau^{ij} n_i)_{\Gamma_\tau}$$

and given displacements

$$v_{i(\Gamma_v)} = (v_i)_{\Gamma_v} \ .$$

The optimization problem is defined as follows, where the objective is to minimize the volume of the component [3,4]:

$$F = \int_\Omega d\Omega \rightarrow \text{Minimum} \tag{3.11}$$

with additional integral expressions as inequality constraints:

$$G_v = \int_\Omega g_v\left(\tau^{ij}, v_i\right) d\Omega < G_v^0 \quad , \ v = 1,2,\dots,N \ , \tag{3.12a}$$

or transformed inequality constraints by using slack variables μ_v^2:

$$G_v = \int_\Omega g_v\left(\tau^{ij}, v_i\right) d\Omega = G_v^0 - \mu_v^2 \ . \tag{3.12b}$$

The basic equations of the linear theory of elasticity (equilibrium conditions, strain-displacement relations, and material law) [17] are considered as equality constraints:

$$\left.\tau^{ij}\right|_{j} + f^{i} = 0 \, , \tag{3.13a}$$

$$\gamma_{ij} = \frac{1}{2}\left(\left.v_{i}\right|_{j} + \left.v_{j}\right|_{i}\right) \, , \tag{3.13b}$$

$$\tau^{ij} = C^{ijkl}\,\gamma_{kl} - \beta^{ij}\,\Theta \tag{3.13c}$$

with C^{ijkl} - components of the isothermal elasticity tensor

$$C^{ijkl} = \frac{E}{2(1+v)}\left(g^{ik}g^{jl} + g^{il}g^{jk} + \frac{2v}{1-2v}g^{ij}g^{kl}\right) \, ,$$

β^{ij} - components of the thermo-elastic tensor

$$\beta^{ij} = \beta g^{ij} = \frac{E\alpha_{T}}{1-2v}g^{ij} \, ,$$

α_{T} - thermal expansion coefficient ,

Θ - temperature difference from the point of reference $\Theta = T - T_{0}$,

$\big|_{j}$ - covariant derivatives due to ξ^{j} .

On the basis of (2.2) together with (3.11) to (3.13) one can formulate the LAGRANGE functional for the solution of convex optimization problems:

$$J = \int_{\Omega} f_{L}\left(\tau^{ij}, v_{i}\right)d\Omega + \sum_{v=1}^{N}\lambda_{v}\left(G_{v}^{0} - \mu_{v}^{2}\right) \, , \tag{3.14}$$

with

$$f_{L} = 1 + \sum_{v=1}^{N}\lambda_{v}g_{v}\left(\tau^{ij}, v_{i}\right) + \psi_{i}\left[\left.\tau^{ij}\right|_{j} + f^{i}\right] + \chi_{ij}\left[\tau^{ij} - C^{ijkl}\gamma_{kl} + \beta^{ij}\Theta\right] \, . \tag{3.15}$$

In (3.15), ψ_{i} and χ_{ij} are adjoint functions as special LAGRANGE-multipliers of the mechanical problem, while the LANGRANGE-multipliers λ_{v} are employed to consider the constraints of the optimization problem.

The adjoint functions ψ_{i} and χ_{ij} can be interpreted as the "*pseudo*" initial displacements and the "*pseudo*" initial stresses of a corresponding body [11, 24]. If the stresses in this body (which is also called *adjoint body*) equal zero, there are already initial displacements, and vice versa. The corresponding body possesses the same dimensions as the original body. The calculation of the adjoint functions, i.e., of the suitable "*pseudo*" initial states, depends on the constraints G_v, and may require a high analytical effort. In the case of so-called *self-adjoint* problems, the corresponding body does not have to be calculated explicitly since the adjoint functions can be determined directly from the state of the original body.

The general optimization problem is expressed by the LAGRANGE-function (3.15). The integrand f_L thus depends on the following vector (without consideration of the volume forces):

$$\mathbf{u} = \left(v^i; \ \tau^{ij}; \ \psi^i; \ \chi^i \ ; \ \lambda_1, \dots \lambda_N \right) .$$

The vector $\mathbf{u}$ consists of 18 components considering the mechanical basic equations, and of N LAGRANGE-multipliers. According to [9], the first variation of (3.14) with respect to the K-th component of $\mathbf{u}$ contains an additional expression that stems from the variation of the variable domain. We thus obtain:

$$\delta J_K = \sum_{i=1}^{3} \sum_{j=1}^{3} \left\{ \int_\Omega \left[\frac{\partial f_L}{\partial u_K} - \left(\frac{\partial f_L}{\partial u_K|_i} \right) \bigg|_j \right] (\delta u_K)_\Gamma \, d\Omega + \int_\Omega \left(\frac{\partial f_L}{\partial u_K|_i} (\delta u_K)_\Gamma \right) \bigg|_j \, d\Omega \right.$$

$$\left. + f_L \delta\Omega \bigg|_{\Omega_{Var}} + 2 \sum_{v=1}^{N} (\lambda_v \mu_v) \delta\mu_v \right\} = 0 \ . \tag{3.16}$$

Here, the variation $(\delta u_K)_\Gamma$ of the undisplaced boundary Γ is linked to the variation δu_K of the displaced boundary via the relation

$$\delta u_K = (\delta u_K)_\Gamma + u_K|_i \delta\xi^i \ . \tag{3.17}$$

This yields for the variation of the single state quantities:

$$\left(\delta\tau^{ij} \right)_\Gamma = \delta\tau^{ij} - \tau^{ij}|_k \delta\xi^k \ , \quad (\delta v_i)_\Gamma = \delta v_i - \gamma_{ij} \delta\xi^j \ .$$

For further calculation one requires the explicit derivatives of the components of $\mathbf{u}$ that can be calculated directly from (3.13b) (symmetry condition: $\gamma_{ij} = v_i|_j = v_j|_i$):

$$\frac{\partial f_L}{\partial \tau^{ij}} = \sum_{v=1}^{N} \lambda_v \frac{\partial g_v}{\partial \tau^{ij}} + \chi_{ij} \ , \qquad\qquad \frac{\partial f_L}{\partial \tau^{ij}|_j} = \psi_i \ ,$$

$$\frac{\partial f_L}{\partial v_i} = \sum_{v=1}^{N} \lambda_v \frac{\partial g_v}{\partial v_i} \ , \qquad\qquad \frac{\partial f_L}{\partial \gamma_{kl}} = - \chi_{ij} C^{ijkl} \ . \tag{3.18}$$

We obtain for f_L in the state of a complete variation:

$$f_L = 1 + \sum_{v=1}^{N} \lambda_v\, g_v \; . \qquad (3.19)$$

The two middle terms in (3.16) can be transformed using GREEN's rule:

$$\iiint_{\Omega} \left. \left(\frac{\partial f_L}{\partial u_K|_i}(\delta u_K)_\Gamma \right) \right|_j \, d\Omega = \iint_{\Gamma} n^j \frac{\partial f_L}{\partial u_K|_i}(\delta u_K)_\Gamma \, d\Gamma \; ,$$

$$f_L\, \delta\Omega \big|_{\Omega_{var}} = \int\int_{\Gamma} n_i f_L \delta\xi^i d\Gamma \; .$$

Based on these transformations we obtain from (3.16) with $\delta J = \sum (\delta J_K)$:

$$\delta J = \int_{\Omega} \left[\left(\sum_{v=1}^{N} \lambda_v \frac{\partial g_v}{\partial \tau^{ij}} + \chi_{ij} \right) - \psi_i \big|_j \right] \left(\delta\tau^{ij} - \tau^{ij} \big|_k \delta\xi^k \right) d\Omega +$$

$$+ \int_{\Gamma} \psi_i n_j \left(\delta\tau^{ij} - \tau^{ij} \big|_k \delta\xi^k \right) d\Gamma +$$

$$+ \int_{\Omega} \left\{ \left(\sum_{v=1}^{N} \lambda_v \frac{\partial g_v}{\partial v^i} \right) - \left(-\chi_{kl} C^{ijkl} \right) \big|_j \right\} \left(\delta v_i - \gamma_{ij} \delta\xi^j \right) d\Omega + \qquad (3.20)$$

$$+ \int_{\Gamma} \left\{ \chi_{kl} C^{ijkl} \right\} n_j \left(\delta v_i - \gamma_{ij} \delta\xi^j \right) d\Gamma +$$

$$+ \int_{\Gamma_{var}} \left(1 + \sum_{v=1}^{N} \lambda_v\, g_v \right) n_i\, \delta\xi^i\, d\Gamma + 2\sum_{v=1}^{N} (\lambda_v \mu_v) \delta\mu_v \; .$$

The first two integrals are obtained by substituting the stress components, and the next two integrals are derived from the displacement components. The variation (3.20) finally leads to the necessary condition of the total problem.

- Evaluation of the EULER equations then yields:

$$\left[\sum_{v=1}^{N}\lambda_v\frac{\partial g_v}{\partial\tau^{ij}}+\chi_{ij}-\psi_i|_j\right]_{\Omega}=0 \quad , \qquad (3.21a)$$

$$\left[\sum_{v=1}^{N}\lambda_v\frac{\partial g_v}{\partial v_i}+\chi_{kl}|_j\,C^{ijkl}\right]_{\Omega}=0 \quad . \qquad (3.21b)$$

- Boundary conditions of the adjoint functions:

$$\left[\psi_i n_j\right]_{\Gamma_v}=0 \quad ,$$

$$\qquad (3.22)$$

$$\left[\chi_{kl}C^{ijkl}\right]_{\Gamma_\tau}=0 \quad .$$

Furthermore, the LAGRANGE-function has to vanish at the variable boundary:

$$\left[1+\sum_{v=1}^{N}\lambda_v g_v\right]_{\Gamma_{var}}=0 \quad . \qquad (3.23)$$

For the product of slack variables and LAGRANGE-multipliers it holds that

$$2\lambda_v\,\mu_v=0, \qquad (3.24)$$

and at the non-variable boundary of the component the following conditions have to be fulfilled:

$$\left[\delta\xi^i\right]_{\Gamma_{\tau,n\,var}}=0 \quad , \qquad\qquad \left[\delta\tau^{ij}n_j\right]_{\Gamma_{\tau,n\,var}}=0 \quad ,$$

$$\qquad (3.25)$$

$$\left[\delta\xi^i\right]_{\Gamma_{v,n\,var}}=0 \quad , \qquad\qquad \left[v_i\right]_{\Gamma_{v,n\,var}}=0 \quad .$$

For a further treatment, the total problem is partitioned into sub-problems, i.e., variation of the stress boundaries δJ_τ, the displacement boundaries δJ_v, and of the unloaded boundaries δJ_0:

$$\delta J = \delta J_v + \delta J_\tau + \delta J_0 = 0 \quad . \qquad (3.26)$$

For the single subvariations it is necessary to separate the corresponding terms from (3.20). We require the integrals over the domain Ω and the variation over Γ_{var} for all partial variations. For the subvariations δJ_τ and δJ_0, one needs the first boundary variation in (3.20), while the second boundary variation is required for δJ_v.

Thus, we can write for the variation over the stress boundaries δJ_τ:

$$\delta J_\tau = \int_{\Gamma_{var}} \left[\psi_i n_j \left(\tau^{ij} - \tau^{ij}\big|_k \delta\xi^k \right) + \left(1 + \sum_{v=1}^N \lambda_v g_v \right) n_k \,\delta\xi^k \right] d\Gamma \ . \qquad (3.27)$$

With the boundary condition at the unloaded boundary

$$t^j(\Gamma_0) = \left(\tau^{ij} n_i \right)_{\Gamma_0} = 0 \ , \qquad (3.28)$$

and because of $\psi_i \tau^{ij}\big|_j n_j + \psi_i\big|_j \tau^{ij} n_j = \left(\psi_i \tau^{ij} \right)\big|_j n_j$ and with $\delta s = n_i \delta\xi^i$, the variation over the unloaded boundaries δJ_0, can be simplified as

$$\boxed{ \delta J_0 = \int_{\Gamma_0} \left[1 + \sum_{v=1}^N \lambda_v g_v - \left(\psi_i \tau^{ij} \right)\big|_j \right] \delta s \, d\Gamma = 0 } \ . \qquad (3.29)$$

The variation expression (3.29) is required for the derivation of an analytical expression for the positioning of a hole (*characteristic function*). At the same time, it is the basis for the variational sensitivity analysis.

3.4 Solution strategy and numerical implementation

Topology optimization via bubble method is based upon a solution concept comprising an iterative positioning of new bubbles followed by a hierarchically secondary shape optimization of the new bubbles together with all variable boundaries.

The optimization process consists of the following steps:

Step 1: For a given topology domain, shape optimization is carried out, considering objective and constraint functionals, after which the structure of the component in its topology class cannot be improved any further.

Step 2: By inserting a hole (transfer to another topology class) one tries to achieve improved results. The coordinates of the optimal position of the new hole (bubble) are required. The positioning is carried out by a so-called positioning criterion to be determined analytically for special objective and constraint functionals (e.g., the global stiffness or the volume) and numerically for general cases.

Step 3:   After positioning, a shape optimization is carried out in order to find the optimal shape of new bubbles and to determine the influence on the other variable boundaries.

Step 4:   Back to Step 2.

After the shape optimization has been completed, a further bubble is inserted by means of the positioning criterion, and is then optimized with respect to its shape. This generates an iterative process, and one obtains a number of possible topologies out of which a suitable variant can be chosen by a criterion still to be developed. This choice, in turn, must follow external demands on the construction (e.g., manufacturing requirements).

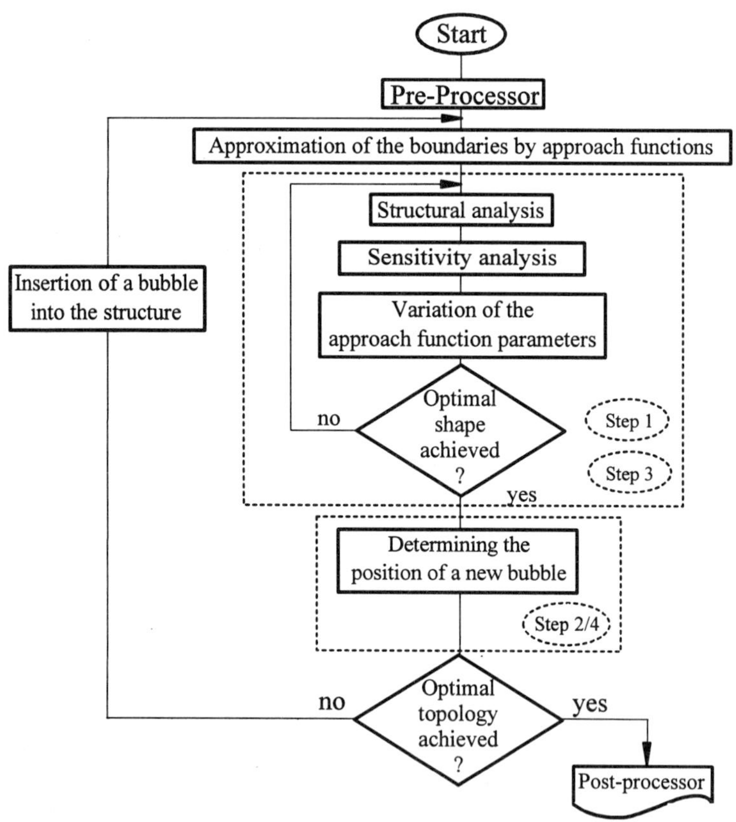

Fig. 3.4: Flow-chart of the bubble method

The bubble method is numerically realized by means of the optimization procedure SAPOP (**S**tructural **A**nalysis **P**rogram and **O**ptimization **P**rocedure) [18, 26]. For that purpose, SAPOP had to be augmented by the corresponding positioning model. Fig. 3.4 presents the flow-chart of the bubble method.

Using FE-analysis during the optimization process, the FE-mesh has to be produced in each iteration by a free mesh generator. If the positioning criterion in the bubble method decides to form an open hole (bubble) at the boundary of the contour, a notch is created. If the hole is positioned within the solid structure, the method generates a closed hole (bubble).The potential of applying the bubble method depends on the works in classical shape optimization. SAPOP offers a large number of different optimization strategies and algorithms which facilitate an effective application of the bubble method [13, 26, 47, 48, 50].

4 Criteria for hole positioning

As we learned in the previous Chapter 3, it is tried to increase the optimal behaviour of a solid body after shape optimization of the outer boundaries by positioning a hole. Therefore, the coordinates of the optimal position of the new hole or the bubble shall be determined now. For complex optimization functionals, the positioning is carried out by means of numerical search procedures, where algorithms of mathematical programming (MP-algorithms) are used.

Considering a simple optimization functional - e.g., mean compliance or volume - an analytical variational term is to be determined to lead to the so-called *characteristic function*. The optimal position of a bubble is found at that point in the structure, where the function has a *minimal* value. By means of (3.29), the explicit variational term of the optimization problem is to be formulated, assuming that the infinitesimal bubble to position is already inserted into the structure.

A solid body has to be designed in such a way that it meets various demands for all operational load cases, where the influence factors may be manifold, e.g., in the case of a complicated material behaviour. Often, these factors render extremely expensive structural analyses necessary. At the example of the mean compliance of a solid body, the essentials of a practice-oriented topology optimization procedure are shown. The respective demands are covered in the form of objective and constraint functionals. A transformation of local functions into global domain functionals has proved to be suitable as the sensitivities of the functionals with respect to the design variables can be described analytically.

4.1 Mean compliance for plane stress problems

By mean compliance one denotes the integral over the product of the boundary stresses or over the volume forces and the corresponding displacements in the state of equilibrium. For a solid body with given boundary stresses $t^j_{(\Gamma_\tau)} = (\tau^{ij} n_i)_{\Gamma_\tau}$ and given boundary displacements $v_{j(\Gamma_v)} = (v_j)_{\Gamma_v}$ according to Fig. 4.1, the *mean compliance* reads

$$G = \int_{\Gamma_\tau} v_j\, t^j_{(\Gamma_\tau)}\, d\Gamma - \int_{\Gamma_v} v_{j(\Gamma_v)} \tau^{ij} n_i d\Gamma + \int_\Omega v_i f^i\, d\Omega \ , \tag{4.1}$$

with f^i as volume forces.

The mean compliance can be described by the total potential Π or by the total complementary potential Π^*, and it is valid for a linear-elastic material behaviour [3,17,49]:

$$G = -2\Pi \quad \text{with} \quad \Pi = \int_\Omega \overline{U}\, d\Omega - \int_\Omega v_i f^i\, d\Omega - \int_{\Gamma_\tau} v_j\, t^j_{(\Gamma_\tau)}\, d\Gamma \tag{4.2}$$

or

$$G = 2\Pi^* \quad \text{with} \quad \Pi^* = \int_\Omega \overline{U}^*\, d\Omega - \int_\Omega v_i f^i\, d\Omega - \int_{\Gamma_v} v_{j(\Gamma_v)} \tau^{ij} n_i d\Gamma \ , \tag{4.3}$$

where $\overline{U}$ and $\overline{U}^*$ denote the specific deformation energy and the specific complementary energy, respectively.

Hence, a reduction of the mean compliance corresponds to an increase of the total potential or of a decrease of the total complementary potential. If no volume forces occur within the structure and if the given displacements at Γ_v equal zero, the compliance minimization problem can be transformed into a minimization of the complementary energy. Thermal influences can easily be considered by augmenting the expression for the

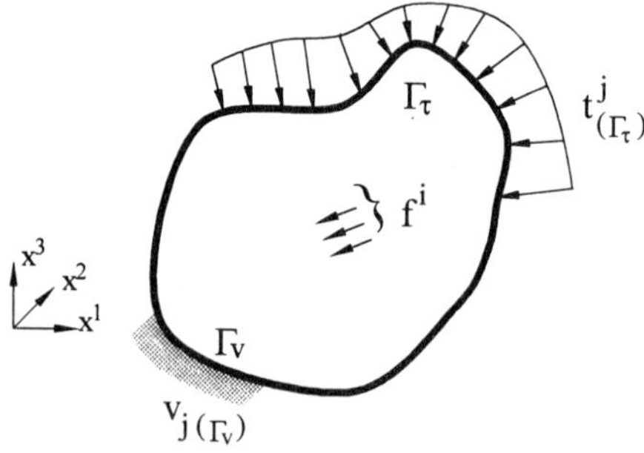

Fig. 4.1: External loads acting on a solid body

complementary energy [17]:

$$G = \int_\Omega \overline{U}^* d\Omega \qquad (4.4a)$$

with

$$\overline{U}^* = \frac{1}{2} D_{ijkl} \tau^{ij} \tau^{kl} + \alpha_T \tau_i^i (T - T_0) \ . \qquad (4.4b)$$

When considering temperature fields an additional term of a stress-independent function occurs in (4.4b). In the present cases it shall be neglected.

In a first step, the complementary energy (4.4b) is investigated for the optimization functional of mean compliance presented in (4.4a):

$$G_1 = \int_\Omega g_1 \, d\Omega \qquad (4.5)$$

with

$$g_1 \mathrel{\hat=} \overline{U}^* = \frac{1}{2} D_{ijkl} \tau^{ij} \tau^{kl} + \alpha_T \tau_i^i (T - T_0) \ .$$

By transforming the mixed tensor τ_i^i into a contravariant tensor $\tau_i^i = g_{ij} \tau^{ij}$, the derivatives required for EULER's equations (3.18) lead to:

$$\frac{\partial g_1}{\partial v_i} = 0 \quad , \qquad (4.6)$$

$$\frac{\partial g_1}{\partial \tau^{ij}} = D_{ijkl} \tau^{kl} + \alpha_T g_{ij} (T - T_0). \qquad (4.7)$$

The differential (4.6) is substituted in (3.21b) to calculate the adjoint functional χ_{ij}:

$$\left[\chi_{kl} \big|_j C^{ijkl} \right]_\Omega = 0 \ . \qquad (4.8)$$

Eq.(4.8) is fulfilled for arbitrary, constant values of the adjoint functions χ_{ij}, since their derivatives equal zero. For further calculation, it holds that

$$\chi_{ij} = 0 \ . \qquad (4.9)$$

The adjoint functions ψ_i are calculated by means of (4.7), (4.9), and EULER's equation (3.21a):

$$\lambda_1 \left[D_{ijkl} \tau^{kl} + \alpha_T g_{ij} (T - T_0) \right] - \psi_i \big|_j = 0 \ . \qquad (4.10a)$$

Introducing the strains and displacements it follows:

$$\lambda_1 \, \gamma_{ij} - \psi_i\big|_j = \lambda_1 \, v_i\big|_j - \psi_i\big|_j = 0 \ . \tag{4.10b}$$

Integration of this term yields for the adjoint functions

$$\psi_i = \lambda_1 v_i + c_i \tag{4.11}$$

with the integration constants c_i. By substituting the adjoint functions ψ_i in (3.29), one obtains for the boundary variation of the bubble $\delta J_\gamma = \delta J_0$:

$$\delta J_\gamma = \int\limits_\gamma \left\{ 1 + \lambda_1 \left(\frac{1}{2} D_{ijkl} \tau^{ij} \tau^{kl} + \alpha_T \tau_i^i (T - T_0) \right) - \left[\left(\lambda_1 v_i + c_i \right) \tau^{ij} \right]\big|_j \right\} \delta s \, d\Gamma = 0.$$

$$\tag{4.12}$$

The application of (4.12) is sufficient, since only the variation of the bubble is of interest for the hole positioning. The variation of the other component boundaries is already carried out by means of shape optimization.

With $\tau^{ij}\big|_j = 0$ and by transforming the differential

$$\left(\lambda_1 v_i \tau^{ij} \right)\big|_j = \lambda_1 \left(\tau^{ij} \gamma_{ij} \right) = \lambda_1 \left[D_{ijkl} \tau^{ij} \tau^{kl} + \alpha_T \tau_i^i (T - T_0) \right] \ ,$$

the functional δJ_γ can be simplified to:

$$\delta J_\gamma = \int\limits_\gamma \left\{ 1 + \lambda_1 \left[\frac{1}{2} D_{ijkl} \tau^{ij} \tau^{kl} + \alpha_T \tau_i^i (T - T_0) \right] - \lambda_1 \left[D_{ijkl} \tau^{ij} \tau^{kl} + \alpha_T \tau_i^i (T - T_0) \right] \right\} \delta s \, d\Gamma = 0. \tag{4.13}$$

For further calculation, the variation of the bubble volume and the variation of the mean compliance are separated:

$$\delta J_\gamma = \int\limits_\gamma \delta s \, d\Gamma - \lambda_1 \int\limits_\gamma \left[\left(\frac{1}{2} D_{ijkl} \tau^{ij} \tau^{kl} \right) - \alpha_T \tau_i^i (T - T_0) \right] \delta s d\Gamma =$$

$$= \delta \Omega - \lambda_1 \int\limits_\gamma \overline{U}^{*'} \, \delta s \, d\Gamma = \delta \Omega - \lambda \left(\delta J_\gamma \right)_1 = 0 \ . \tag{4.14}$$

The first term is constant, because the inserted bubble possesses an arbitrary but constant volume variation $\delta\Omega$. For positioning, the variation of the complementary energy $(\delta J_\gamma)_1$ is employed:

$$\left(\delta J_\gamma\right)_1 = \int_\gamma \overline{U}^{*'}\,\delta s\,d\Gamma = \int_\gamma \left[\left(\frac{1}{2}D_{ijkl}\tau^{ij}\tau^{kl}\right) + \alpha_T\tau_i^i(T-T_0)\right]\delta s\,d\Gamma = 0 \quad . \qquad (4.15)$$

To calculate the complementary energy, the local stresses at the boundary of the bubble in dependence on the global state of stress are required.

4.1.1 Positioning of circular holes

Many design components are plane load-bearing structures (disks, plates) or they consist of them. Disks are subjected to loads in their plane. In the following, positioning criteria with regard to different hole shapes are presented for such disks.

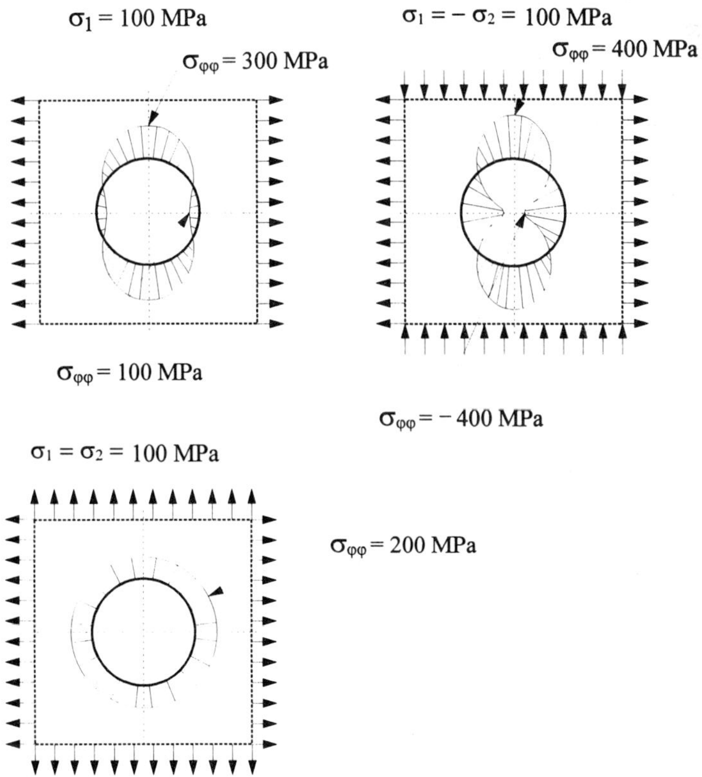

Fig. 4.2: Stresses at the boundary of the circular hole in dependence on different global principal stresses

In order to calculate the stresses at the boundary of a circular hole in a disk (plane stress state) without temperature effects, one can determine the stresses at the circular hole in dependence on the global principal stresses σ_1 and σ_2 at the boundary of the evaluation domain (see Fig. 4.2) [17]

$$\sigma_{\varphi\varphi} = (\sigma_1 + \sigma_2) - 2(\sigma_1 - \sigma_2)\cos 2\varphi . \tag{4.16}$$

Since the boundary of the hole is unstressed, the radial stress elements vanish:

$$\sigma_{rr} = \tau_{r\varphi} = 0 . \tag{4.17}$$

From the boundary stresses, the complementary energy $\overline{U}^{*'}$ yields in polar coordinates:

$$\overline{U}^{*'} = \frac{1}{2E}\left[\left(\sigma_{rr} + \sigma_{\varphi\varphi}\right)^2 - 2(1+\nu)\left(\sigma_{rr}\sigma_{\varphi\varphi} - \tau_{r\varphi}^2\right) \right] . \tag{4.18}$$

Inserting of (4.16) and (4.17) the term for $\overline{U}^{*'}$ along the boundary of the hole in dependence on the principal stresses σ_1 and σ_2 then reads:

$$\overline{U}^{*'} = \frac{1}{2E}\left[(\sigma_1 + \sigma_2)^2 + 2(\sigma_1 - \sigma_2)^2 - 4\left(\sigma_1^2 - \sigma_2^2\right)\cos 2\varphi + 2(\sigma_1 - \sigma_2)^2 \cos 4\varphi \right] .$$

$$\tag{4.19}$$

Eq. (4.19) must be substituted into the functional (4.14) and integrated along the boundary of the hole $d\gamma = r_B d\varphi$ to obtain the characteristic function for positioning the bubble:

$$\left(\delta J_\gamma\right)_{1CD} = \int_0^{2\pi}\left\{ \frac{1}{2E}\left[(\sigma_1 + \sigma_2)^2 + 2(\sigma_1 - \sigma_2)^2 - 4\left(\sigma_1^2 - \sigma_2^2\right)\cos 2\varphi + \right.\right.$$
$$\left.\left. + 2(\sigma_1 - \sigma_2)^2 \cos 4\varphi \right] \right\} r_B \, t \, d\varphi \, \delta s . \tag{4.20}$$

The index 1CD denotes the application of a circular bubble C in disks D with respect to the complementary energy 1.

Integration yields:

$$\left(\delta J_\gamma\right)_{1CD} = \frac{\pi}{E}\left[(\sigma_1 + \sigma_2)^2 + 2(\sigma_1 - \sigma_2)^2 \right] r_B \, t \, \delta s . \tag{4.21}$$

For further considerations the functional is split into two terms:

$$\left(\delta J_\gamma\right)_{1SK} = \delta\Omega\,\Phi_{1CD}(\sigma_1,\sigma_2) \quad \text{with} \quad \delta\Omega = 2\pi\,r_B\,t\,\delta s \qquad (4.22)$$

and

$$\boxed{\Phi_{1CD}(\sigma_1,\sigma_2) = \frac{1}{2E}\left[(\sigma_1+\sigma_2)^2 + 2(\sigma_1-\sigma_2)^2\right]} \qquad (4.23)$$

In (4.23), $\delta\Omega$ is a virtual, infinitesimally small value. But, due to (3.24) and (3.29), the variation over the boundary of the hole $\delta J_\gamma = \delta J_0$ as well as the variation over the boundary of the stress functional δJ_τ and the boundary of the displacement functional δJ_v must equal zero. This indicates that the function $\Phi_{1CD}(\sigma_1,\sigma_2)$ depending on the coordinates x^1 and x^2 must become zero. Since in most cases this does not occur, the bubble is positioned at the point in the disk, where the *characteristic function* takes the minimal value.

The entire disk can be analysed to find the minimum because of the simple structure of the *characteristic function*. No further optimization step is necessary. In case of using a finite-element-model, all knots existing in the topology domain are analysed, which finally results in the optimal positioning vector $\mathbf{r}^*(x,y)$.

4.1.2 Positioning of non-circular holes

The positioning of circular holes does not allow to derive any information on the orientation and the optimal shape of the bubble. However, this information in particular is highly suitable for a subsequent shape optimization, as one may derive appropriate approach functions for the boundary description, and an efficient initial design.

In order to obtain a complete expression for the characteristic function, the stress distribution is determined at an arbitrary boundary shape of a hole, using the Method of Complex Stress Functions according to KOLOSOV [23, 36, 44]. In this method, an elasticity problem is solved by determining two complex functions $\Phi(z)$ and $\Psi(z)$ of the complex variables $z = x^1 + ix^2$. The two complex stress functions then yield the stresses σ_{11}, σ_{22} and τ_{12} and the displacements u and v (in Cartesian coordinates) from the so-called KOLOSOV-formulas:

$$\sigma_{11} + \sigma_{22} = 2\left[\Phi'(z) + \overline{\Phi}'(\bar{z})\right] = 4\,\mathrm{Re}\{\Phi'(z)\}\ ,$$

$$\sigma_{11} - \sigma_{22} + 2i\tau_{12} = -2\left[z\overline{\Phi}''(\bar{z}) + \overline{\Psi}'(\bar{z})\right]\ , \qquad (4.24)$$

$$2G(u+iv) = \kappa\,\Phi(z) - z\overline{\Phi}'(\bar{z}) - \overline{\Psi}(\bar{z})$$

with $\qquad \kappa = \dfrac{3-\nu}{1+\nu}$ $\qquad$ for the state of plane stress (SPStress) ,

$\qquad\qquad \kappa = 3 - 4\nu$ $\qquad$ for the state of plane strain (SPStrain) .

$\overline{\Phi}(\overline{z}) = \overline{\Phi(z)}$ and $\overline{\Psi}(\overline{z}) = \overline{\Psi(z)}$ are the conjugate-complex functions to Φ and Ψ .

When calculating the stresses at the boundaries of a hole it is sensible to transfer the KOLOSOV-formulas onto a coordinate system locally defined at the boundary. By this, the components of the stresses and displacements σ_{rr}, v_r have an orientation normal to the boundary of the hole. With respect to the r,φ -coordinates, the KOLOSOV-formulas yield with $z = r\,e^{i\varphi}$, $\overline{z} = r\,e^{-i\varphi}$ and $(v_r + iv_\varphi) = e^{-i\varphi}(u + iv)$:

$$\sigma_{rr} + \sigma_{\varphi\varphi} = 2\left[\Phi'(z) + \overline{\Phi}'(\overline{z})\right] = 4\,\mathrm{Re}\{\Phi'(z)\} ,$$

$$\sigma_{rr} - \sigma_{\varphi\varphi} + 2i\tau_{r\varphi} = -2\left[\overline{z}\,\overline{\Phi}''(\overline{z}) + \frac{\overline{z}}{z}\,\overline{\Psi}'(\overline{z})\right] , \qquad (4.25)$$

$$2G(v_r + iv_\varphi) = e^{-i\varphi}\left[\kappa\,\Phi(z) - z\overline{\Phi}'(\overline{z}) - \overline{\Psi}(\overline{z})\right] .$$

If we proceed from an unloaded hole in an infinitely extended disk it is required

$$\Psi(z) = -\overline{\Phi}(\overline{z}) - \overline{z}\Phi'(z)$$

as an additional equation for the boundary conditions.

Conformal mapping allows to consider arbitrary shapes of a hole. Using the polynomial approach [23]

$$z(\zeta) = \frac{c_{-1}}{\zeta} + \sum_{k=1}^{n} c_k \zeta^k , \qquad (4.26)$$

the outer domain of the arbitrarily perforated z-plane can conformally be mapped onto the inner domain of the unit circle of the ζ-plane $\zeta = re^{i\varphi}, |\zeta| \leq 1$ (Fig. 4.3), where c_{-1} and c_k are real constants. In the case of complicated hole shapes, the formulation of suitable mapping functions is difficult. Elliptical and triangular holes are the type of holes most frequently used in examples of application.

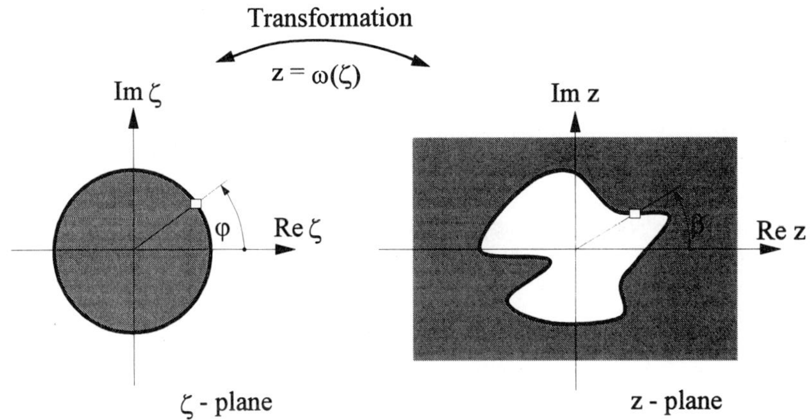

Fig. 4.3: Conformal mapping of an arbitrary hole contour onto a unit circle

For the transformed stress functions

$$\Phi(z), \Psi(z) \xleftrightarrow{\ \text{Mapp.}\ } \Phi[z(\zeta)] = \Phi(\zeta),\ \Psi[z(\zeta)] = \Psi(\zeta)\ , \qquad (4.27)$$

we can state, with the required derivatives

$$\Phi'(\zeta) = \frac{d\Phi}{d\zeta} = \frac{d\Phi}{dz}\frac{dz}{d\zeta} = \Phi'(z) \cdot z'(\zeta)\ , \qquad (4.28)$$

the *transformed KOLOSOV-formulas* as:

$$\sigma_{rr} + \sigma_{\varphi\varphi} = 2\left[\Omega(\zeta) + \overline{\Omega}(\overline{\zeta})\right] = 4\,\mathrm{Re}\left\{\Omega(\overline{\zeta})\right\}\ ,$$

$$\sigma_{rr} - \sigma_{\varphi\varphi} + 2i\tau_{r\varphi} = -2\,\frac{\overline{\zeta}}{\xi z'(\zeta)}\left[z(\zeta)\overline{\Omega}'(\overline{\zeta}) + \overline{\Psi}'(\overline{\zeta})\right]\ , \qquad (4.29)$$

$$2G(v_r + iv_\varphi) = \frac{\overline{\xi}\overline{z}'(\overline{\zeta})}{r\left|\overline{z}'(\overline{\zeta})\right|}\left[\kappa\,\Phi(\zeta) - z(\zeta)\overline{\Omega}(\overline{\zeta}) - \overline{\Psi}(\overline{\zeta})\right]$$

with the abbreviation $\Omega(\xi) = \dfrac{\Phi'(\zeta)}{z'(\zeta)}$.

The problem is then solved for a suitable function $z(\zeta)$ by expanding the two stress functions in TAYLOR-series, and by adapting the coefficients in such a way that the stresses in the infinite ($z \rightarrow \infty$) remain finite.

Examples:

a) *Positioning of an elliptical hole*

The *optimal position and orientation* of a hole can relatively easily be shown for an elliptical hole (Fig. 4.4). The parametrical description of an ellipse referring to a coordinate system in the middle of the hole reads:

$$x^1 = a\cos\varphi = R(1 + m)\cos\varphi,$$
$$x^2 = b\sin\varphi = R(1 - m)\sin\varphi \qquad (4.30)$$

with $\quad m = \dfrac{a-b}{a+b}\quad,\quad R = \dfrac{a+b}{2}\quad$ (middle semi-axis).

The x^1/x^2-ratio yields the relation between an angle φ in the ζ-plane and the real angle of projection β in the z-plane:

$$\tan\beta \;=\; \frac{x^2}{x^1} = \frac{1-m}{1+m}\tan\varphi$$

The factor m denotes the excentricity of the ellipse, i.e., its deviation from the circle. Special cases are $\quad m = 0\;\to\;$ ellipse becomes a circle.
$\qquad\qquad\qquad\qquad m = 1\;\to\;$ ellipse takes the form of a slot with the length 4R.

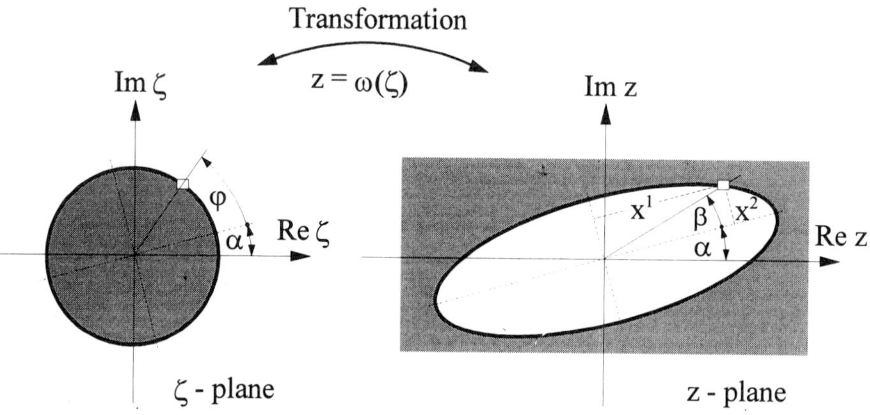

Fig. 4.4: Geometrical description of an elliptical hole

If an infinite plane with an elliptical hole is to be mapped onto the inner region of the unit circle, the mapping function (4.26) reads as follows:

$$z(\zeta) = R\left(\frac{1}{\zeta} + m\zeta\right) \tag{4.31}$$

with the middle semi-axis $R = \dfrac{a+b}{2}$.

In order to calculate the stresses, the stress function approaches $\Phi(\zeta), \Psi(\zeta)$ and the mapping function are substituted into (4.25) which fulfills the boundary conditions. The stress function $\Phi(\zeta)$ for the first global principal stress then reads:

$$\Phi_1(\zeta) = \frac{R}{4}\sigma_1\left[\frac{1}{\zeta} + \left(2e^{2i\alpha} - m\right)\zeta\right] \tag{4.32}$$

with the angle of orientation α between the main axis of the ellipse and the global principal stress.

Since no external loading acts at the boundary of the hole, only a tangential stress $\sigma_{\varphi\varphi}$ occurs:

$$(\sigma_{\varphi\varphi})_1 = 4\mathrm{Re}\{\Omega(\zeta)\} = 4\mathrm{Re}\left\{\frac{\Phi(\zeta)}{z'(\zeta)}\right\} = \sigma_1\frac{(1-m^2) + 2(m\cos 2\alpha - \cos 2(\alpha+\varphi))}{m^2 - 2m\cos 2\varphi + 1} . \tag{4.33}$$

The second global principal stress causes the following boundary stress:

$$(\sigma_{\varphi\varphi})_2 = \sigma_2\frac{(1-m^2) - 2(m\cos 2\alpha - \cos 2(\alpha+\varphi))}{m^2 - 2m\cos 2\varphi + 1} . \tag{4.34}$$

Superposition of these stresses then yields the total stress at the boundary of the hole:

$$\sigma_{\varphi\varphi} = \frac{(\sigma_1 + \sigma_2)(1-m^2) + 2(\sigma_1 - \sigma_2)\left[m\cos 2\alpha - \cos 2(\alpha+\varphi)\right]}{m^2 - 2m\cos 2\varphi + 1} . \tag{4.35}$$

Fig. 4.5 shows the results for the stresses at the boundary of the ellipse with $m=1/2$ (corresponding to an axes ratio of $a/b=3$). The results are drawn for four different load cases.

The complementary energy expression corresponding to (4.4b)

$$\overline{U}^{*'} = \frac{1}{2}D_{\alpha\beta\gamma\delta}\tau^{\alpha\beta}\tau^{\gamma\delta}$$

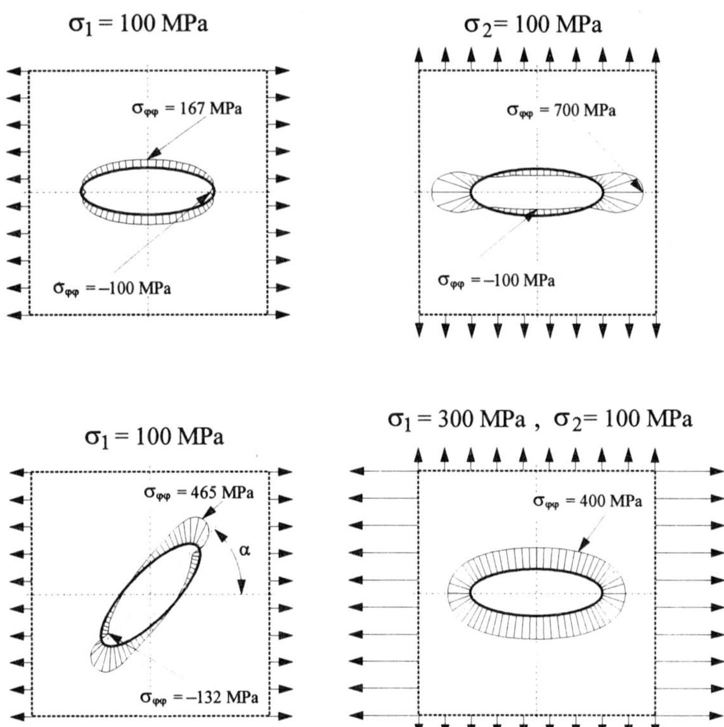

Fig. 4.5: Stresses at the boundary of the ellipse with m=1/2

in elliptic-hyperbolical coordinates

$$x^1 = c\,\cosh\xi^1\cos\xi^2$$
$$x^2 = c\,\sinh\xi^1\sin\xi^2$$

is determined by the metric tensor [17]

$$(g_{\alpha\beta}) = \begin{bmatrix} 1 & 0 \\ 0 & 1 \end{bmatrix} c^2\,\cosh^2\xi^1\cos^2\xi^2,$$

where c denotes the distance between the focus of the ellipse and the coordinate origin. Thus, we obtain

$$\overline{U}^{*'} = \frac{c^4}{2E}\left(\cosh^2\xi^1 - \cos^2\xi^2\right)^2 \cdot \left[(\tau^{11} + \tau^{22})^2 - 2(1+\nu)(\tau^{11}\tau^{22} - \tau^{12}\tau^{12})\right]. \quad (4.36)$$

The expression is simplified by the boundary conditions of the hole $\tau^{11} = 0,\ \tau^{12} = 0$.

Substituting the stress τ^{22} by its physical component

$$\sigma_{\varphi\varphi} = \tau^{22} c^2 (\cosh^2 \xi^1 - \cos^2 \xi^2),$$

the specific complementary energy $\overline{U}^{*\prime}$ follows to:

$$\overline{U}^{*\prime} = \frac{\sigma_{\varphi\varphi}^2}{2E} = \frac{1}{2E}\left[\frac{(\sigma_1 + \sigma_2)(1-m^2) + 2(\sigma_1 - \sigma_2)(m\cos 2\alpha - \cos 2(\alpha + \varphi))}{m^2 - 2m\cos 2\varphi + 1} \right]^2 . \quad (4.36)$$

For the variation expression (4.10), the integration of $\overline{U}^*$ over the boundary of the hole yields

$$(\delta J_\gamma)_{1ED} = \int_0^{2\pi} \frac{1}{2E}\left[(\sigma_1 + \sigma_2)(1-m^2) + \right. \qquad\qquad\qquad (4.37)$$

$$\left. +2(\sigma_1 - \sigma_2)(m\ \cos 2\alpha - \cos 2(\alpha + \varphi)) \right] g(\varphi, m)\, \delta s\, t\, d\beta$$

with

$$g(\varphi, m) = \frac{\sqrt{1 + 2m\cos 2\varphi + m^2}}{(m^2 - 2m\cos 2\varphi + 1)^2} \qquad \text{and} \qquad \beta = \arctan\left(\frac{1-m}{1+m} \tan\varphi \right) .$$

The index 1ED denotes the use of an elliptical hole E in disks D when considering the complementary energy.

The integral, which is dependent on φ and m, is solved numerically. The present integrals of different type are approximated by polynomial approaches:

$$I_a = \int_0^{2\pi} g(\varphi, m) d\beta \qquad\qquad := \sum_{i=0}^{n} a_i m^i ,$$

$$I_b = \int_0^{2\pi} \cos 2\varphi\, g(\varphi, m) d\beta \quad := \sum_{i=0}^{n} b_i m^i ,$$

$$I_c = \int_0^{2\pi} \sin 2\varphi\, g(\varphi, m) d\beta \quad := \sum_{i=0}^{n} c_i m^i = 0 ,$$

$$I_d = \int_0^{2\pi} \cos 4\varphi\, g(\varphi,m)\,d\beta \;:=\; \sum_{i=0}^{n} d_i m^i \;,$$

$$I_e = \int_0^{2\pi} \sin 4\varphi\, g(\varphi,m)\,d\beta \;:=\; \sum_{i=0}^{n} e_i m^i = 0 \;.$$

The polynomial approaches yield the following integral values for different excentricities of the ellipse (polynomium of 10th degree, $n=10$):

m	0	0.5	0.8
I_a	6.2829	10.0519	44.5820
I_b	-0.2E-3	5.1221	43.1306
I_d	-0.3E-3	7.2944	40.8223

The *characteristic function* for the ellipse-shaped hole is calculated in the same manner as for the positioning of a circular hole (see 4.2.1). Again, it proves suitable to split the functional into two expressions:

$$\left(\delta J_\gamma\right)_{1ED} = \delta\Omega\, \Phi_{1SE}(\sigma_1,\sigma_2) \quad \text{with} \quad \delta\Omega = 2\pi\, R_b\, t\, \delta s \tag{4.38a}$$

and

$$\begin{aligned}
\Phi_{1ED}(\sigma_1,\sigma_2,\alpha) = \frac{K_E}{4\pi E}\Big[&(\sigma_1+\sigma_2)^2(1-m^2)^2 I_a + 2(\sigma_1-\sigma_2)^2(I_a - I_d) + \\[4pt]
&+4(\sigma_1^2 - \sigma_2^2)(1-m^2)^2 \cos 2\alpha\ (mI_a - I_b) + \\[4pt]
&+4(\sigma_1 - \sigma_2)^2(1-m^2)^2 \cos 2\alpha\ (m^2 I_a - 2mI_b + \tfrac{1}{2} I_d) \Big]
\end{aligned} \tag{4.38b}$$

The factor K_E has been introduced in order to compare the positioning of an elliptical hole with that of a circular hole, where the volume variation $\delta\Omega$ shall be equal for both cases. Thus, it is possible to state the characteristic equation independently of the size of a hole. If the area of the elliptical hole shall be set equal to the area of the circular hole ($r_B = 1$), the middle semi-axis of the ellipse

$$R_B = \frac{1}{2}\left(\sqrt{\frac{1-m}{1+m}} + \sqrt{\frac{1+m}{1-m}} \right)$$

must be included into the calculations. With the approximate calculation of the circumference of the ellipse we obtain the factor K_E :

$$K_E = 2\pi \, R_B \, \frac{64 - 3m^4}{64 - 16m^2} \quad .$$

When positioning a hole with prescribed shape (m = const) this factor does not have to be calculated explicitly since it is constant and thus it does not gain influence on the position of the hole.

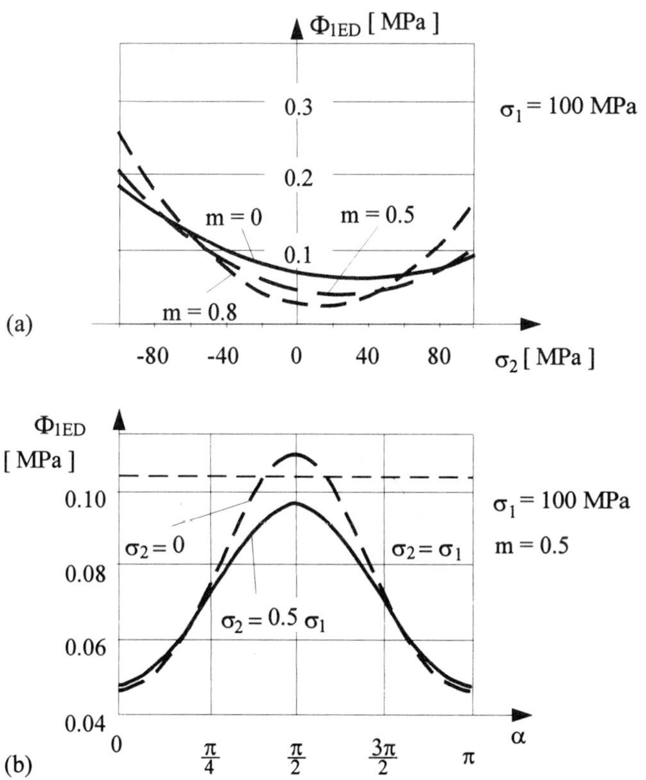

Bild 4.6: Characteristic function for the positioning of an elliptical hole

    (a) in dependence on the acting global stresses,

    (b) in dependence on the angle between principal axis and the direction of the first principal stress

Fig. 4.6 presents the characteristic function (4.38b) in dependence on the different influence quantities (principal stresses, orientation angle, and measure of excentricity of the ellipse). Herewith, statements can be made concerning the optimal orientation and the optimal shape. The characteristic function contains an angle α between the principal axis of the ellipse and the direction of the first global principal stress. A minimum of this function can be determined at the point(s) where

$$\frac{\partial \Phi_{1ED}}{\partial \alpha} = 0 \ , \qquad \frac{\partial^2 \Phi_{1ED}}{\partial \alpha^2} > 0 \ . \qquad\qquad (4.39)$$

For $\alpha = 0$ one obtains a relative minimum. At the optimum, the directions of the largest principal stress and of the principal axis are equal.

In addition to the optimal position and orientation, the derived characteristic function yields the optimal axis ratio of the ellipse.

b) Positioning of a triangular hole

Truss structures can be generated using triangular bubbles, where rounded edges are employed in order to reduce notch effects (Fig. 4.7).

The geometry of an equilateral triangle can sufficiently be described by a two-term conformal mapping (4.26) [44]:

$$z(\zeta) = R\left(\frac{1}{\zeta} + \frac{1}{3}\zeta^2\right) \ . \qquad\qquad (4.40)$$

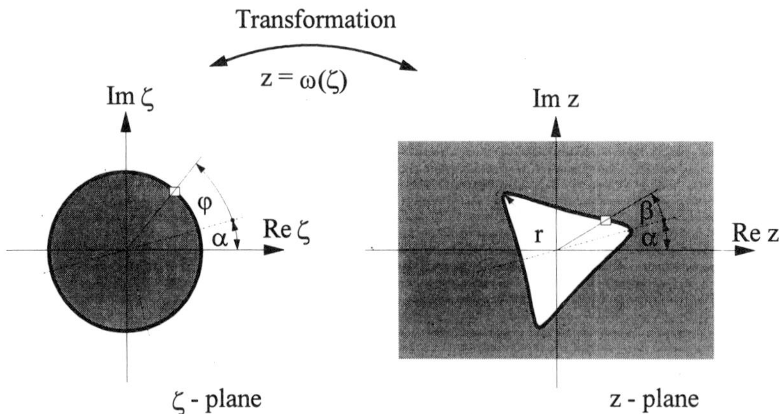

Fig. 4.7: Geometrical description of a triangular hole

By separating (4.40) into a real and an imaginary part ($\zeta = R\,e^{i\varphi}$, $R = 1$), one obtains the equations of the considered boundary of the hole as well as the radius of curvature at the edges:

$$x^1 = R(\cos\varphi + \tfrac{1}{3}\cos 2\varphi)\ ,$$

$$x^2 = -R(\sin\varphi - \tfrac{1}{3}\sin 2\varphi)\ , \qquad r = \tfrac{1}{21}R = \tfrac{1}{14}H\ ,$$

where H denotes the height of the equilateral boundary of the hole. The angle α describes the orientation of the hole (Fig. 4.7).

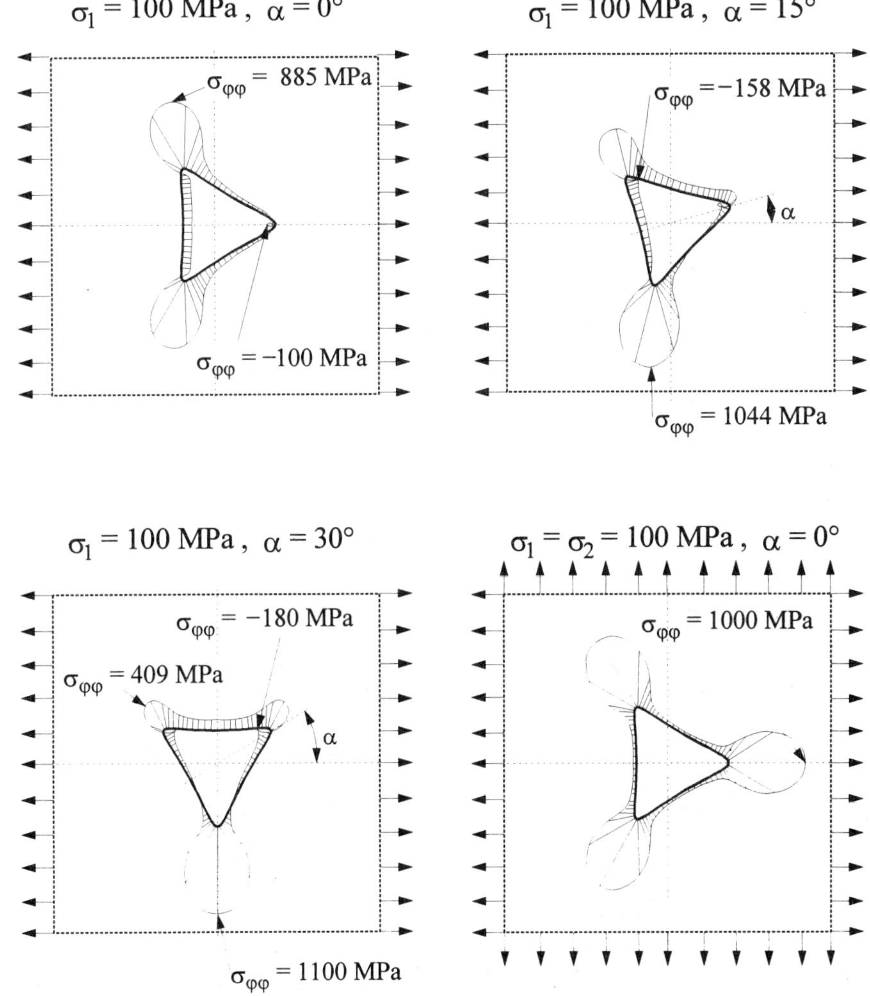

Fig. 4.8: Stresses at the boundaries of a triangular hole

Analogous to the elliptical hole, the stresses at the triangular hole are determined by means of the complex stress function. As no loading occurs at the boundary of the hole, only the tangential stress $\sigma_{\varphi\varphi}$ must be considered.

$$(\sigma_{\varphi\varphi})_1 = 4\mathrm{Re}\{\Omega(\xi)\} = \sigma_1 \frac{5 - 18\cos(2(\alpha + \varphi)) + 12\cos(2\alpha - \varphi)}{13 - 12\cos 3\varphi} . \tag{4.41}$$

The second global principal stress causes the following boundary stress:

$$(\sigma_{\varphi\varphi})_2 = \sigma_2 \frac{5 + 18\cos(2(\alpha + \varphi)) - 12\cos(2\alpha - \varphi)}{13 - 12\cos 3\varphi} . \tag{4.42}$$

Superposition of the two elements yields the total stress at the boundary of the hole (Fig. 4.8):

$$\sigma_{\varphi\varphi} = \frac{5 \cdot (\sigma_1 + \sigma_2) + [-18\cos(2(\alpha + \varphi)) + 12\cos(2\alpha - \varphi)] \cdot (\sigma_1 - \sigma_2)}{13 - 12\cos 3\varphi} . \tag{4.43}$$

In analogy with the calculation of the elliptical hole, the integration of $\overline{U}^{*'}$ over the boundary is carried out with polynomial approaches. The *characteristic function* for positioning of triangular holes then reads:

$$\boxed{\Phi_{1\mathrm{TD}}(\sigma_1, \sigma_2) = \frac{1}{2E}\left[c_1 \cdot (\sigma_1 + \sigma_2)^2 + c_2 \cdot (\sigma_1 - \sigma_2)^2\right]} \tag{4.44}$$

with $c_1 = 7.000$ and $c_1 = 14.472$.

The characteristic function is independent of the angle of orientation. Owing to the strong notch effects at the edges of the triangle, the level of disturbances is substantially higher than for a circular hole (4.18), and the sensitivity of a triangular hole to stress differences $\sigma_1 - \sigma_2$ is slightly stronger. The optimal position of the triangular hole is practically the same as that of the circular hole.

4.2 Mean compliance for plate problems

Proceeding from the variation expression (4.10), a characteristic function for the positioning of holes can also be determined for plate problems. For that purpose, the stresses at the boundaries of the inserted holes have to be calculated for different load cases. Then, the characteristic function can be determined analogously to the plane stress problem (disks, see 4.1). The calculations can be simplified substantially by confining the considerations to the special case of a shear-rigid plate (KIRCHHOFF´s plate theory),

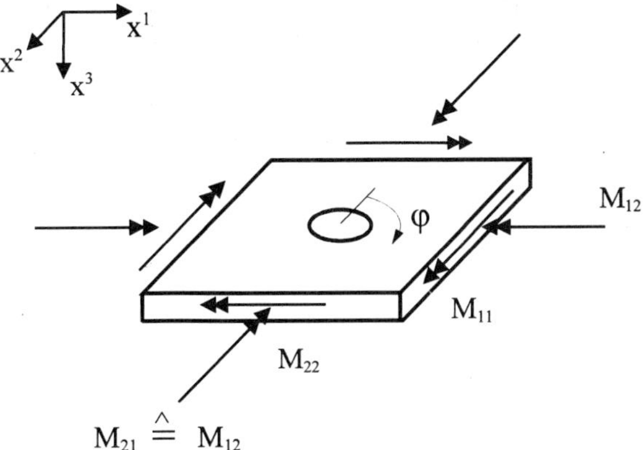

Fig. 4.9: Resulting moments at the evaluation domain

where no shear deformations occur in transverse direction, i.e., $\gamma_{23} = \gamma_{13} = 0$. This simplification requires the introduction of substitute transverse forces and their corresponding edge forces in order to form the equilibrium [17], however, edge forces and plate deformations are not coupled by the constitutive law, and the deformations and stresses in a shear-rigid plate can thus be completely described by two bending moments (M_{11}, M_{22}) and one torsional moment (M_{12}) (see Fig. 4.9).

When calculating the perforated plate we furthermore assume the stress evaluation to be carried out in an evaluation domain that is much smaller than the overall dimensions of the plate and in which the stress peaks decay completely. The occuring stresses at the boundary of the hole are determined by superposing two bending moments and one torsional moment.

4.2.1 Positioning of circular holes

Stress calculation

The single moments at the boundary of a circular hole read [44]:

− Moments at the boundary due to a bending moment M_{11} :

$$M_{\varphi\varphi}^{(1)} = M_{11}\left[1 - \frac{2(1+\nu)}{3+\nu}\cos 2\varphi\right]. \tag{4.45}$$

− Moments at the boundary due to a bending moment M_{22} :

$$M_{\varphi\varphi}^{(2)} = M_{22}\left[1 - \frac{2(1+\nu)}{3+\nu}\cos(2\varphi - \pi)\right] = M_{22}\left[1 + \frac{2(1+\nu)}{3+\nu}\cos 2\varphi\right]. \tag{4.46}$$

– Moments at the boundary due to a torsional moment M_{12} :

$$M_{\varphi\varphi}^{(3)} = M_{12} \frac{4(1+\nu)}{3+\nu} \sin 2\varphi \ . \tag{4.47}$$

By superposing the single solutions one obtains the resulting bending moment at an arbitrary point of the circular hole as:

$$M_{\varphi\varphi} = M_{11}\left[1 - \frac{2(1+\nu)}{3+\nu}\cos 2\varphi\right] + M_{22}\left[1 + \frac{2(1+\nu)}{3+\nu}\cos 2\varphi\right] + $$
$$+ M_{12}\frac{4(1+\nu)}{3+\nu}\sin 2\varphi \ , \tag{4.48}$$

and thus the stresses occuring at the boundary of the hole:

$$\sigma_{\varphi\varphi} = 12\frac{M_{\varphi\varphi}}{t^3}x^3 \ , \quad \sigma_{rr} = 0 \ , \quad \tau_{r\varphi} = 0 \ , \quad \tau_{rz} = 0 \ , \quad \tau_{\varphi z} = 0 \ .$$

For the specific complementary energy $\overline{U}^{*'}$ holds that

$$\overline{U}^{*'} = \frac{\sigma_{\varphi\varphi}^2}{2E} = \frac{1}{2E}\left(\frac{12x^3}{t^3}\right)^2\left\{M_{11}\left[1 - \frac{2(1+\nu)}{3+\nu}\cos 2\varphi\right] + M_{22}\left[1 + \frac{2(1+\nu)}{3+\nu}\cos 2\varphi\right] + \right.$$
$$\left. + M_{12}\frac{4(1+\nu)}{3+\nu}\sin 2\varphi\right\}^2 \ . \tag{4.49}$$

For the variation expression (4.10) we integrate $\overline{U}^{*'}$ over the boundary from $\varphi = 0$ to $\varphi = 2\pi$:

$$(\delta J_\gamma)_{ICP} = \frac{12\pi}{Et^3}r_B\,\delta s\left\{(M_{11} + M_{22})^2 + \left(\frac{2(1+\nu)}{3+\nu}\right)^2\left[(M_{11} - M_{22})^2 + 4M_{12}^2\right]\right\}. \tag{4.50}$$

In order to determine the characteristic function, the functional is split again into two terms in accordance with (4.18):

$$(\delta J_\gamma)_{ICP} = \delta\Omega \ \Phi_{ICP}(M_{11}, M_{22}, M_{12}) \quad \text{with} \quad \delta\Omega = 2\pi r_b t\delta s \tag{4.51a}$$

and

$$\boxed{\Phi_{ICP} = \frac{1}{6E}\left\{(M_{11} + M_{22})^2 + \left(\frac{2(1+\nu)}{3+\nu}\right)^2\left[(M_{11} - M_{22})^2 + 4M_{12}^2\right]\right\}} \ . \tag{4.51b}$$

Alternatively, the *characteristic function* can be expressed by the two global principal stresses on the surface of the plate:

$$\left.\begin{matrix}\sigma_1\\\sigma_2\end{matrix}\right\} = \frac{6}{t^2}\left[\frac{M_{11}+M_{22}}{2}\pm\sqrt{\left(\frac{M_{11}+M_{22}}{2}\right)^2+M_{12}^2}\,\right].$$

The function then reads:

$$\Phi_{ICP}(\sigma_1,\sigma_2)=\frac{1}{6E}\left[(\sigma_1+\sigma_2)^2+\left(\frac{2(1+v)}{3+v}\right)^2(\sigma_1-\sigma_2)^2\right]. \tag{4.51c}$$

The bubble is positioned at that point of the structure where the characteristic function attains a minimum value. While the difference of principal stresses $\sigma_1-\sigma_2$ in the case of disk problems gains an influence that is twice as high as the stress sum $\sigma_1+\sigma_2$, the opposite feature occurs at plate problems. For a POISSON ratio of $v = 0.3$ one here obtains a factor of 0.621 for the stress difference.

4.2.2 Positioning of non-circular holes

The *positioning of triangular holes* in plates is carried out in analogy with the calculations at disk problems. The moments at the boundary of the hole caused by a bending moment M_{11} read in dependence on the angle of orientation α [44]:

$$M_\varphi^{(1)} = M_{11}\left[\frac{2}{3+v}+\frac{1+v}{3(3+v)}\,\frac{5-18\cos(2\varphi+2\alpha)+12\cos(\varphi-2\alpha)}{13-12\cos 3\varphi}\right]. \tag{4.52}$$

Superposition of the given loads, calculation of $\overline{U}^{*'}$, and integration over the boundary of the hole then yield the characteristic function in dependence on the global principal stresses on the surface of the plate:

$$\Phi_{IPD}(\sigma_{10},\sigma_{20})=\frac{1}{18E}\frac{1}{(3+v)^2}\left[(84.0+16.0(1+v)+\right.$$
$$\left.+2.1(1+v)^2)(\sigma_{10}+\sigma_{20})^2+43.4(1+v)^2(\sigma_{10}-\sigma_{20})^2\right], \tag{4.52}$$

where σ_1,σ_2 are the principal stresses on the upper surface of the plate. The coefficients have been calculated numerically by means of polynomium approaches.

Result:

As in disk problems (4.1.2), the characteristic function for the positioning of triangular holes in plates is independent of the angle of orientation. The sensitivity towards differences of the principal stresses $\sigma_1-\sigma_2$ is slightly higher than in the case of a circular hole.

4.3 Mean compliance for three-dimensional problems

For treating three-dimensional (spatial) bodies, it is necessary to position a three-dimensional hollow bubble (sphere). Here, the characteristic function is derived in the same manner as for disks and plates.

The stresses for the boundary of the sphere (boundary of the bubble) are analogous to the plane solutions, and they are to be determined in dependence on the three principal stresses. Based on procedures for an explicit solution of partial differential equations, NEUBER (1985) [37] gives general solution methods for the calculation of stresses in bodies with spheres. The special case of a sphere in an infinite body subjected to uni-axial tension dates back to LEON (1908) [32]. The stresses at the boundary of the sphere read for the first global principal stress σ_1 (Fig. 4.10):

$$\sigma_{\varphi\varphi 1} = \frac{\sigma_1}{14 - 10\nu}\left[30\sin^2\varphi - 3 - 15\nu\right] , \tag{4.54}$$

$$\sigma_{\vartheta\vartheta 1} = \frac{\sigma_1}{14 - 10\nu}\left[30\nu\sin^2\varphi - 3 - 15\nu\right] . \tag{4.55}$$

The remaining stress components on the surface of the sphere are equal to zero. It can be observed that the stress increase caused by the three-dimensional bubble is lower than in the corresponding plane case. By superposing three global principal stresses perpendicular to each other, one obtains:

$$\sigma_{\varphi\varphi} = \frac{1}{14 - 10\nu}\left[\sigma_1(30\sin^2\varphi_1 - c) + \sigma_2(30\sin^2\varphi_2 - c) + \sigma_3(30\sin^2\varphi_3 - c)\right], \tag{4.56}$$

$$\sigma_{\vartheta\vartheta} = \frac{1}{14 - 10\nu}\left[\sigma_1(30\nu\sin^2\varphi_1 - c) + \sigma_2(30\nu\sin^2\varphi_2 - c) + \sigma_3(30\nu\sin^2\varphi_3 - c)\right] \tag{4.57}$$

with

$$\varphi_1 = \arccos\frac{x^1}{r} = \varphi, \qquad \varphi_2 = \arccos\frac{x^2}{r} = \arccos(\sin\varphi\cos\vartheta),$$

$$\varphi_3 = \arccos\frac{x^3}{r} = \arccos(\sin\varphi\sin\vartheta), \qquad c = 3 + 15\nu.$$

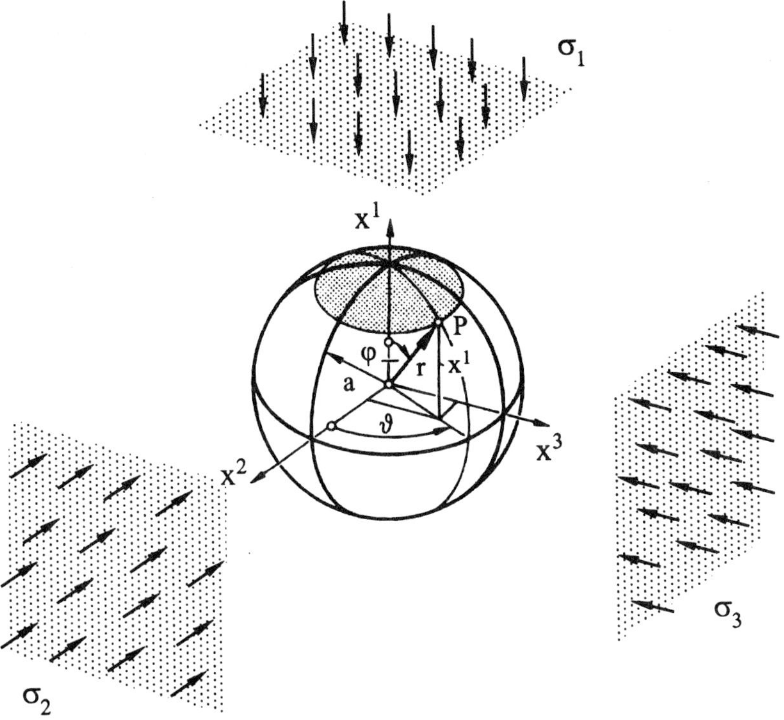

Fig. 4.10: Sphere in an infinitely extended body

The variational functional (4.14) leads to the following form:

$$(\delta J_\gamma)_{1SV} = \frac{8r^2}{2E} \int\limits_0^{\pi/2} \int\limits_0^{\pi/2} \left(\sigma_{\varphi\varphi}^2 + \sigma_{\vartheta\vartheta}^2 - 2\nu\sigma_{\varphi\varphi}\sigma_{\vartheta\vartheta}\right) \sin\varphi \, d\vartheta \, d\varphi \, \delta s. \qquad (4.58)$$

The index 1SV denotes the application of a sphere S in a three-dimensional space (volume) with respect to the complementary energy 1.

After integration one obtains the functional in dependence on the three principal stresses:

$$(\delta J_\gamma)_{1SV} = \frac{8r^2}{2E(14 - 10\nu)} \left[c_a\left(\sigma_1^2 + \sigma_2^2 + \sigma_3^2\right) + c_b\left(\sigma_1\sigma_2 + \sigma_1\sigma_3 + \sigma_2\sigma_3\right)\right] \delta s. \qquad (4.59)$$

Hence, the characteristic function for the positioning of a sphere reads:

$$\Phi_{\text{ISV}}(\sigma_1,\sigma_2,\sigma_3) = \frac{1}{2E(14-10v)}\left[c_a(\sigma_1^2+\sigma_2^2+\sigma_3^2)+c_b(\sigma_1\sigma_2+\sigma_1\sigma_3+\sigma_2\sigma_3)\right]$$

$$(4.60)$$

with

$$c_a = 1920\pi(1-v^2)-160\pi c(1-v^2)+8\pi c^2(1-v),$$

$$c_b = 1440\pi(1-3v^2)+160\pi c(2v^2+v-1)+8\pi c^2(1-2v).$$

The coefficients c_a and c_b have been calculated numerically using 10-th degree polynomium approaches.

5   Applications

5.1 Cantilever disk

The task for this example is to find a best-possible initial design for a prescribed topology domain (Fig. 5.1a). The disk is clamped on the left-hand side and is subjected to a load on the right-hand side. The optimization problem consists of a minimization of the complementary energy of this component, i.e., a minimization of the mean compliance, while considering a volume constraint as equality condition. A triangular disk is chosen as initial design in the half-occupied topology domain. In this case, the problem reads as follows:

$$\underset{\mathbf{x}\in\mathfrak{R}^n}{\text{Min}}\left\{U^*(\mathbf{x})\mid V(\mathbf{x}) = \frac{1}{2}V_{\text{domain}}\ ,\ \mathbf{h}(\mathbf{x}) = \mathbf{0}\right\} \qquad (5.1)$$

with $U^*(x)$ complementary energy,

$\quad\qquad V(x)$ volume of the structure,

$\quad\qquad \mathbf{h}(\mathbf{x})$ vector of q equality constraints.

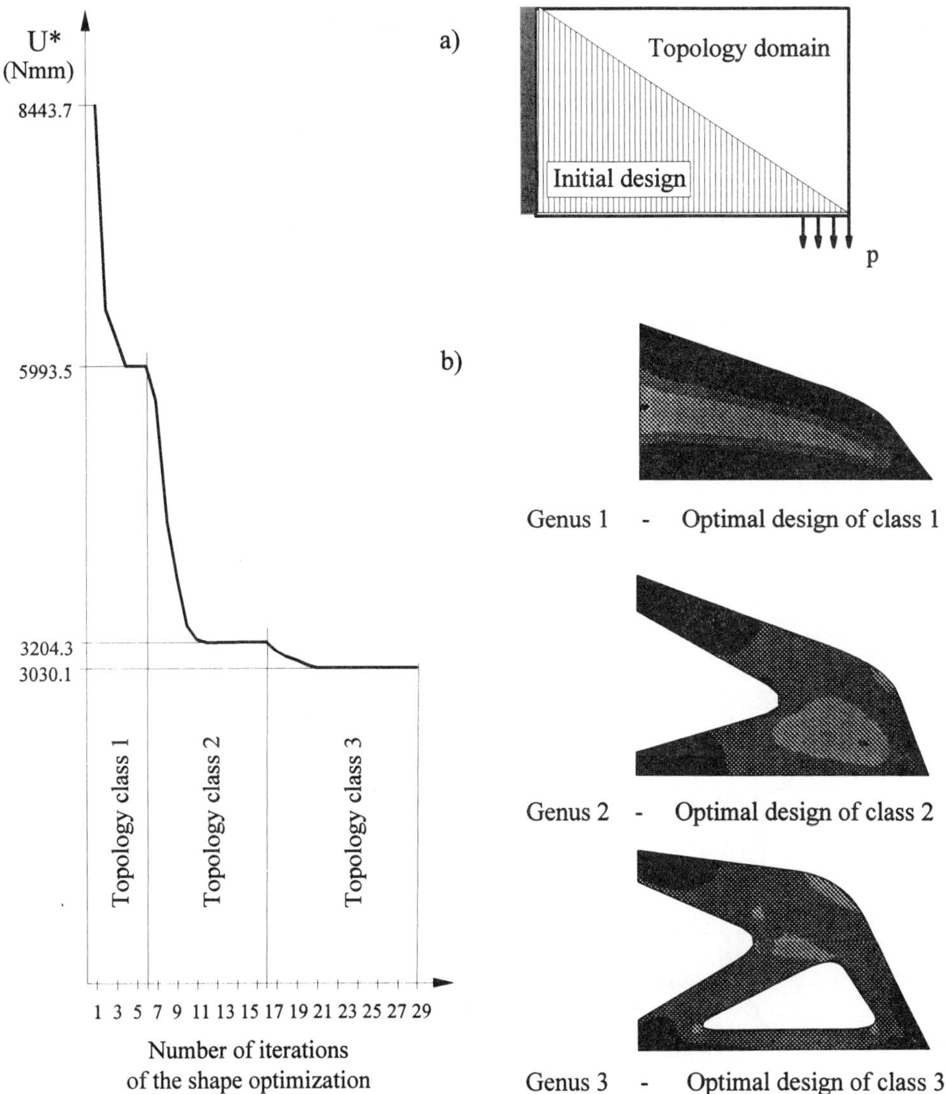

Fig. 5.1: Topology optimization of a cantilever disk

Remarks on the solution procedure:

If a shape optimization is carried out on this structure, a shape as presented in *genus 1* is achieved (Fig. 5.1b). If a bubble is positioned at the point of the minimum of the characteristic function (in the middle of the supported edge) and a new shape optimization is carried out, we achieve *genus 2*. By positioning the next bubble at the point of the minimum of the characteristic function, *genus 3* is achieved. In the present example, the optimization is finished at this point.

In the current state of development, the computation time for the optimization is still substantial, since this procedure carries out a numerical sensitivity analysis. The efficiency of the bubble method may be exploited to its full potential if an analytical sensitivity can be achieved.

5.2 Panel truss structure of a radiotelescope

In the following, we will treat an example from the field of the design of radio telescopes. Radio-astronomical observations in the millimeter- and submillimeter-range require highly accurate telescopes. The reflector of such a type of telescope consists of a large number of single panels mounted on a supporting back-up structure (Fig. 5.2). The panel

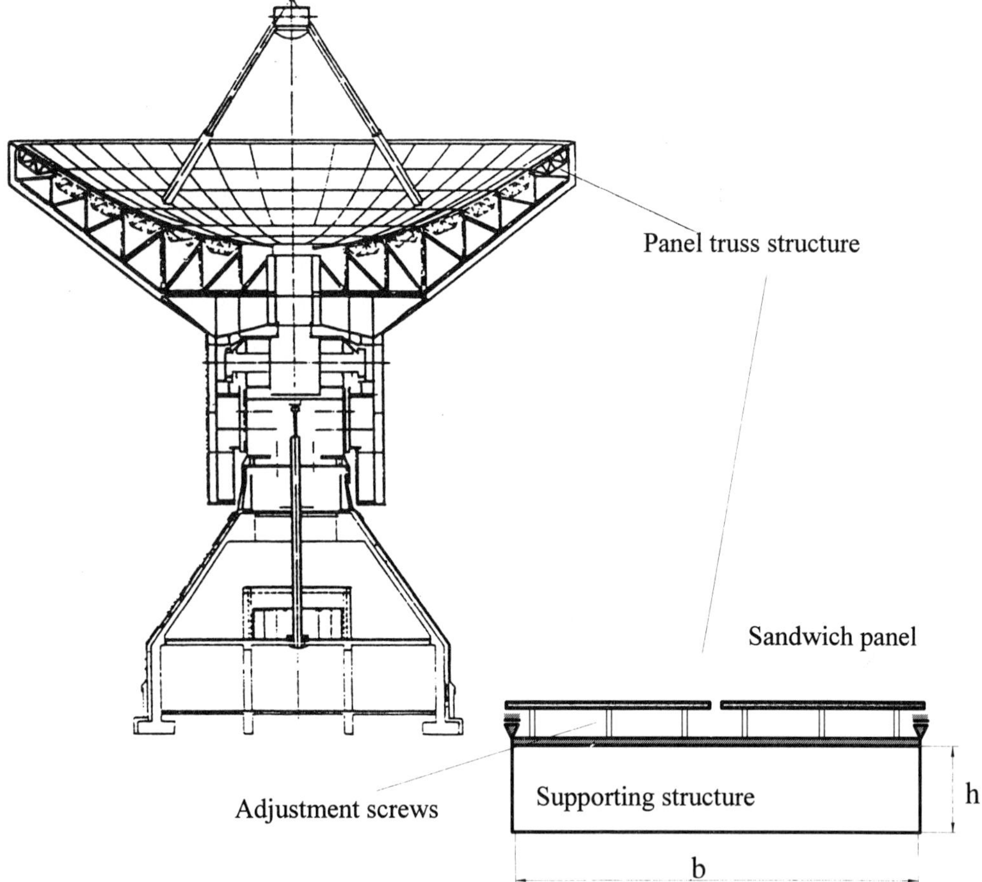

Fig. 5.2: Panel truss structure of the 30m-Millimeter-radiotelescope MRT [12]

truss structures have a substantial influence on the structural derivations, since they deform due to weight and wind, and due to temperature influences. In [12] it is shown how a best-possible panel truss structure can be determined by means of parameter studies. However, these investigations are normally very time-consuming. When the bubble method is applied, the development process can be carried out more efficiently.

In this example a two-dimensional substitute model (Fig. 5.3) is dealt with. For this purpose, two topology domains are chosen to determine best-possible initial designs for two given topology domains. The first topology domain has a depth of 150 mm (version 1), and the second one has a depth of 250 mm (version 2). In both versions, the topology domain has a width of $b = 1000$ mm. This structure is subjected to a loading that is composed of deadweight and wind load. In a first step, the topology optimization shall be performed with a vertical surface load as illustrated in Fig. 5.3.

As a measure of accuracy, we take the mean displacement of the surface. Provided that no temperature influence occurs, the displacement is linearly dependent on the complementary energy (see Chapter 4). The volume of the panel structure must not be larger than the volume of the structure with a rectangular base of 100 mm. In order to treat this example, the standardized approach functions described in 3.2 are augmented with respect to their flexibility.

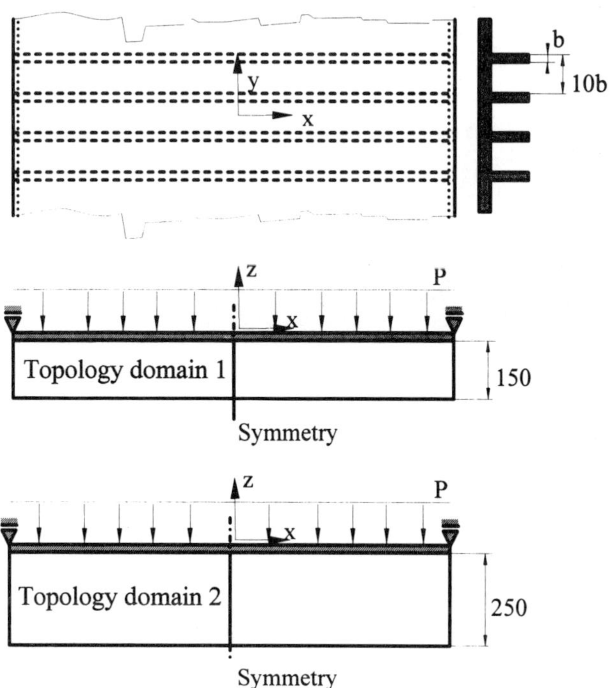

Fig. 5.3: Two-dimensional substitute model for different topology domains

Results:

The results for the topology optimization of version 1 are shown in Fig. 5.4. The minimum of the characteristic function for positioning a triangular hole can be found in the lower corner of the structure. Thus, one obtains a shape optimization of the outer boundary in the same topology class.The subsequent holes are positioned at the marked points. This leads to notches as the points are positioned at a non-variable boundary of the structure. The topology optimization is limited when the next following hole is positioned at the boundary of the original bubble. Further optimization then requires a refinement of the approach function of the original bubble not be described in this paper.

Fig. 5.5 illustrates the results of the topology optimization for version 2, which yields substantially improved results for higher topology classes. Here, the suitable choice of the topology domain gains a decisive influence on the optimal structure of the component.

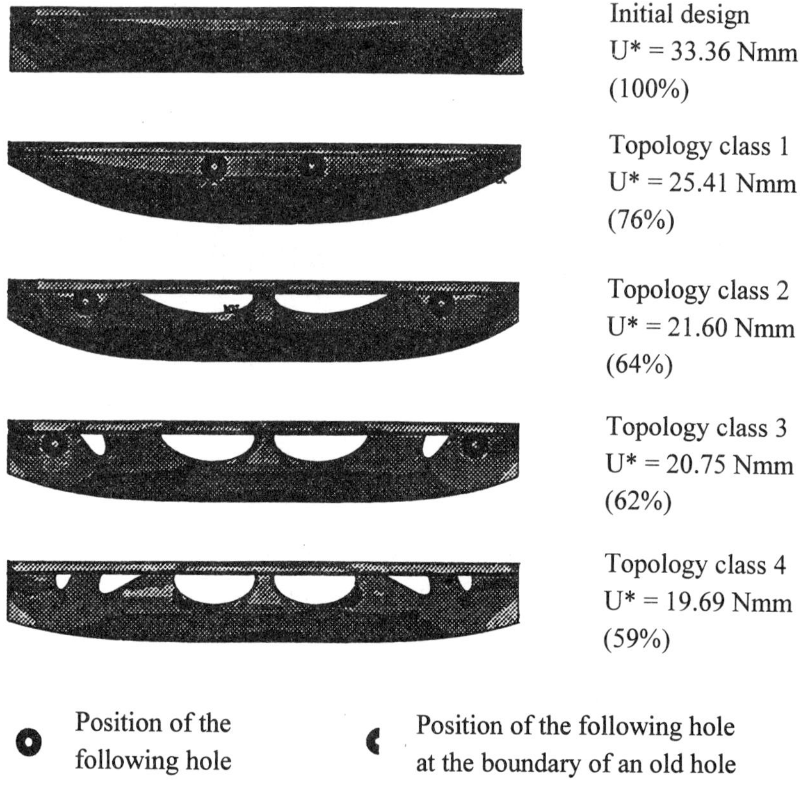

Fig. 5.4:   Topology optimization of version 1

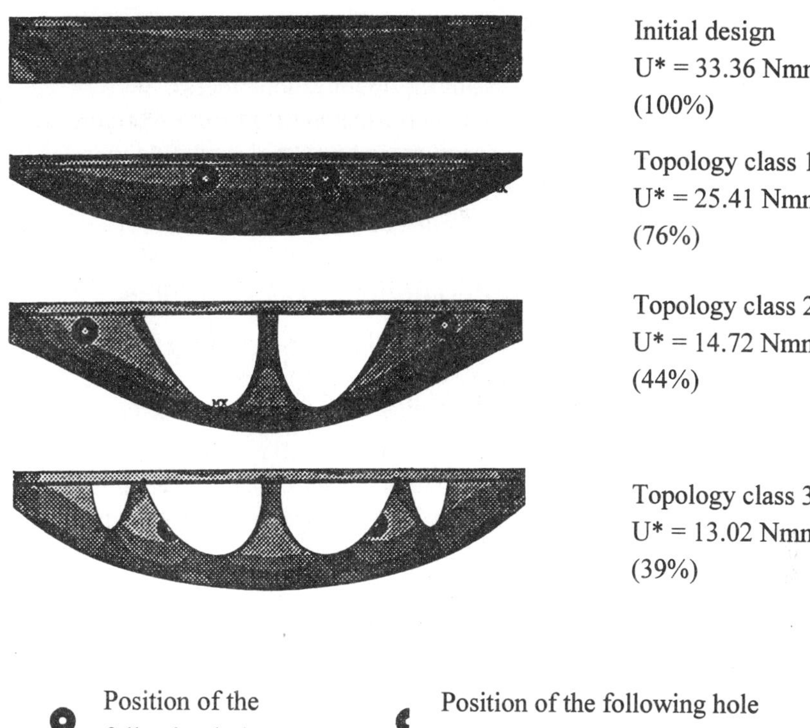

Fig. 5.5: Topology optimization of the version 2

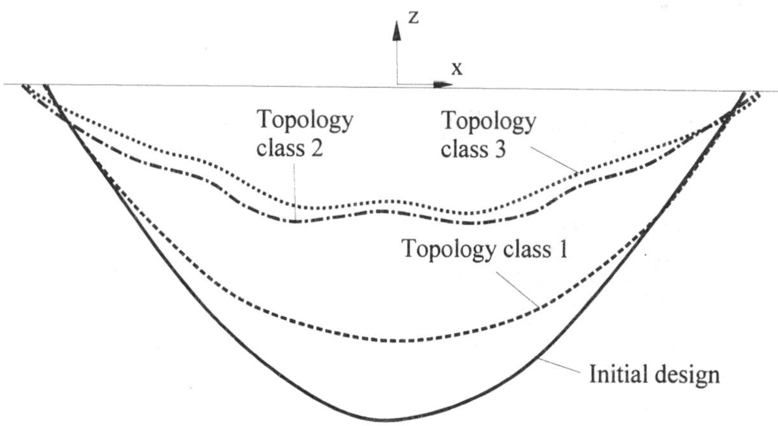

*Fig. 5.6: Deformations of the intermediate results of version 2
(magnifying factor 2000)*

Fig. 5.6 shows the curves of deformations for the intermediate results of version 2. It becomes obvious that a substantial reduction of the deformations in the middle of the panel surface can be achieved in the course of the optimization process.

The results found by means of the bubble method confirm the parameter study in [12]. Comparing the computing times of both procedures, the computing time for the parameter study took several weeks, whereas the time required for one optimization run amount to about four hours, using a DEC-AXP-3000 workstation.

5.3 Quadratic plate with different boundary conditions

Fig 5.7a presents an example for the positioning of circular holes in quadratic plates with a single force in the centre. One of the plates is simply supported at its boundaries (Fig. 5.7a). The other sketch in Fig. 5.7a shows a plate with clamped edges.

In Fig. 5.7b, the calculation results of the characteristic function (4.51) is presented for both plates. Fig. 5.8a,b illustrates the optimal position for single holes. For the simply supported plate, the holes are set in the middle of the plate edge, and for the clamped plate the holes are positioned at the corner of the plate. The shape optimization is carried out for quarter-circles.

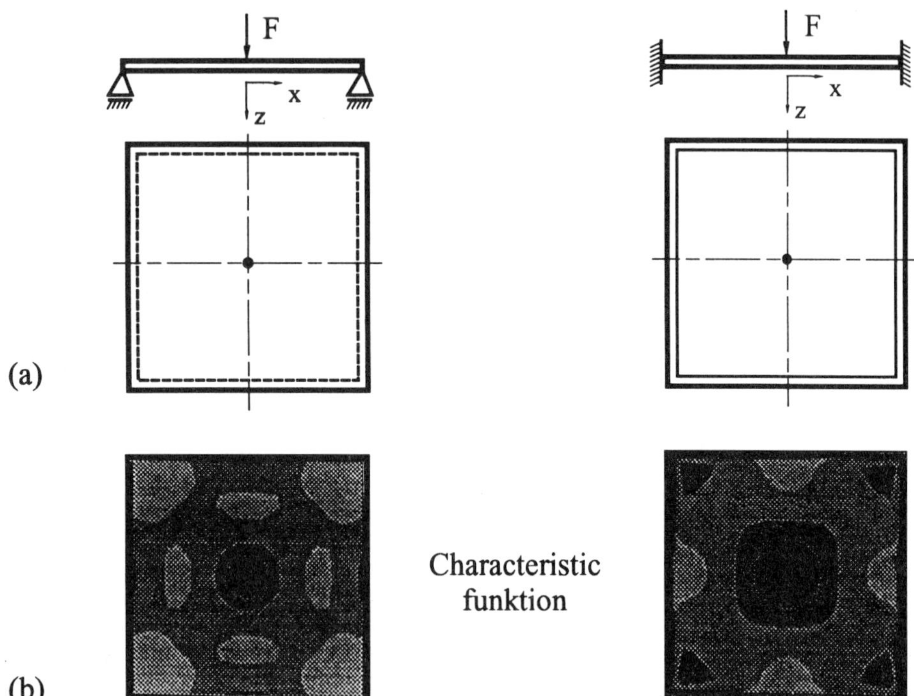

Fig. 5.7: a) Quadratic plate under single force with different boundary conditions
b) Evaluation of the characteristic function

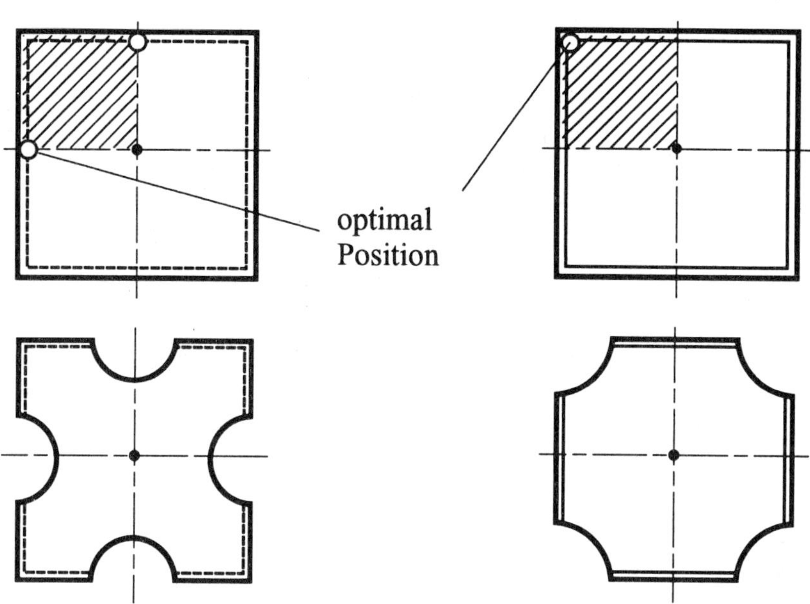

Fig. 5.8: Determination of optimal hole positioning with subsequent shape optimization

Interpretation of the results:

- In order to reduce compliance, a *simply supported plate* requires stiffened corners. As both edges meet in the corner, the corner point can almost be treated as a clamped point. This fact yields substantial corner forces which strongly influence the stiffness of the plate.
- In the case of the *clamped plate* with its corner supports, the shortest distance between the application point of the force F and the clamping forms a supporting joint. The corner forces are not required to increase the stiffness of the plate.

5.4 Casing of a handsaw-grip

The bubble method was practically applied to the topology optimization of the casing of a handsaw grip , especially its internal stiffeners. In order to avoid stress peaks, a reduction of special parts of the cross-sections with local stress peaks in a crash shall be avoided. In a first approach, we are here using again the complementary engergy for the optimization. By minimizing the complementary energy in the grip, the stresses are equalized and stress peaks can be reduced. For the computation, the grip is clamped at the motor suspension, and is subjected to a horizontal load at the rear end of the grip. The structural model is shown in Fig. 5.9.

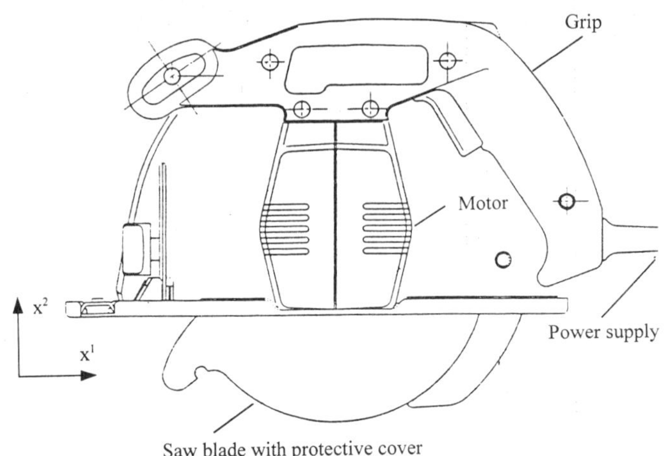

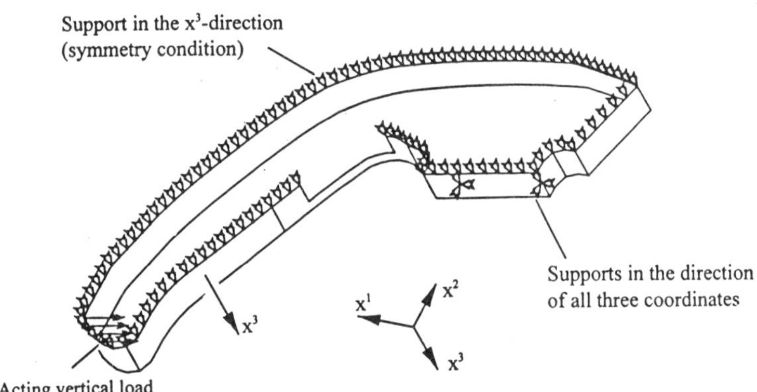

Fig. 5.9: Principle sketch and structural model of a handsaw grip

In the optimization, holes at the inner side of the grip are positioned and optimized with regard to shape. In order to ensure an effective optimization calculation, a special geometry definition is employed for the bubbles that are to be inserted. The coordinates of the corner points of the bubbles are described by means of two curvilinear coordinate systems (move variables on the outer hull of the grip), and the size of the corresponding bubbles is kept variable by scaling factors. Furthermore, they are additional design variables in the optimization process.

The step-wise positioning and shape optimization of new bubbles provide helpful information on the optimal position and arrangement of stiffeners. Further details on the optimization process can be found in [47]. The optimization history with the final optimal design is presented in Fig. 5.10.

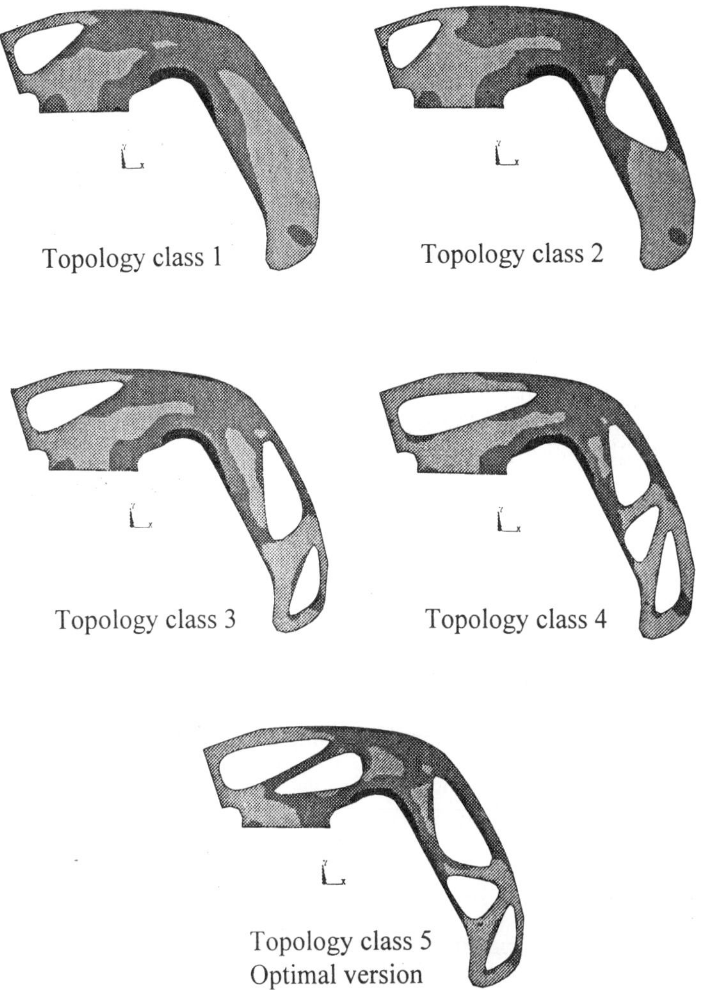

Topology class 1   Topology class 2

Topology class 3   Topology class 4

Topology class 5
Optimal version

Fig. 5.10: Step-wise positioning and shape optimization in single topology classes

5.5 Topology optimization of a wing rib

Wing ribs stabilize the profile of the wing of an airplane (see Fig. 5.11a), and they connect the spars and the plating of the supporting structure. The ribs are clamped at the spars at 15% and 75% of the wing depth. The plating is mounted on their outer sides. For reasons of assemblage, the wing ribs must have recesses in the inner part of a wing. These notches shall be optimized by means of a shape and topology optimization.

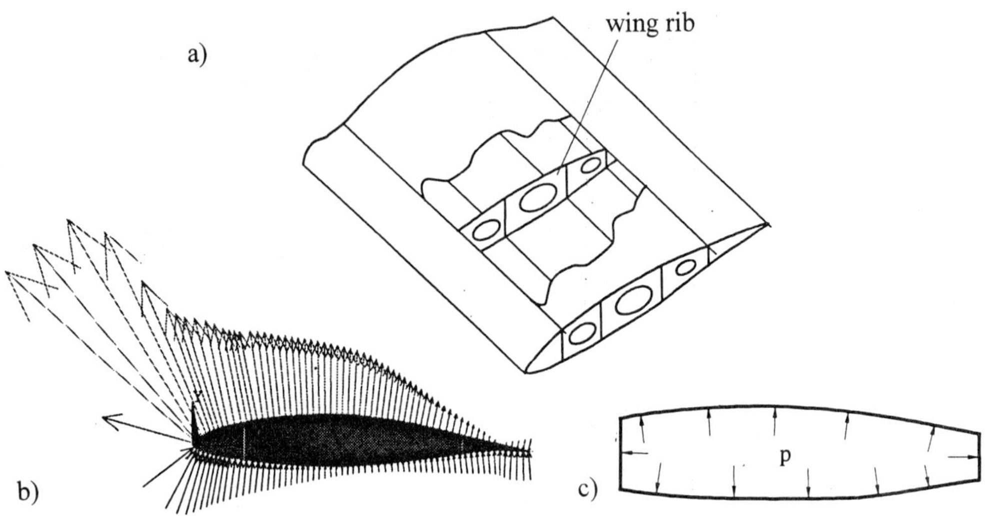

Fig. 5.11:   a) Principle sketch of a wing rib of an airplane
b) Load case 1: Pull-out after gliding fight
b) Load case 2: Tank pressure (2 bar)

The following load cases are dealt with in this example (see Fig. 5.11b,c).

1. Air-induced forces along the rib length:
   a. Pull-out after gliding flight (Fig. 5.11b)
   b. Vertical dive

2. Tank pressure (2 bar) (Fig. 5.11c)

The loads are defined on a parameterized geometry model that is independent of the Finite-Element mesh. The outer cover (wall-thickness 5mm) and the spars (wall-thickness 5mm) are simulated by means of beam elements. As an example, calculation is carried out for a stiffener thickness of 5mm and a stiffener distance of 1000mm.

The geometric centre of gravity of the inserted triangular bubble is placed at a point determined by the positioning criterion. The side of the triangle closest to the boundary shall be parallel to this boundary. The coordinates of the corner points of the bubble are guided along the curvilinear coordinate systems of the stiffener boundaries. The sizes of the respective bubbles are kept variable by means of scaling factors, and they are used as design variables in the optimization process.

For the first optimization we consider load case 1a (pull-out after gliding flight) [40]. The optimization steps are presented in Fig. 5.12. With unchanged volume, the complementary energy (as a measure of mean compliance) could be reduced from 61818.5 Nmm (class 1) to 60697.6 Nmm (class 3).

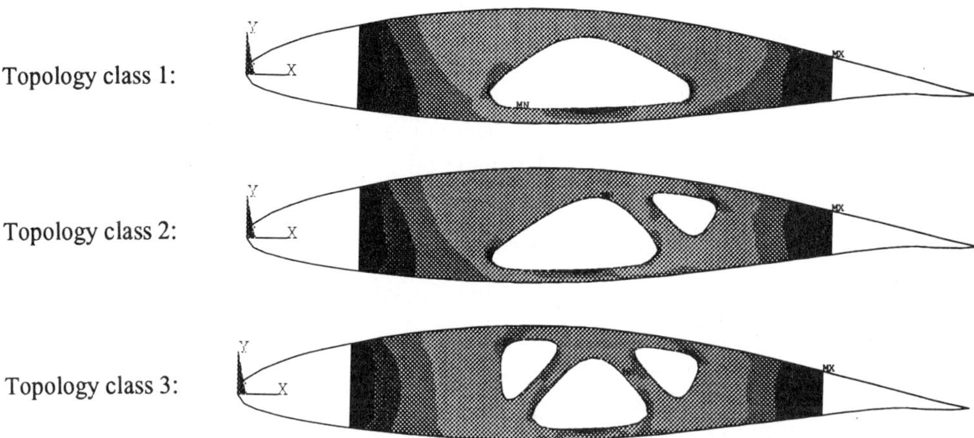

Fig. 5.12: *Step-wise positioning and shape optimization of a wing rib in three topology classes (Load case 1: Pull-out after gliding flight)*

Load case 2 (tank pressure) is treated in a second optimization process. The single steps are presented in Fig. 5.13. With unchanged volume, the complementary energy could be reduced from 57725.6 Nmm (class 1) to 17264.6 Nmm (class 2).

Future investigations deal with further single and multiple load cases. In the present example, superposition of the pull-out load and of the tank pressure load leads to the results for the pull-out-load, as the sensitivities of the hole positions and their shapes are very low in the load case tank pressure.

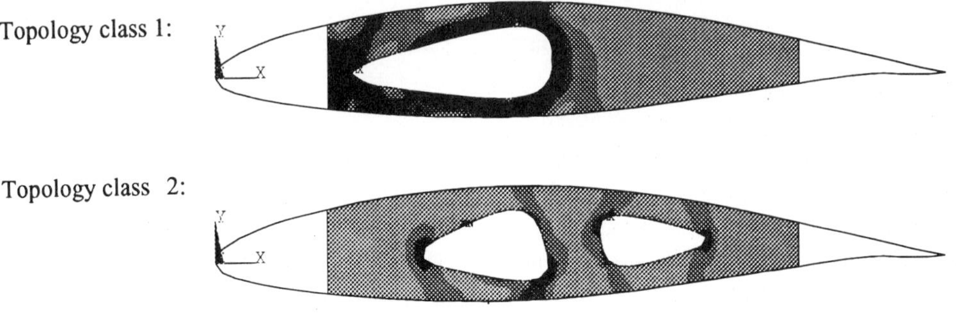

Fig. 5.13: *Step-wise positioning and shape optimization of a wing rib in two topology classes (Load case 2: Tank pressure)*

6 Final remarks and future works

In this paper, the theoretical aspects and the potential of applications of the bubble method are considered. As we learned, the solution concept of the bubble method admits interaction during finding optimal layouts for structures. Besides other topology procedures like the homogenization method, the bubble method can also be viewed as a valuable contribution to the design process. Some of the special characteristics of the bubble method are:

- It determines the optimal position by inserting a small bubble into the structure.

- The bubble is moved proceeding from different initial positions by means of optimization algorithms.

- Shape and topology are combined in the solution procedure.

- Because of the smooth component boundaries in each stage of the optimization, no post-smoothing is required.

- The use of smooth boundaries allows to consider local effects in the topology optimization.

Further investigations are being in progress, especially with respect to the size of a hole, the employed initial points, and the applied optimization algorithms. The augmentation of the optimization functionals for determining characteristic functions for further applications has to be strongly dealt with. Therefore, for the future the following problems are to be solved:

- Treatment of local displacements and stresses in structures [51];

- Structural responses of dynamic systems [33];

- Introduction of local failure criteria [14];

- Establishing of augmented cost functions by use of advanced materials [14];

- Consideration of material behaviour (brittle and composite materials) and manufacturing requirements [14].

Acknowledgement:

The authors would like to thank the German Research Association DFG for sponsoring the research project *Topologieoptimierung von Bauteilstrukturen unter Verwendung von Lochpositionierungskriterien - Bubble Method* (DFG Es 53/8-1,2). Our thanks are also due to M. Wengenroth for preparing the manuscript in its final form.

References

1. ALLAIRE, G., KOHN, R.V.: Optimal design for minimum weight and compliance in plane stress using extremal microstructures, Eur. J. Mech., A/Solids, 12, No.6 1993, 839-878.

2. ATREK, E.: SHAPE: A Program for Shape Optimization of Continuum Structures. Proc. First Int. Conf.: Opti'89. Comp. Mechanics Publications, Springer Verlag, Berlin, 1989, 135-144.

3. BANICHUK, N.V.: Introduction to Optimization of Structures. Springer-Verlag, New York 1990.

4. BANICHUK, N.V.: Shape Design Sensitivity Analysis for Optimization Problems with Local and Global Functionals, Mech. Struct. & Mach., 21(3) (1993), 375-397.

5. BENDSØE, M.P., KIKUCHI, N.: Generating optimal topologies in structural design using a homogenization method, Comp. Meth. Appl. Mech. Eng. 71 (1988), 197-224.

6. BENDSØE, M.P., DIAZ, A., KIKUCHI, N.: Topology and Generalized Layout Optimization of Elastic Structures. In: M.P. BENDSØE, C.A. MOTA SOARES (eds.): Topology Design of Structures. Kluwer Academic Publishers, Dordrecht, Netherlands 1993, 159-205.

7. BENDSØE, M.P.: Methods for the Optimization of Structural Topology, Shape and Material. Springer-Verlag, Berlin 1994.

8. BOURGAT, J.F.: Numerical Experiments of the Homogenisation Method for Operators with Periodic Coefficients, Lecture Notes in Mathematics 704, Springer-Verlag, Berlin 1973, 330-356.

9. COURANT, R., HILBERT, D.: Methods of Mathematical Physics I. Interscience Publishers, Inc., New York 1968.

10. DE BOOR, C.: On Calculation with B-Splines, J. of Approx. Theory 6 (1972), 50-62.

11. DEMS, K.: First and second-order shape sensitivity analysis of structures, J. Structural Optimization 3 (1991), 79-88.

12. ESCHENAUER, H.A., GATZLAFF, H., KIEDROWSKI, H.W.: Entwicklung und Optimierung hochgenauer Paneeltragstrukturen. Techn. Mitt. Krupp, Forsch.-Ber., Band 38, H.1. In: H.A. ESCHENAUER: Numerical and Experimental Investigations on Structural Optimization of Engineering Design. Institute of Mechanics and Control Engineering, University of Siegen 1980, 153-167.

13. ESCHENAUER, H.A., KOSKI, J., OSYSCZKA, A.: Multicriteria Design Optimization - Procedures and Applications. Springer-Verlag Berlin, Heidelberg 1990.

14. ESCHENAUER, H.A., SCHUMACHER, A., VIETOR, T.: Decision makings for initial designs made of advanced materials, In: M.P. BENDSØE, C.A. MOTA SOARES (eds.): Topology Design of Structures. Kluwer Academic Publishers, Netherlands 1993, 469-480.

15. ESCHENAUER, H.A., SCHUMACHER, A.: Possibilities of Applying Various Procedures of Topology Optimization to Components subject to Mechanical Loads, ZAMM-Z. angew. Math. Mech. 73 (1993), T 392-T394.

16. ESCHENAUER, H.A., SCHUMACHER, A.: Bubble Method: A special strategy for finding best possible initial designs, Proc. of the 1993 ASME Design Technical Conference-19th Design Automation Conference, Albuquerque, New Mexiko, Sept. 19-22, 1993, Vol.65-2, 437-443.

17. ESCHENAUER, H.A.; OLHOFF, N.; SCHNELL, W.: Applied Structural Mechanics, Berlin, Heidelberg, New York, Springer-Verlag 1996.

18. ESCHENAUER, H.A., GEILEN, J., WAHL, H.J.: SAPOP - An Optimization Procedure for Multicriteria Structural Design. In: H. HÖRNLEIN, K. SCHITKOWSKI: Numerical Methods in FE-based Structural Optimization Systems, Int. Series of Num. Math., Birkhäuser-Verlag, Basel 1993, 207-227.

19. ESCHENAUER, H.A., KOBELEV, V.V., SCHUMACHER, A.: Bubble method for topology and shape optimization of structures, J. Structural Optimization 8 (1994), 42-51.

20. FARIN, G.: Splines in CAD/CAM, Surveys on Mathematics for industry, Springer-Verlag, Austria 1991, 39-73.

21. GILL, P.E., MURRAY, W., WRIGHT, M.H.: Practical Optimization. Academic Press, London 1981.

22. GREIG, D.M.: Optimisation. Longman Group, London 1980.

23. HAHN, H.G.: Bruchmechanik: Einführung in die theoretischen Grundlagen. Teubner, Stuttgart 1976.

24. HAUG, E.J., CHOI, K.K., KOMKOV, V.: Design Sensitivity Analysis of Structural Systems. Academic Press, Orlando, Florida 1985.

25. HIMMELBLAU, D.M.: Applied Nonlinear Programming. Mc Graw-Hill, New York 1972.

26. HÖRNLEIN, H.R.E.M., SCHITTKOWSKI, K.: Numerical Methods in FE-based Structural Optimization Systems. Int. Series of Num. Math., Birkhäuser-Verlag, Basel 1993.

27. JÄGER, J.: Elementare Topologie. Schöningh-Verlag, Paderborn 1980.

28. JÄNICH, K.: Topologie. Springer-Verlag, Berlin, Heidelberg 1980.

29. KIRSCH, U.: On the Relationship between Optimum Structural Topologies and Geometries, J. Struct. Opt. 2 (1990), 39-45.

30. KORYCKI, R., ESCHENAUER, H.A., SCHUMACHER, A.: Incorporation of Adjoint Method Sensitivity Analysis with the Optimization Procedure SAPOP, Proc. of 11th Polish Conference on Computer Methods in Mechanics, 11-15 May 1993, Kielce, Poland 1993.

31. KORYCKI, R.: Multiparametric shape optimal design of disks, J. Structural Optimization 9 (1995), 25-32.

32. LEON, A.: Über die Störungen der Spannungsverteilung, die in elastischen Körpern durch Bohrungen und Bläschen entstehen, Österr. Wochenschrift für den öffentl. Baudienst, Nr.9 (1908), 163-168.

33. MA, Z.D., KIKUCHI, N., CHENG, H.C.: Topology and Shape Optimization Methods for Structural Dynamic Problems. Proc. of IUTAM-Symposium on Optimal Design with Advanced Materials, Lyngby, Denmark, Aug. 18-20, 1992, 247-261.

34. MAXWELL, J.C.: On Reciprocal Figures, Frames and Diagrams of Forces, Scientific Papers, Vol.2 (1989), 160-207.

35. MICHELL, A.G.M.: The Limits of Economy of Materials in Frame Structures, Philosophical Magazine, Series 6, Vol. 8, No. 47 (1904), 589-597.

36. MUSSCHELISCHWILI, N.I.: Einige Grundaufgaben zur mathematischen Elastizitätstheorie. VEB Fachbuchverlag, Leipzig 1971.

37. NEUBER, H.: Kerbspannungslehre. Springer-Verlag, Berlin, Heidelberg 1985.

38. PRAGER, W.: Optimality Criteria Derived from Classical Extremum Principles, SM Studies Series, Solid Mechanics Division, University of Waterloo, Ontario 1969.

39. PRAGER, W.: A note on discretized Michell structures, Computer Methods in Applied Mechanics and Engineering 3 (1974), 349-355.

40. RIEGELS, F.W.: Aerodynamische Profile. R. Oldenbourg-Verlag, München 1958.

41. ROSEN, D.W., GROSSE, I.R.: A Feature Based Shape Optimization Technique for the Configuration and Parametric Design of Flat Plates, Eng. with. Comp. 8 (1992), 81-91.

42. ROZVANY, G.I.N., ZHOU, M., ROTTHAUS, M., GOLLUB, W., SPENGEMANN, F.: Continuum - type optimality criteria methods for large Finite Element Systems with displacement constraints, Part I + II, J, Structural Optimization 1 (1989), 47-72.

43. SANKARANARYANAN, S., HAFTKA, R.T., KAPANIA, R.K.: Truss Topology Optimization with stress and displacement constraints. In: M.P. BENDSØE, C.A. MOTA SOARES (eds.): Topology Design of Structures. Kluwer Academic Publishers, Netherlands 1993, 71-78.

44. SAWIN, G.N.: Spannungserhöhung am Rande von Löchern. VEB Verlag Technik, Berlin 1956.

45. SCHMIT, L.A., MALLET, R.H.: Structural Synthesis and Design Parameters, Hierarchy Journal of the Structural Division, Proceedings of the American Society of Civil Engineers, Vol. 89, No.4 (1963), 269-299.

46. SCHRAMM, U., PILKEY, W.D.: The Coupling of Geometric Descriptions and Finite Elements using NURBS - A Study in Shape Optimization, Finite Elements in Analysis and Design, 15 (1993), 11-34.

47. SCHUMACHER, A.: Topologieoptimierung von Bauteilstrukturen unter Verwendung von Lochpositionierungskriterien, Dr.-Thesis, Universität-GH Siegen, TIM-Bericht Nr. T09-01.96.

48. SWANSON, J.A.: ANSYS User's Manual. Swanson Analysis System Inc., P. O. Box 65, Houston, PA 1994.

49. WASIUTYNSKI, Z.: On the Equivalence of Design Principles: Minimal Potential - Constant Volume and Minimum Volume - Constant Potential, Bulletin de l'Academie Polonaise des Sciences, Serie des sciences techniques, Vol. XIV, No. 9 (1966), 537-539.

50. WEINERT, M.: Sequentielle und parallele Strategien zur Auslegung komplexer Rotationsschalen, Dr.-Thesis, . Universität-GH Siegen, FOMAAS, TIM-Bericht Nr. T 05-05.94.

51. WHEELER, L.T.: On the role of constant-stress surfaces in the problem of minimizing elastic stress concentration, Int. J. Solids and Structures 12 (1976), 779-789.

52. WIEGHARDT, K.: Ein Konzept zur interaktiven Formoptimierung kontinuierlicher Strukturen. Dr.-Thesis, Institut für konstruktiven Ingenieurbau, Mitteilung Nr. 95-3, Ruhr-Univ.-Bochum 1995.

REDUCTION AND EXPANSION PROCESSES
IN TOPOLOGY OPTIMIZATION

U. Kirsch

Israel Institute of Technology, Haifa, Israel

ABSTRACT

A two-stage layout optimization procedure, consisting of reduction and expansion processes, is presented. The object in developing this procedure is to use the advantages of both processes. In the reduction process, a reduced structure with a limited number of members and joints is established by solving large scale idealized problems. An expansion process is then employed, where members and joints are added to the reduced structure. At this stage, small problems are solved, considering general variables, all relevant constraints and the real objective function.

1. INTRODUCTION

Most of the work that has been done on optimum structural design is related to optimization of cross sections. Much less effort has been devoted to optimization of the layout (geometry and topology). It is recognized, however, that optimization of the structural layout can greatly improve the design [1 - 5]. Because of the complexity in simultaneous optimization of the geometry, the topology and the cross sections, two classes of problems are often considered in this type of optimization [6]:

a. Topological optimization, where the spatial sequence of members and joints is optimized.
b. Geometrical optimization, where joint coordinates and cross-sectional sizes are optimized. The solution of each problem affects indirectly the other one. That is, the geometry is affected by elimination of members during topological optimization whereas the topology might be changed due to zero cross sections or the coalescence of joints during geometrical optimization.

In this paper layout optimization is viewed as a two-stage procedure, consisting of reduction and expansion processes (Table 1). The object in developing this procedure is to use the advantages of both processes. Topological optimization is usually based on a reduction process, where members and joints are eliminated from an initial highly connected ground structure. In a typical reduction process, solution of a large scale idealized problem is achieved by assuming various simplifications. In the approach presented in this study, the object at this stage is to establish an initial reduced structure (IRS) with a limited number of members and joints, using available analytical and numerical methods. An expansion process, where members and joints are added to an initial structure, is uncommon due to the lack of effective systematic procedures. In the procedure presented, relatively small problems are solved during the expansion process, considering general variables, all relevant constraints and the real objective function. The object at this stage is to find the final optimum by adding successively members and joints to the IRS. For each candidate topology the geometry is optimized and all intermediate solutions are feasible.

Table 1 Reduction and expansion processes

Process	Reduction	Expansion
Problem size	Large	Small
Formulation	Idealized-simplified	Real-practical
Solution stages	- Exact LB (analytical) - Approximate LB (discretized) - Establishing an IRS	- Selection of geometrical variables - Geometrical optimization - Adding members and joints
Difficulty	Computational effort	Selection of variables
Advantage	Automated process	Improved feasible solutions

2. THE REDUCTION PROCESS

2.1 General considerations

A reduction process is characterized by elimination of members and joints from an initial structural topology. The number of members in the latter topology might be indefinitely large (in exact-analytical formulations), very large (in approximate-discretized formulations) or a reduced one. In a typical reduction process, a large scale structure with numerous members is often solved. Although most of the work on layout optimization is related to reduction processes, the problem solved is usually highly idealized. Various simplifications are often assumed in the problem formulation due to some basic difficulties involved in the solution.

One problem is that, unlike common optimization problems, the structural model is itself allowed to vary during the design process. Another difficulty is that the number of possible element-joint connectivities in the initial structure is very large. In addition, the problem can have singular global optima that cannot be reached by assuming a continuous set of variables. These and other difficulties make the topological optimization problem perhaps the most challenging of the structural optimization tasks. The various simplifications assumed in the problem formulation include consideration of simplified sizing variables (e.g. a single cross-sectional variable per member), only certain constraints (stress constraints), simplified objective function (weight or compliance), simple structural systems (trusses), approximate analysis models (rigid plastic) and a limited number of loading conditions.

In the approach presented, the object of the reduction process is to establish an initial reduced structure (IRS), consisting of a limited number of members and joints. Selection of the IRS is based on available analytical and numerical solutions of simplified problems.

Since an idealized problem is solved, the solution can be viewed as a lower bound on the optimum. To obtain a practical topology for the IRS, the lower bound solution can be modified by eliminating or adding members and joints.

2.2 Exact and approximate lower bounds

Some analytical and numerical solutions used to establish the IRS are briefly described subsequently. Similar to most studies on layout optimization, truss structures will be considered; however, reduction concepts are applicable also in other types of structure. Exact-analytical solutions provide a theoretical lower bound on the weight of the structure. The general concept of the structural layout can often be established by the classical Michell theory. Michell structures can also be used as reference solutions for assessment of the efficiency of practical configurations. However, they are seldom suitable for direct use in practical design due to several reasons:
- A fundamental limitation of this approach is that it is not general. General variables, constraints, objective function and loading conditions are not considered.
- The solutions usually consist of an indefinitely large number of infinitesimal members, and the resulting structures are potentially unstable if alternative loads are applied.
- Michell layouts have only been determined for a few simple loading conditions, and there is no systematic procedure to construct a structure for an arbitrary set of loads.

Michell's early work was further developed by others (see reviews [2, 4, 5]). Techniques for assessing the efficiency of near optimal trusses have been presented, some practical aspects have been studied, and solutions for several alternative load conditions have been demonstrated. Prager and Rozvany [7] developed a layout theory as a generalization of Michell's effort. It deals with the layout of low density grid-type structures, called gridlike continua. Recently, layout theory for high density structures has been developed [8]. Approximate-discretized solutions are usually based on the ground structure approach. Member areas are allowed to reach zero and hence can be deleted automatically from the structure. While the displacement method is the prevalent structural analysis tool in current computational practice, the force method formulation is adopted in many topological optimization problems. The main reason is that a linear programming (LP) formulation is obtained under certain assumptions [2 - 5]. The main advantages of the LP formulation is that the global optimum is reached in a finite number of steps and large structures can efficiently be solved. The LP solution satisfies the equilibrium and stress constraints, but it might not satisfy the compatibility conditions or may represent unstable configuration under a general loading. For structures subjected to a single loading condition the optimum represents a statically determinate or an unstable structure. In cases where the optimal LP solution represents a statically indeterminate structure the compatibility conditions might not be satisfied, but a certain deviation from elastic force distribution is often allowed on account of the inelastic behavior. Another approach to achieve an approximate lower bound, based on optimality criteria, has been used successfully in recent years [9, 10].

2.3 Establishing an IRS

Optimal layouts obtained by solution of idealized problems are often impractical. They might represent an unstable structure (a mechanism) or consist of too many members and joints. To achieve a feasible practical design for the IRS, the lower bound solutions are modified by eliminating or adding members and joints. Denoting the exact lower bound on the optimum achieved by analytical solutions as Z_L, the approximate lower bound achieved by discretized solutions as Z_A, and the solution corresponding the selected IRS modified topology as Z_{IRS} , then

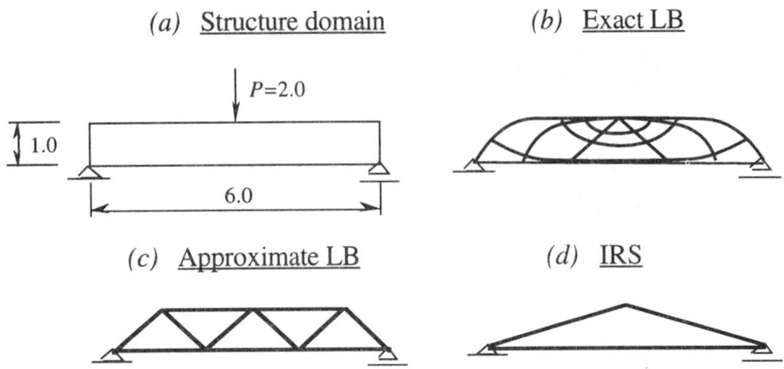

Fig. 1 Reduction process - simply-supported structure

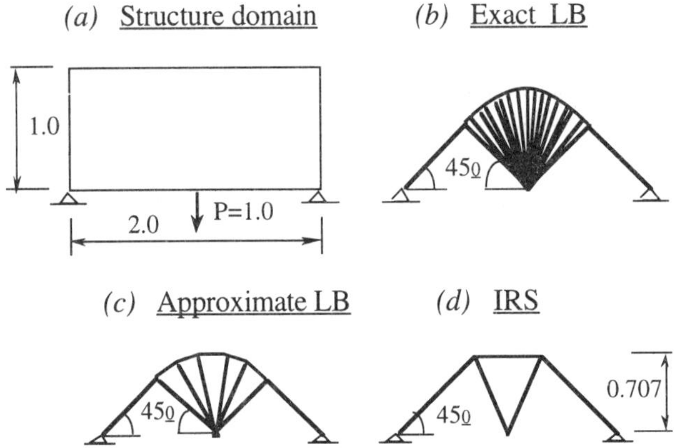

Fig. 2 Reduction process - fixed-supported structure

$$Z_L \leq Z_A \leq Z_{IRS}$$

It will be shown subsequently that the difference between the above solutions achieved during the reduction process is often insignificant. Moreover, the effect of geometrical optimization on the optimum might be larger than that of topological changes. In the following three typical examples of reduction (the results are summarized in Table 2), the objective function is the volume of material, the constraints are related only to stresses, and the allowable stress is 1.0 (arbitrary units have been assumed):

a. Simply-supported structure, shown in Fig. 1*a* . The lower bound on the optimum is a Michell structure shown in Fig. 1*b*, an approximate lower bound achieved by LP is shown in Fig. 1*c* and a three-bar truss selected as an IRS is shown in Fig. 1*d* .

b. Fixed-supported structure, shown in 2*a*. The lower bound on the optimum is a Michell structure shown in Fig. 2*b*, an approximate lower bound achieved by an optimality criteria method is shown in Fig. 2*c* [10] and a five bar truss selected as an IRS is shown in Fig 2*d*. It should be noted that all the topologies shown in Fig. 2 are unstable.

c. Cantilever structure (Fig. 3*a)*. The lower bound, an approximate lower bound and the selected six bar IRS are shown in Figs 3*b*, 3*c* , 3*d,* respectively.
It can be seen that the difference in weight between highly idealized lower bound solutions (with an indefinitely large number of members) and simplified structures consisting of 5 - 6 members might be only 3-4%.

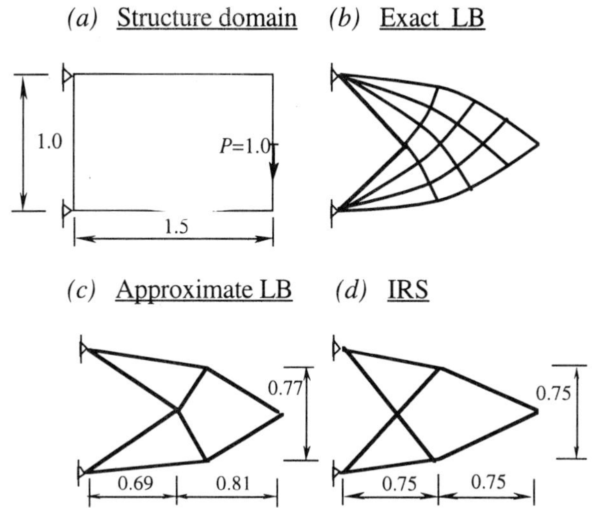

Fig. 3 Reduction process - cantilever structure

Table 2 Results, reduction process

Structure	Figure		Z_L	Z_A	Z_{IRS}
Simply-supported	1	Number of members *(n)*	∞	11	3
		Minimum weight	28.2	$1.06\,Z_L$	$1.35\,Z_L$
Fixed-supported	2	Number of members *(n)*	∞	13	5
		Minimum weight	2.57	$1.01\,Z_L$	$1.03\,Z_L$
Cantilever	3	Number of members *(n)*	∞	8	6
		Minimum weight	4.50	$1.02\,Z_L$	$1.04\,Z_L$

3. THE EXPANSION PROCESS

3.1 General considerations

Following the reduction process, an expansion process is employed, characterized by addition of members and joints to an initial structural topology. The object at this stage is to find the final optimum by adding successively members and joints to the IRS and optimizing the real problem for each candidate topology. The expansion process consists of the following main stages:
a. Selecting geometrical variables for the given topology.
b. Optimizing the geometry and cross sections.
c. Modifying the structural topology by adding members and joints.

To obtain an upper bound on the optimum Z_U, the IRS is first optimized considering geometrical and sizing variables, all relevant constraints and the real objective function. Introducing successively improved feasible designs by adding members and joints and optimizing the geometry of the resulting topologies, Z_U is improved and the final optimal design Z_{OPT} is approached from the interior side of the feasible region. The optimum is in between Z_U and the theoretical lower bound on the optimum Z_L

$$Z_L \leq Z_{OPT} \leq Z_U$$

The expansion process is characterized by the following features:
a. The structures optimized are simple, consisting of a limited number of members and joints. Therefore, the computational effort involved in the solution process is significantly reduced.
b. Since at each iteration the real problem is solved, all intermediate solutions are feasible designs satisfying all the constraints.
c. Problems of singular optima that might be encountered in the common reduction process are eliminated since we start with reduced structures having a small number of members.

3.2 Selecting the geometrical variables

During geometrical optimization the design variables are assumed to be continuous. Effective selection of these variables is most important due to the following reasons:
a. Poor selection of the variables might lead to non optimal or singular solutions.
b. A large number of variables increases significantly the computational effort whereas a small number of effective variables might be adequate to achieve a near optimal solution.
Modification of the topology during geometrical optimization might occur due to deletion of zero size members, obtained for certain geometries, or deletion of *non zero size members* due to the coalescence of joints. Elimination of zero length or parallel members, in cases where some joints tend to coalesce during geometrical optimization will change the topology and the resulting structures might represent singular optima [11]. To illustrate this phenomenon, assume the eleven-bar truss shown in Fig. 4 , the optimal depth $Y = 4.24$ and only a single geometrical variable X . Variation of Z with X is shown in Fig. 5. It can be seen that the global optimum is at point GS ($X = 3.0$, $Z = 17.0$) which is a singular point in the design space, representing a three-bar truss. However, the solution process converges to point L ($X = 1.0$, $Z = 31.1$), which is a local optimum. That is, the solution reached by optimizing X is heavier than the global optimum by 83%! The singular optimum is the result of changes in the topology of the structure. Specifically, as X approaches 3.0, the joints *A-B*, *C-D*, *E-F-G*, and the members 6-7-8, 9-10-11 tend to coalesce. Just before that, the forces in members 6, 8, 9, 11 are identical, while the forces in members 7, 10 have the same magnitude but with opposite direction. At the limit ($X = 3.0$) the three members 6, 7, 8 (and 9, 10, 11) become a single member. The result is a reduction of 2/3 of the weight of the diagonal members, leading to the singular global optimum at GS.

To illustrate the effect of alternative selections of variables, consider the following cases for the eleven-bar truss shown in Fig. 4:
a. A uniform depth with a single vertical variable Y.
b. A uniform depth with a vertical variable Y and a horizontal variable X.
c. A non uniform depth with two vertical variables: Y_F (joint F) and Y_E (joints E and G).

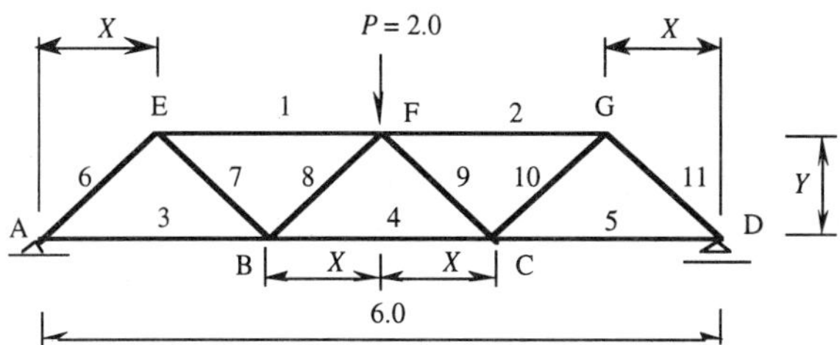

Fig. 4   Eleven-bar truss

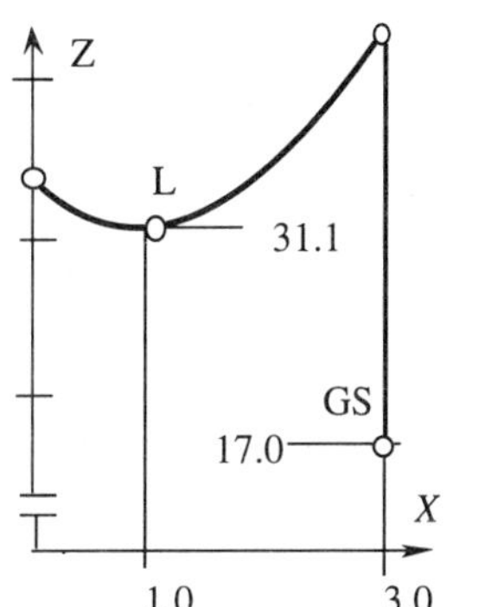

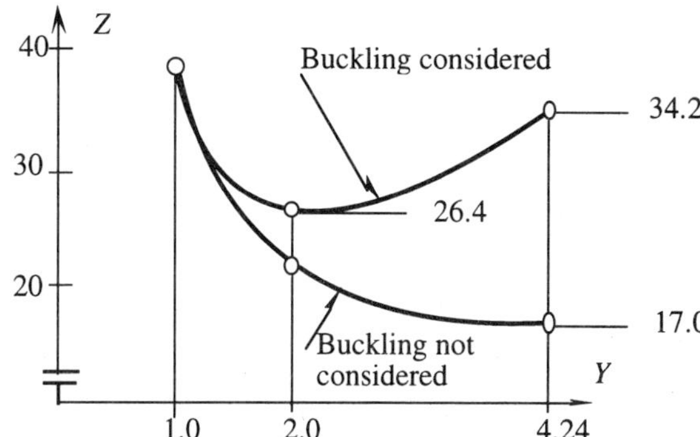

Fig. 5 Z versus X   **Fig. 6** Effect of buckling, three-bar truss

Table 3 Eleven-bar truss, various geometrical variables

Case	Variables	Optimal geometry	Z	Type of optimum
a	Y	$Y = 2.0$	24.0	Global
b	Y , X	$Y = 2.0$, $X = 1.0$	24.0	Local
c	Y_F, Y_E	$Y_F = 4.24$, $Y_E = 1.41$	17.0	Global

The optimal solutions are summarized in Table 3. It can be seen that in case a the global optimum is obtained, but better solutions could be achieved if more vertical variables are assumed (case c below). In case b a local optimum is obtained, since the true singular optimum could not be achieved by the optimization process. In case c the global optimum is achieved (the topology is changed into the three-bar truss). In conclusion, poor selection of geometrical variables might lead to non optimal or singular solutions. In the example presented, selection of two vertical variables (Y_F and Y_E) is most effective, since the global optimum is reached even in cases of changes in the topology during optimization.

As noted earlier, some important constraints are often neglected in the reduction process in order to simplify the solution. These constraints can readily be considered in the expansion process, since the structures optimized are relatively small and simple. To illustrate the effect of Euler buckling constraints, assume tubular members with a predetermined diameter-to-thickness ratio. The allowable buckling stress can be expressed as $\sigma_E = -cX/L^2$, where c is a constant depending on the modulus of elasticity, L is the member length, and X is the cross-sectional area. Assuming $c = 4.0$, then the variation of the optimal objective function value with the depth Y for the three-bar simply-supported truss (IRS) of Fig. 1d is shown in Fig. 6. It can be seen that the buckling constraints affect significantly both the optimal geometry and the objective function value. Specifically, if buckling is not considered the optimum is $Y = 4.24$, $Z_L = 17.0$. If buckling is considered for this geometry then $Z_U = 34.2$. Optimizing the geometry with buckling constraints gives the near optimal solution $Y = 2.0$, $Z = 26.4$.

3.3 Adding members and joints

During the expansion process, modified structures are introduced successively by adding members and joints. Although several expansion approaches have been proposed in the past, it is usually difficult to carry out this stage automatically and there is no single best method in terms of efficiency, reliability, and ease-of. implementation. The following approaches have been considered in this study:

a. Adding a single joint and members connecting it with existing joints. The initial position of the joint is arbitrary, whereas its final location is determined by optimizing the geometry. The advantage is that this procedure can readily be automated, but in cases of multiple local optima the solution might be affected by the initial joint position.

b. Adding multiple joints and members. This possibility is more general and also might be suitable for automated implementation. However, experience has shown that it is not very .effective since some of the new members might not be needed and the final solution could be a local optimum.

c. Adding a limited number of joints and members. This approach usually involves interactive decisions, but the combination of automated optimization and interactive design might prove useful.

To illustrate the latter approach, consider the three bar truss (Fig 7a) as an IRS. Assuming only some regular layouts, the solution process involves the following steps (Fig. 7):

a. The geometry of the IRS (Fig. 7a) is optimized with a single geometrical variable Y_1.

b. Adding four internal members and defining the new geometrical variables Y_1 and Y_2 shown in Fig. 7b, the resulting eleven-bar truss is optimized.

c. Adding four internal members and defining the new geometrical variables Y_1, Y_2 and Y_3 shown in Fig. 7c, the resulting nineteen-bar truss is optimized.

From the results shown in Fig. 8 and Table 3 it can be seen that the optimal layout for all cases of upper limit on the depth (Y_U) is the eleven-bar truss. In addition, the effect of the depth is more significant than the effect of the number of members. In particular, the difference in weight between the eleven-bar and the nineteen-bar topologies is small.

(a) IRS=three-bar truss   *(b)* Eleven-bar truss   *(c)* Nineteen-bar truss

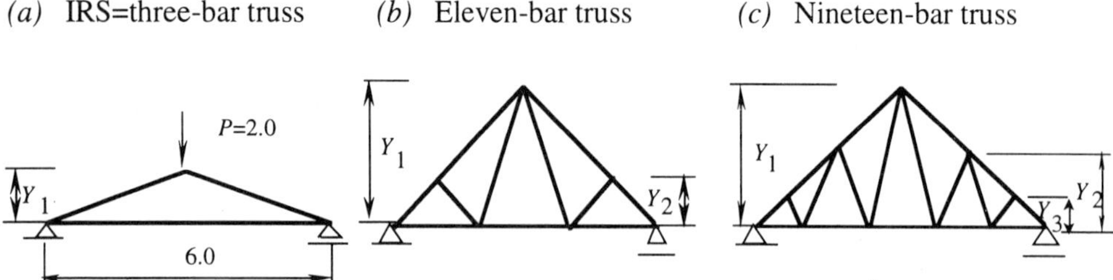

Fig. 7 A typical expansion process

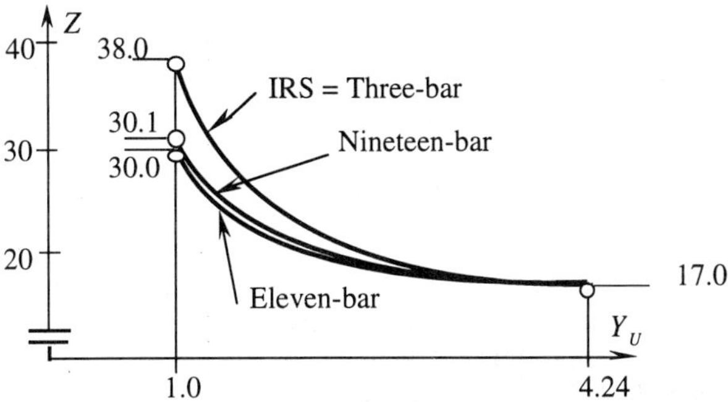

Fig. 8 Results, expansion process

Table 4 Results, expansion process

	Three-bar		Eleven-bar			Nineteen-bar			
Y_U	Z	Y_1	Z	Y_1	Y_2	Z	Y_1	Y_2	Y_3
1.0	38.0	1.0	30.0	1.0	1.0	30.1	1.0	0.8	0.6
2.0	22.0	2.0	20.7	2.0	1.0	21.6	2.0	1.4	0.8
4.24	17.0	4.24	17.0	4.24	1.21	17.0	4.24	2.55	0.85

4. CONCLUDING REMARKS

The common approach in topological optimization is based on a reduction process, where members and joints are eliminated from an initial highly connected structure. Since the problem solved is usually large, idealized simplified formulations are often assumed. An expansion process, where members and joints are added to an initial reduced structure, is not common due to the lack of automated systematic procedures. The approach presented in this paper is based on integration of the two processes into a general design procedure. In the reduction process, several methods are used to establish the IRS. It has been shown that the difference in weight between highly idealized lower bounds on the optimum (with an indefinitely large number of infinitesimal members) and simplified structures consisting of 5 - 6 members might be only 3-4 %. In addition, the effect of geometrical optimization on the optimum might be larger than that of topological changes.

During the expansion process, problems with general variables, constraints and objective function are solved successively. The main advantages at this stage are:

a. The structures optimized are simple and small, therefore the computational effort involved in the solution process is considerably reduced.

b. At each iteration the optimum is a feasible design satisfying all practical constraints.

c. Problems of singular optima that might be encountered in the reduction process are eliminated.

It has been shown that poor selection of the geometrical variables might lead to local or singular optima that cannot be reached by numerical optimization. The procedure presented in this paper can find such optima at early stages while optimizing reduced structures with a small number of members. Finally, development of a systematic expansion approach for adding members and joints is still a challenge.

ACKNOWLEDGMENT

The author is indebted to the Alexander von Humboldt Foundation for supporting this work.

REFERENCES

1. Bendsoe, M. P. and Mota Soares, C. (Eds.): Proceedings of NATO ARW on Topology design of structures, Kluwer Academic Publishers, Dordrecht 1992.
2. Kirsch U.: Optimal topologies of structures, Appl. Mech. Rev., 42 (1989), 223-239.
3. Kirsch, U.: Structural Optimizations, Springer-Verlag, Heidelbrg 1993.
4. Rozvany, G.I.N., Bendsoe, M. P. and Kirsch, U.: Layout optimization of structures, Appl. Mech. Rev., 48 (1995), 41-119.
5. Topping B.H.V.: Shape optimization of skeletal structures: a review, J. of Structural Engineering, ASCE, 109 (1983), 1933-1951.
6. Kirsch, U.: On the relationship between optimum structural geometries and topologies, Structural Optimization , 2 (1990), 39-45.
7. Prager, W. and Rozvany, G.I.N.: Optimal layout of grillages, J. Struct. Mech., 5 (1977), 1-18.
8. Bendsoe, M. P. and Kikuchi, N.: Generating optimal topologies in structural design using the homogenization method, Comp. Methods in Appl. Mech. and Engrg., 71 (1988), 197-224.
9. Rozvany, G.I.N.: Structural Design via Optimality Criteria, Kluwer, Dordrecht 1989.
10. Zhou, M. and Rozvany, G.I.N.: Iterative continuum-type optimality criteria methods in structural optimization, Research Report, Essen University 1990.
11. Kirsch, U.: Singular optima in geometrical optimization of structures, AIAA J. 33 (1995), 1165-1167.

SINGULAR AND LOCAL OPTIMA IN LAYOUT OPTIMIZATION

U. Kirsch

Israel Institute of Technology, Haifa, Israel

ABSTRACT

The major difficulty in problems having singular and local optima is that the solution process might converge to a non optimal design. In this paper, the special circumstances that might lead to these situations are discussed. Singular and local optima, encountered in layout optimization of cross-sectional variables and geometrical variables, are demonstrated. It is shown that the type of the optimum depends on the chosen design variables and preassigned parameters. In addition, both types of optima are associated with different load paths or changes in the topology of the structure. The effect of various preassigned parameters (objective function coefficients, cross-sectional dimensions, geometrical parameters, allowable stresses, limits on the design variables and external loadings) on the optimum is illustrated.

1. INTRODUCTION

In this paper, some structural optimization problems having singular and local optima, are discussed. The objective is to illustrate the special circumstances that might lead to these situations. Once the phenomena of singular and local optima are understood, more effective solution approaches could be developed, to deal with structural optimization problems in general, and topological optimization in particular.

A point is said to be *a local (relative) minimum* if it has the least function value in its neighborhood, but not necessarily the least function value for all the feasible region. Relative minima may occur due to the nature of either the objective function or the constraints, or of both. In structural optimization, relative minima usually occur due to the form of the constraints [1, 2]. Problems of multiple relative minimum points, encountered in cross-sectional optimization, are often associated with different load paths or changes in the topology of the structure, as will be shown subsequently.

A point is said to be *a singular optimum* if it has the least function value in its neighborhood, but it is connected to the feasible region by a reduced (degenerate) space, formed by assuming certain variables as zero. Singular optima are usually associated with changes in the topology of the structure. If the optimal solution is a singular point in the design space, it might be difficult or even impossible to arrive at the true optimum by numerical search algorithms.

Most topological optimization studies are based on the assumption of an initial ground structure that contains many joints and members connecting them. Member areas are allowed to reach zero and hence can be deleted automatically from the structure. This permits elimination of uneconomical members during the optimization process [3 - 5]. The singularity of the optimal topology in cross-sectional optimization of truss structures was first shown by Sved and Ginos [6]. Singular optima of grillages [7, 8] and some properties of singular optimal topologies [9, 10] have been studied later. In particular, problems of continuous constraint functions have been discussed, and limiting stresses obtained in cases of elimination of members have been defined [10].

In simultaneous optimization of geometrical and cross-sectional variables, it is possible that some joints tend to coalesce during the solution process. The result of elimination of members that coincide with others due to coalescing of joints is a reduced optimal structure. It will be shown in this article that under certain circumstances, such reduced optimal structures represent singular optima that cannot be reached by a simple numerical optimization. To overcome this difficulty, it might be necessary to adopt new layout optimization approaches, where members and joints are added to an initial structure [11].

In the structural optimization problem considered in this study, $\mathbf{X}$ is the vector of design variables and $\mathbf{C}$ is a matrix of preassigned parameters, consisting of the following quantities:

$$\mathbf{C} = \{\mathbf{S}, \mathbf{P}, \mathbf{T}, \mathbf{X}^L, \mathbf{X}^U, \sigma^L, \sigma^U\} \tag{1}$$

where $\mathbf{S}$ is the vector of structural parameters; $\mathbf{P}$ is the external load vector; $\mathbf{T}$ is a vector representing the objective function coefficients; and $\mathbf{X}^L$, $\mathbf{X}^U$, σ^L, σ^U are vectors of lower and upper limits on the variables $\mathbf{X}$, and on the stresses σ, respectively. The preassigned parameters affect directly the feasible region, the objective function contours, and the type of optimal solution, as will be demonstrated later in section 3. For simplicity of presentation the objective function is assumed to be a linear function of $\mathbf{X}$, that is

$$Z = \ell(\mathbf{T})^T \mathbf{X} \to \min \tag{2}$$

where ℓ is, for example, the vector of members' lengths. The constraints are related to limits on stresses σ and design variables $\mathbf{X}$

$$\sigma^L \leq \sigma \leq \sigma^U \tag{3a}$$

$$\mathbf{X}^L \leq \mathbf{X} \leq \mathbf{X}^U \tag{3b}$$

Other constraints, related to displacements, buckling strength, etc. could be considered; but it has been found that these constraints do not affect directly the class of particular optima discussed in this article and, therefore, they are not included in the problem formulation.

Singular and local optima that occur due to the type of the chosen design variables are discussed in section 2, and the effect of the preassigned parameters on the nature and the type of optimum is illustrated in section 3.

2. EFFECT OF DESIGN VARIABLES

To illustrate singular and local optima in optimization of cross-sectional variables, consider the three beam grillage shown in Fig. 1 and subjected to two concentrated loads $P = 1$ at the intersections (arbitrary dimensions are assumed in all examples). The cross-sectional variables for the longitudinal and transverse beams are denoted as X_1 and X_2, respectively, the allowable stress is $\sigma^U = 10$, and the members' lengths are $\ell_x = 10.0$ and $\ell_y = 14.0$. The following relations are assumed for the cross-sectional area, the modulus of section and the moment of inertia, respectively, of the two cross sections

$$A_i = \alpha X_i^k \qquad W_i = \beta X_i^m \qquad I_i = \gamma X_i^n \qquad (i = 1, 2) \qquad (4)$$

in which α, β, γ, k, m and n are given constants. Neglecting torsional rigidity of the elements, assuming only stress constraints and the volume of material as an objective function, it will be shown subsequently that, depending on the choice of the variables, the optimal solution might be either a singular point or a local optimum.

The following notation will be used for the various optimum design points : G = global optimum; L = local optimum; GS = global singular optimum; LS = local singular optimum.

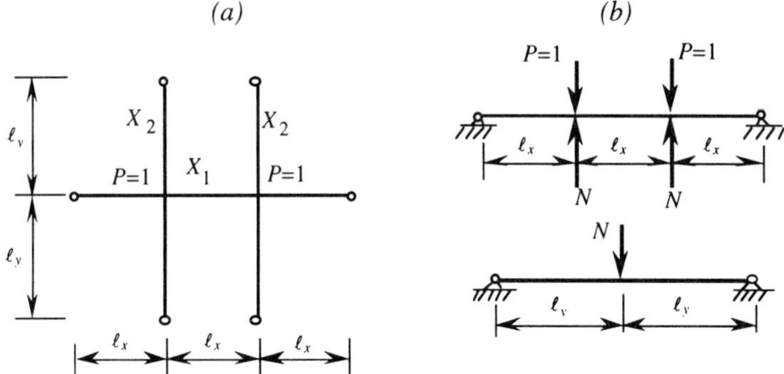

Fig. 1 Grillage

2.1 Singular optima

Assuming constant depth ($X_p = h = 2.45$) and the width of the cross sections as design variables, the design space is shown in Fig. 2a and the solution converges to point A. However, since $X_2 = 0$ at this point, the two transverse members and the constraint $\sigma_2 \le \sigma^U$

are eliminated. Thus, the resulting global singular optimum is at point GS (X_1 =1.0, X_2= 0, Z =73.5) and not at point A (X_1 =1.275, X_2= 0, Z =93.7). It can be observed that the line segment A - GS is a reduced (degenerate) part of the feasible region. Also, both points A and GS represent a topology of only the longitudinal beam. Although the optimum is a singular point in the design space, there is a single feasible region. Two or more separated regions might exist if upper or lower limits on design variables are considered [10].

Changes in the coefficients of the objective function might result in different points of convergence and optima. Assuming, for example, the objective function

$$Z = 2.45(TX_1 + 56X_2) \rightarrow \min \tag{5}$$

where T is a predetermined parameter, then the various situations that might occur are summarized in Table 1 and shown in Fig. 2. It can be seen that convergence to the true optimum might be achieved only in cases where $T \geq 39.2$. The singularity phenomenon occurs also in alternative formulations of the problem, such as simultaneous analysis and design (SAND). It is instructive to note that singular optima do not occur in plastic design, where the compatibility conditions are neglected.

Table 1 Effect of objective function coefficients

Value of T	Point of convergence	True optimum
$T < 30.75$	A	GS
$T = 30.75$	Along A - B	GS
$30.75 < T < 39.2$	B	GS
$T = 39.2$	B	B and GS
$T > 39.2$	B	B

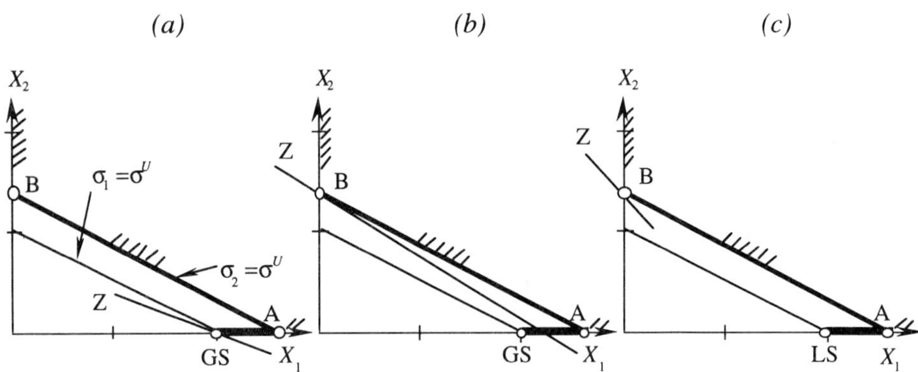

Fig. 2 Effect of T: a. T < 30.75. b. 30.75 < T < 39.2. c. 39.2 < T .

2.2. Local optima

Assuming constant width ($X_p = b = 1.2$) and the depth of the cross sections as design variables, then three relative optima, representing three different topologies, can be identified (Fig. 3a): G ($X_1 = 2.24$, $X_2 = 0$, $Z = 780.6$) , L1 ($X_1 = 1.63$, $X_2 = 1.28$, $Z = 144.7$) , L2 ($X_1 = 0$, $X_2 = 1.87$, $Z = 125.7$). Neglecting the compatibility conditions, both stress constraints will be active and the expanded feasible region (Fig. 3a) is bounded by the curved line G - L1 - L2. It can be seen that the number of local optima has been reduced to two (points G and L2). In addition, since both points represent statically determinate structures, the compatibility conditions are automatically satisfied.

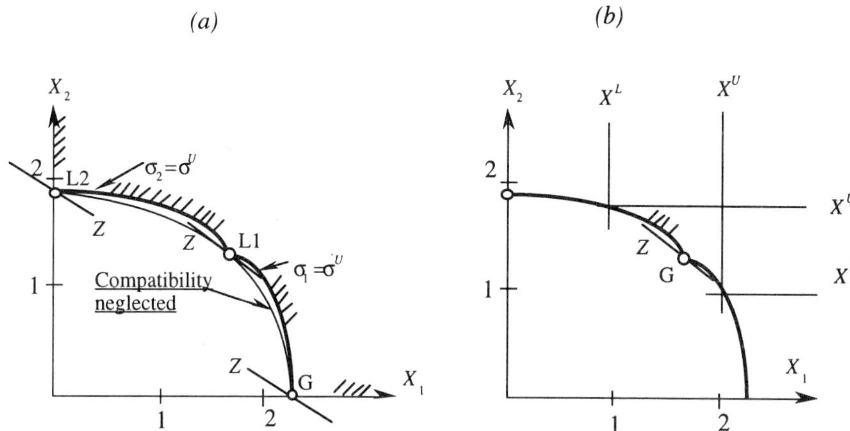

Fig. 3 Local optima: a. No limits on design variables. b. Limits on variables.

3. EFFECT OF PREASSIGNED PARAMETERS

3.1 Structural parameters

Consider again the grillage example shown in Fig. 1. It has been noted earlier that, depending on the choice of the cross-sectional parameters, the optimal solution might be either a singular point or a local optimum. It will be shown in this section that these types of optima exist only for certain values of the structural parameters.

Assuming constant depth of the cross sections and optimizing the grillage for various geometrical parameters ($\ell_y = 14.0$, $\ell_y = 60.0$ and $\ell_y = 6.0$), the results are shown in Fig. 4 and in Table 2. As noted earlier, for $\ell_y = 14.0$ the singular global optimum is at point GS (Fig. 4a). For larger ℓ_y (e.g. 60.0) the optimum is at point G ($X_2 = 0$, Fig. 4b), and for smaller ℓ_y (e.g. 6.0) the optimum is at point G ($X_1 = 0$, Fig. 4c), both being nonsingular global optima. Similarly, it can be shown that the nature and the number of local optima also depend on the structural parameters.

Table 2 Effect of structural parameters

ℓ_y	Point	X_1	X_2	Z
14	A	1.275	0	93.7
	GS	1.0	0	73.5
60	G	1.0	0	73.5
	A	0	43.2	25396
6	A	69.4	0	5103
	G	0	3.0	176.4

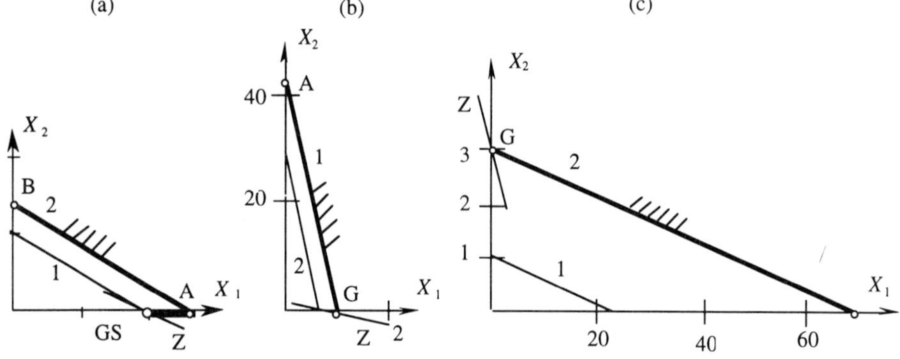

Fig. 4 Effect of ly on singular optima of a grillage: a. ℓ_y =14 b. ℓ_y =60 c. ℓ_y =6

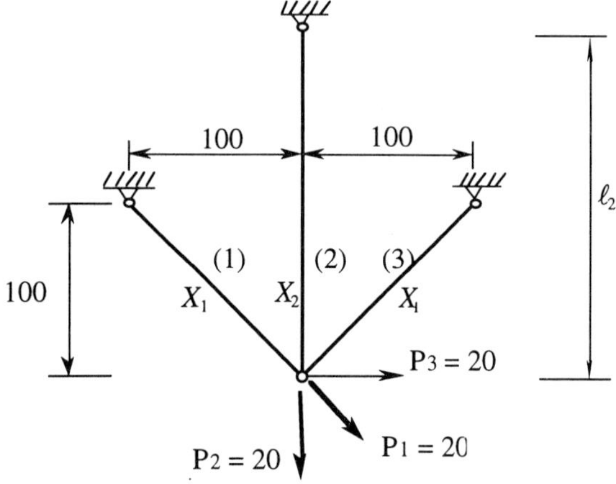

Fig. 5 Three-bar truss

3.2 Allowable stresses

Consider the symmetric three-bar truss shown in Fig. 5, subjected only to the load P_1. The two design variables are the cross-sectional areas X_1 and X_2, and the objective function represents the volume of material. To illustrate the effect of the allowable stresses on the optimum, assume $(\sigma_2^U = \sigma_3^U = 20)$ and the following three cases of allowable stresses:

   a. $\sigma_1^U = 20$,   b. $\sigma_1^U = 25$,   c. $\sigma_1^U = 50$.

The results (Fig. 6 and Table 3) show that a local singular optimum and a global singular optimum are obtained for $\sigma_1^U = 25$ and $\sigma_1^U = 50$, respectively.

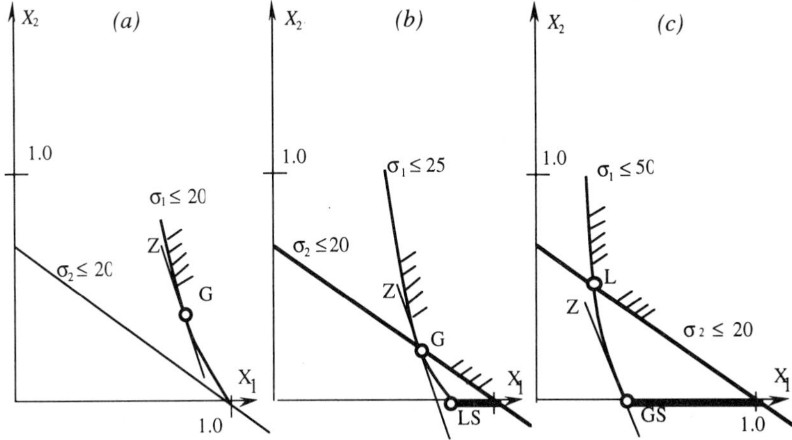

Fig. 6 Effect of allowable stress on the optimum, P1: a. $\sigma_1 \leq 20$ b. $\sigma_1 \leq 25$ c. $\sigma_{31} \leq 50$.

Table 3 Optimum points, three-bar truss (load P_1), effect of allowable stress

Case	σ_1^U	Point	X_1	X_2	. Z
Fig. 6a	20	G	0.79	0.41	264
Fig. 6b	25	G	0.67	0.23	212
		LS	0.80	0	226
Fig. 6c	50	GS	0.40	0	113
		L	0.25	0.53	124

3.3 Limits on design variables

Modifying the limits on **X**, might change the feasible region, including elimination of some singular or local optima. To illustrate the effect of these parameters, consider again the grillage example of Fig. 1. Assuming the depth of the cross sections as design variables and the lower limit $X^L = 1.0$, then the optima at points G and L2 (Fig. 3a) are eliminated and there is only a single global optimum at point G (Fig 3b). The upper limits, shown also in Fig. 3b, affect the feasible region in a similar manner.

3.4 Loadings

Changes in **P** are equivalent to modifications in the coefficients of the stress constraint expressions . If **P** is changed to μ**P**, where μ is a scalar multiplier, this modification is equivalent to linear transformation, or scaling, of the stress constraints. That is, in the case of only stress constraints, the feasible region is scaled.

To illustrate the effect of changes in the direction of the loading, consider again the three-bar truss shown in Fig. 5 and subjected separately to a vertical load (P_2), a 45° load (P_1), and a horizontal load (P_3). The results (Fig. 7) show that the type of the optimum and the corresponding topology are highly dependent on the direction of the load. Specifically, for the vertical load (Fig 7a) the global optimum is at point G (0, 1.0); for the 45° load (Fig 7b) it is at point G (0.79, 0.41); and for the horizontal load (Fig 7c) it is at point G (0.707, 0).

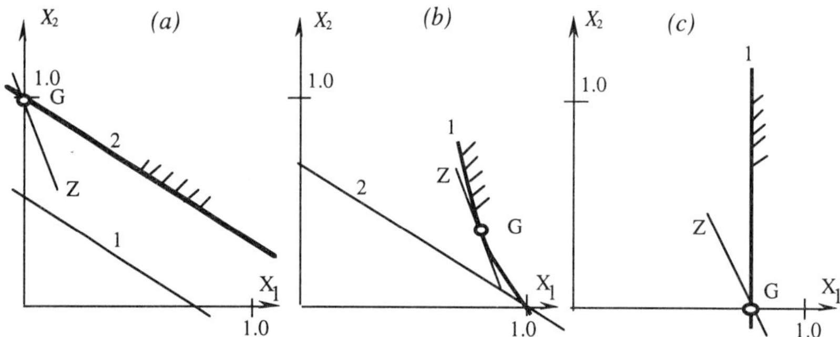

Fig. 7 Effect of direction of load on the optimum: a. vertical load b. 45 degrees load c. horizontal load.

4. CONCLUDING REMARKS

The major difficulty in problems having singular and local optima is that the solution process might converge to a non optimal design. It has been shown that the type of the optimum depends on the chosen design variables and on the selected preassigned parameters. The effect of various design variables and parameters (objective function coefficients, structural parameters, allowable stresses, limits on the design variables and external loadings) on the optimum is demonstrated.

Problems of multiple relative minimum points in cross-sectional optimization are often associated with different load paths or changes in the topology of the structure. Singular optima in either cross-sectional optimization or geometrical optimization are usually associated with changes in the topology. It has been shown [11] that singular optima in geometrical optimization may exist even in simple statically determinate structures, where joints and members coalesce. If the optimal solution is a singular point in the design space, it might be difficult or even impossible to arrive at the true optimum by numerical search algorithms.

The examples illustrate singular optima in various problems of topological optimization, where the common ground structure approach is applied and members are eliminated from the structure during the solution process. These optima are independent of the problem formulation and occur either in the common formulation in the design variables space or in the SAND formulation. To overcome the difficulty involved in the solution process it might be necessary to adopt new layout optimization approaches, where members and joints are added to an initial reduced structure. Such an approach will be discussed in a future article.

ACKNOWLEDGMENT

The author is indebted to the Alexander von Humboldt Foundation for supporting this work.

REFERENCES

1. Kirsch, U.,:*Structural optimizations, fundamentals and applications,* Springer-Verlag, Heidelbrg, 1993.
2 Moses, F. and Onoda, S.: Minimum weight design of structures with application to elastic grillages, Int. J. Num. Meth. Eng., 7, (1973), 125-136
3. Dorn, W.S., Gomory, R.E., and Greenberg, H.J.: Automatic design of optimal structures. Journal de Mecanique 3, (1964), 25-52
4. Hemp, W.S.: Optimum structures. Oxford, U K.: Clarendon Press 1973
5. Reinschmidt, K.F., and Russell, A.D.: Applications of linear programming in structural layout and optimization. Computers and Structures 4, (1974), 855-869
6. Sved, G., and Ginos, L.: Structural optimization under multiple loading. Int. J. Mech. Sci. 10, (1968), 803-805
7. Kirsch, U., and Taye, S.: On optimal topology of grillage structures. Engineering with Computers 1, (1986), 229-243
8. Kirsch, U.: Optimal topologies of flexural systems. Engineering Optimization 11, (1987), 141-149
9. Kirsch, U.: On singular topologies in optimum structural design. Structural Optimization 2, (1990), 133-142
10. Cheng, G., Jiang, Z.: Study on topology optimization with stress constraints, Engineering Optimization, 20, (1992), 129-148
11. Kirsch, U.: Singular optima in geometrical optimization of structures, AIAA J. 33 (1995), 1165-1167.

REANALYSIS MODELS FOR TOPOLOGY OPTIMIZATION
PART 1 - CONCEPTS

U. Kirsch

Israel Institute of Technology, Haifa, Israel

ABSTRACT

In part 1 of this paper, efficient reanalysis method for topological optimization of structures is presented. The method is based on combining the computed terms of a series expansion, used as high quality basis vectors, and coefficients of a reduced basis expression. The advantage is that the efficiency of local approximations and the improved quality of global approximations are combined to obtain an effective solution procedure.

The method is based on results of a single exact analysis and it can be used with a general finite element program. It is suitable for different types of structure, such as trusses, frames, grillages, etc. Calculation of derivatives is not required, and the errors involved in the approximations can readily be evaluated.

In part 2, several numerical examples illustrate the effectiveness of the solution procedure. It is shown that high quality results can be achieved with a small computational effort for various changes in the topology and the geometry of the structure.

1. INTRODUCTION

1.1 Reanalysis and topological optimization

In general, optimization of the structural topology can greatly improve the design. That is, potential savings affected by the topology are usually more significant than those resulting from cross-sections optimization. However, topological optimization did not enjoy the same degree of progress as fixed-layout optimization due to some difficulties involved in the solution process. These difficulties make the problem perhaps the most challenging of the structural optimization tasks [1, 2]. One basic problem is that the structural model is itself allowed to vary during the design process. Discrete structures are generally characterized by the fact that the finite element model of the structure is not modified during the optimization process. In topological design, however, since members are deleted or added during the design process, both the finite element model and the set of design variables change.

In most structural optimization problems the implicit behavior constraints must be evaluated for successive modifications in the design. For each trial design the analysis equations must be solved and the multiple repeated analyses usually involve extensive computational effort. This difficulty motivated extensive studies on explicit approximations of the structural behavior in terms of the design variables [3]. The various methods can be divided into the following classes:

a. *Global approximations* (called also *multipoint approximations*), such as a polynomial fitting or the reduced basis method [4 - 7]. The approximations are obtained by analyzing the structure at a number of design points, and they are valid for the whole design space (or, at least, large regions of it). However, global approximations may require much computational effort in problems with a large number of design variables.

b. *Local approximations* (called also *single-point approximations*), such as the first-order Taylor series expansion or the binomial series expansion about a given design. Local approximations are based on information calculated at a single design point. These methods are most efficient but they are effective only in cases of small changes in the design variables. For large changes in the design the accuracy of the approximations often deteriorates and they may become meaningless. That is, the approximations are only valid in the vicinity of a point in the design space. To improve the quality of the results, second-order approximations [8] can be used. Another possibility is to assume the reciprocal cross-sectional areas as design variables [9, 10]. A hybrid form of the direct and reciprocal approximations which is more conservative than either can be introduced [11]. This approximation has the advantage of being convex [12]. More accurate convex approximations can be introduced by the method of moving asymptotes [13]. One problem in using intermediate variables, such as the reciprocal variables, is that it might be difficult to select appropriate variables for cases of general optimization where geometrical, topological or shape design variables are considered.

c. *Combined approximations*, which attempt to give global qualities to local approximations. One approach to introduce combined approximations is to scale the initial stiffness matrix such that the changes in the design variables are reduced [14, 15]. The advantage of this approach is that, similar to local approximations, the solution is based on results of a single exact analysis. It has been shown that the scaling procedure is useful for various types of design variables and behavior functions. In particular, simplified approximations can be achieved for homogeneous functions [16]. Several criteria for selecting the scaling multiplier have been proposed. These include geometrical considerations [14] and mathematical criteria [15, 16]. The concept of scaling has been extended recently to include also the approximate displacements, in addition to the initial stiffness matrix, thereby allowing improved results [17, 18]. The effectiveness of combined approximations in problems of cross-sections optimization as well as geometrical optimization has been demonstrated elsewhere [19, 20].

In this paper combined approximations, suitable for topological optimization, are presented. The drawbacks of local and global approximations motivated combination of the two approaches to achieve an improved solution procedure. The presented method is based on combining the computed terms of a series expansion, used as high quality basis vectors, and coefficients of a reduced basis expression. The latter coefficients can readily be determined by solving a reduced set of the analysis equations. The advantage is that the efficiency of local approximations and the improved quality of global approximations are combined to achieve an effective solution procedure. It is shown that high quality approximations can be achieved with a small computational effort for topological changes: deletion as well as addition of members and joints from an initial ground structure. The method is based on results of a single exact analysis and can be used with a general finite element program. It is suitable for different types of structure, such as trusses, frames, grillages, etc. Calculation of derivatives is not required, and the errors involved in the approximations can readily be evaluated. It is shown that the presented procedure is a powerful tool to achieve efficient and high quality approximations of the structural behavior in topological optimization where changes in the design are very large.

1.2 Topological changes

In layout optimization, members and joints are added or deleted during the solution process and the geometry of the structure is modified due to changes in joint coordinates. Developing a reanalysis procedure for general layout modifications is most challenging, particularly in cases where the number of DOF is modified and the structural behavior is significantly changed. For such modifications, approximate reanalysis methods are usually not suitable, since they provide inadequate or even meaningless results.

Considering a general layout optimization problem, the various modifications in the structure can be classified as follows:

a. *Deletion of members and joints,* where both the design variable vector and the number of DOF are reduced. If only members are deleted, the value of some design variables becomes zero and can be eliminated from the set of variables, but the analysis model is unchanged.

b. *Addition of members and joints,* where both the design variable vector and the number of DOF are increased. If only members are added, then the vector of design variables is expanded, but the number of DOF is unchanged.

c. *Modification in the geometry,* where there is no change in the number of variables and in the number of DOF. In this case only the numerical values of the variables are modified.

The approximate method presented in this paper can deal with all types of layout modifications.

2. PROBLEM FORMULATION

Assume an initial design variables vector $\mathbf{X}_0$, the corresponding stiffness matrix $\mathbf{K}_0$ and the displacement vector $\mathbf{r}_0$, computed by the equilibrium equations

$$\mathbf{K}_0\,\mathbf{r}_0 = \mathbf{R} \tag{1}$$

The elements of the load vector $\mathbf{R}$ are assumed to be independent of the design variables. However, the approach presented herein is suitable also for cases of changes in the load vector. The stiffness matrix $\mathbf{K}_0$ is often given from the initial analysis in the decomposed form

$$\mathbf{K}_0 = \mathbf{U}_0{}^T\mathbf{U}_0 \tag{2}$$

where $\mathbf{U}_0$ is an upper triangular matrix.

Most reanalysis methods developed in the past are suitable for the relatively simple case where the structural model (or the number of DOF), and the size of $\mathbf{K}_0$, $\mathbf{r}_0$ and $\mathbf{R}$, are unchanged. The reanalysis method developed in this study is intended for problems where

the size of the above quantities is changed due to changes in the number of DOF. In the formulation presented in this section, a distinction is made between the following two cases:
a. The common case where the number of DOF is not increased.
b. The more challenging case considered in this paper, where the number of DOF is increased.

2.1 The number of DOF is not increased

Assume a change $\Delta \mathbf{X}$ in the design variables so that the modified design is

$$\mathbf{X} = \mathbf{X}_0 + \Delta \mathbf{X} \tag{3}$$

and the corresponding stiffness matrix is

$$\mathbf{K} = \mathbf{K}_0 + \Delta \mathbf{K} \tag{4}$$

where $\Delta \mathbf{K}$ is the change in the stiffness matrix due to the change $\Delta \mathbf{X}$.

The problem under consideration can be formulated as follows. Given $\mathbf{K}_0$ and $\mathbf{r}_0$, the object is to find efficient and high quality approximations of the modified displacements $\mathbf{r}$ due to various changes in the design variables $\Delta \mathbf{X}$, without solving the modified analysis equations

$$\mathbf{K}\,\mathbf{r} = (\mathbf{K}_0 + \Delta \mathbf{K})\mathbf{r} = \mathbf{R} \tag{5}$$

The elements of the stiffness matrix are not restricted to certain forms and can be general functions of the design variables. That is, the design variables $\mathbf{X}$ may represent coordinates of joints, the structural shape, members' cross sections, etc.

The reanalysis model presented herein is intended to replace the implicit analysis equations (5). Once the displacements $\mathbf{r}$ are evaluated, the stresses σ can be calculated by the explicit stress-displacement relations

$$\sigma = \mathbf{S}\,\mathbf{r} \tag{6}$$

in which $\mathbf{S}$ is the stress-transformation matrix. The elements of matrices $\mathbf{K}$ and $\mathbf{S}$ are some explicit functions of the design variables. Since the stresses are explicit functions of $\mathbf{S}$ and $\mathbf{r}$, they can readily be evaluated.

2.2 The number of DOF is increased

Consider the case where a new joint and some members connecting this joint to existing joints are added to the initial structure such that the number of DOF is increased. The modified analysis equations are

$$\mathbf{K}_M\,\mathbf{r}_M = \mathbf{R}_M \tag{7}$$

in which subscript M denotes quantities related to the modified design. If the new joint is not loaded, then the modified load vector can be expressed as

$$\mathbf{R}_M = \left\{ \begin{array}{c} \mathbf{R} \\ \mathbf{0} \end{array} \right\} \tag{8}$$

In addition, the modified stiffness matrix can be partitioned to obtain

$$\mathbf{K}_M = \left[\begin{array}{cc} \mathbf{K}_{00} & \mathbf{K}_{0M} \\ \mathbf{K}_{M0} & \mathbf{K}_{MM} \end{array} \right] \tag{9}$$

where $\mathbf{K}_{00} = \mathbf{K}_0$ is a sub matrix of the original DOF and $\mathbf{K}_{MM}$ is a sub matrix of stiffness coefficients of the new joint.

The object is to find the modified displacements $\mathbf{r}_M$ due to addition of the new joint and members, without solving the modified analysis equations (7). Evidently, developing a reanalysis method in this case is more challenging since both the size and the numerical values of the elements of the displacement vector are changed.

In the next chapter, the approach proposed to solve this problem is presented. For completeness of presentation, the approach developed in previous studies for the common case, where the number of DOF is not increased, is first described. Then, the method developed in this study, for the case where the number of DOF is increased, is presented.

3. THE SOLUTION APPROACH

3.1 The number of DOF is not increased

Approximate solution The Combined Approximations (CA) approach, has previously been demonstrated for problems with unchanged numbers of design variables and DOF [19]. The solution procedure is based on combining the reduced basis method and the first terms of a series expansion. Assuming the reduced basis method and considering second-order approximations, the displacements are expressed as

$$\mathbf{r} = y_0\, \mathbf{r}_0 + y_1\, \mathbf{r}_1 + y_2\, \mathbf{r}_2 = \mathbf{r}_B\, \mathbf{y} \tag{10}$$

where $\mathbf{r}_0$, $\mathbf{r}_1$ and $\mathbf{r}_2$ are the first three terms of the series. The matrix $\mathbf{r}_B$ and the vector $\mathbf{y}$ of coefficients to be determined are defined as

$$\mathbf{r}_B = \{\mathbf{r}_0, \mathbf{r}_1, \mathbf{r}_2\}$$
$$\mathbf{y}^T = \{y_0, y_1, y_2\} \tag{11}$$

In previous studies [17 - 20] either the Taylor series or the binomial series have been used. The advantage of using the latter series is that, unlike the Taylor series, calculation of derivatives is not required. This makes the method most attractive in layout optimization problems where derivatives are often not available.

The binomial series approximations can be obtained by rearranging (5) to read

$$\mathbf{K}_0\, \mathbf{r} = \mathbf{R} - \Delta\mathbf{K}\, \mathbf{r} \tag{12}$$

Writing this equation as the recurrence relation

$$\mathbf{K}_0 \, \mathbf{r}^{(k+1)} = \mathbf{R} - \Delta\mathbf{K} \, \mathbf{r}^{(k)} \tag{13}$$

where $\mathbf{r}^{(k+1)}$ is the value of $\mathbf{r}$ after the kth cycle, and assuming the initial value $\mathbf{r}^{(1)} = \mathbf{r}_0$, the following series is obtained

$$\mathbf{r} = (\mathbf{I} - \mathbf{B} + \mathbf{B}^2 - \dots) \, \mathbf{r}_0 \tag{14}$$

In this equation matrix $\mathbf{B}$ is defined by

$$\mathbf{B} \equiv \mathbf{K}_0^{-1} \, \Delta\mathbf{K} \tag{15}$$

Thus, the terms for $\mathbf{r}_1$ and $\mathbf{r}_2$ are given by

$$\mathbf{r}_1 = -\mathbf{B} \, \mathbf{r}_0$$
$$\mathbf{r}_2 = -\mathbf{B} \, \mathbf{r}_1 \tag{16}$$

Given $\mathbf{K}_0$ and $\mathbf{r}_0$, the following procedure is carried out to evaluate the displacements and the stresses for any change $\Delta\mathbf{K}$ in the stiffness matrix.

a. The modified matrix $\mathbf{K} = \mathbf{K}_0 + \Delta\mathbf{K}$ and the basis vectors $\mathbf{r}_1$ and $\mathbf{r}_2$ [Eq. (16)] are introduced. In cases where the load vector is modified due to changes in the design, this modification can be considered in calculation of the basis vectors.

b. The reduced (3×3) matrix $\mathbf{K}_R$ and the reduced (3×1) vector $\mathbf{R}_R$ are calculated by the expressions

$$\mathbf{K}_R = \mathbf{r}_B^T \, \mathbf{K} \, \mathbf{r}_B \qquad\qquad \mathbf{R}_R = \mathbf{r}_B^T \, \mathbf{R} \tag{17}$$

c. The coefficients $\mathbf{y}$ are calculated by solving the set of (3×3) equations

$$\mathbf{K}_R \, \mathbf{y} = \mathbf{R}_R \tag{18}$$

d. The final displacements and stresses are evaluated by Eqs. (10) and (6), respectively.

In the procedure described above second-order series approximations have been assumed, therefore it is called Combined Approximations of order 2 (CA2). If first-order approximations are assumed (only two basis vectors), the coefficients $\mathbf{y}$ in step *c* are calculated by solving a set of (2×2) equations and the procedure is called accordingly Combined Approximations of order 1 (CA1). The effectiveness of this approach has been demonstrated in several studies [17 - 20]. It has been found that high quality approximations can be achieved for very large changes in the design.

It should be noted that calculation of the basis vectors [Eq. (16)] involves only forward and backward substitutions if $\mathbf{K}_0$ is given in the decomposed form (2). The calculation of $\mathbf{r}_1$, for example, is carried out by means of this equation

$$\mathbf{K}_0 \mathbf{r}_1 = -\Delta\mathbf{K} \, \mathbf{r}_0 \tag{19}$$

We first solve for $\mathbf{t}$ by the forward substitution

$$\mathbf{U}_0^{T}\mathbf{t} = -\Delta\mathbf{K}\ \mathbf{r}_0 \qquad (20)$$

then $\mathbf{r}_2$ is calculated by the backward substitution

$$\mathbf{U}_0\mathbf{r}_1 = \mathbf{t} \qquad (21)$$

Similarly, $\mathbf{r}_2$ is calculated from

$$\mathbf{K}_0\mathbf{r}_2 = -\Delta\mathbf{K}\ \mathbf{r}_1 \qquad (22)$$

Evaluation of the Results To evaluate the quality of the results, we substitute the approximate displacements [Eq. (10)] into the modified analysis equations (5). To find the errors in satisfying the latter equations we may define *a fictitious load vector* $\mathbf{R}_a$ by

$$\mathbf{R}_a \equiv \mathbf{K}\ \mathbf{r} \qquad (23)$$

If $\mathbf{r}$ are the exact displacements, then $\mathbf{R}_a = \mathbf{R}$ (the actual given loading). That is, the approximate displacements can be viewed as the exact displacements for $\mathbf{R}_a$. The difference between the fictitious loading and the real loading

$$\Delta\mathbf{R} = \mathbf{R}_a - \mathbf{R} = \mathbf{K}\ \mathbf{r} - \mathbf{R} \qquad (24)$$

indicates the discrepancy in satisfying the modified equilibrium conditions due to the approximate solution. Thus, $\Delta\mathbf{R}$ can be used to evaluate the quality of the approximations. Let us define the common measure of smallness of $\Delta\mathbf{R}$ by the quadratic form

$$q = \Delta\mathbf{R}^T\Delta\mathbf{R} \qquad (25)$$

A possible criterion for acceptable approximations is

$$q \leq q^U \qquad (26)$$

in which q^U is a predetermined upper bound on q. Alternatively, we may use the criteria

$$\Delta\mathbf{R}^L \leq \Delta\mathbf{R} \leq \Delta\mathbf{R}^U \qquad (27)$$

where $\Delta\mathbf{R}^L$ and $\Delta\mathbf{R}^U$ are the predetermined acceptable bounds on $\Delta\mathbf{R}$.

Exact solution It has been shown [21] that for cross-sectional changes in truss structures, the exact solution is achieved if for each changed member a corresponding basis vector is assumed. Specifically, for simultaneous changes in m members, the exact solution is given by

$$\mathbf{r} = \mathbf{r}_0 + \sum_{i=1}^{m} y_i\ \mathbf{r}_i \qquad (28)$$

in which the basis vectors are defined as

$$\mathbf{r}_0 = \text{initial displacements}$$

$$\mathbf{r}_i = \mathbf{K}_0^{-1}\Delta\mathbf{K}_i\mathbf{r}_0 \quad i = 1, ..., m \tag{29}$$

and $\Delta\mathbf{K}_i$ is the contribution of the ith member to $\Delta\mathbf{K}$. This procedure is efficient in cases where a change has been made in a limited number of members. If some of the basis vectors are linearly dependent, the exact solution can be achieved for a smaller number of basis vectors. Exact solutions can be achieved also in other cases such as scaling of cross sections or scaling of the geometry.

3.2 The number of DOF is increased

Establishing a Modified Initial Design (MID) Adding a joint to the structure, the number of DOF is increased. Therefore, it is necessary first to expand the basis vectors and to introduce a *modified initial design* (MID), such that the new degrees of freedom are included in the analysis model. The MID can be selected such that reanalysis will be convenient and not necessarily a particular modified design. Any requested design can be analyzed at a later stage by corresponding changes. Once the MID is introduced, it is then possible to analyze conveniently modified structures due to addition or deletion of members, keeping the number of degrees of freedom unchanged.

The MID and the expanded basis vectors can be established in several ways. A simple and convenient procedure for introducing and reanalyzing the MID is demonstrated in this section. Using this approach, the MID is formed by adding a new joint and horizontal and vertical members connecting this joint with existing joints in the structure. In cases where these members are not needed, they can be eliminated from the structure later, while applying changes to the MID. In addition, a convenient temporary location for the new joint can be selected. Then, it is possible to modify the joint coordinates by the procedure described later.

Assuming a given initial design and adding a new joint and horizontal and vertical members, the equilibrium equations of the resulting MID are given by Eq. (7). Considering the modified stiffness matrix of Eq. (9), its inverse is a flexibility matrix, $\mathbf{F}_M$, consisting of the corresponding sub matrices

$$\mathbf{K}_M^{-1} = \mathbf{F}_M = \begin{bmatrix} \mathbf{F}_{00} & \mathbf{F}_{0M} \\ \mathbf{F}_{M0} & \mathbf{F}_{MM} \end{bmatrix} \tag{30}$$

In general, calculation of $\mathbf{K}_0^{-1}$ is not needed [22]. However, since only a horizontal member and a vertical member are connected to the new joint, the various sub matrices are readily available. Specifically, it can be observed that

$$\mathbf{F}_{00} = \mathbf{K}_0^{-1}$$

$$\mathbf{F}_{0M} = \mathbf{F}_{M0}^T = \text{corresponding rows and columns of } \mathbf{K}_0^{-1} \tag{31}$$

$$\mathbf{F}_{MM} = (\text{flexibility matrix of the new members, } \mathbf{K}_{MM}^{-1}) + (\text{corresponding elements of } \mathbf{K}_0^{-1})$$

Noting that $\mathbf{K}_0^{-1}$ is given from the initial analysis, all sub matrices in Eq. (26) can be determined in a simple way.

The modified initial displacement vector is given by

$$\mathbf{r}_M = \left\{ \begin{array}{c} \mathbf{r}_0 \\ \Delta\mathbf{r}_0 \end{array} \right\} \tag{32}$$

in which $\Delta\mathbf{r}_0$ is the sub vector of displacements of the new joint. Since there is no force in the members connected to the added joint, the elements of $\Delta\mathbf{r}_0$ are simply the corresponding displacements of the existing joints, as will be demonstrated later by some numerical examples.

Further changes Once the MID has been introduced, evaluation of the displacements for further changes in the topology and the geometry of the structure is straightforward. Considering the given initial value of the inverse $\mathbf{K}_M^{-1}$ [Eq. (30)] and the initial displacements $\mathbf{r}_M$ [Eq. (32)], the modified basis vectors are determined for any assumed $\Delta\mathbf{K}$ by

$$\mathbf{r}_{1M} = -\mathbf{K}_M^{-1} \Delta\mathbf{K}\, \mathbf{r}_M$$

$$\mathbf{r}_{2M} = -\mathbf{K}_M^{-1} \Delta\mathbf{K}\, \mathbf{r}_{1M} \tag{33}$$

Members can be deleted from the structure by assuming zero cross sections. It should be noted that a joint is automatically eliminated if all members connected to the joint are deleted from the structure. Addition of members is considered by assuming certain cross sections for such members. As noted earlier, the exact solution can be achieved if the basis vectors are determined by Eq. (29). Since the number of basis vectors is $m + 1$, this procedure is efficient only in the case of changes in a limited number of members.

The exact solution can be achieved also for geometrical modifications, as follows. In the case of changes in a small number of members, the exact solution can efficiently be obtained by viewing these changes as corresponding topological modifications. Modifying, for example, the coordinates of a single joint, it is possible to obtain the exact solution for the new design by viewing the change in the geometry as the following two successive changes in the topology:
a. All members connected to that joint are eliminated.
b. New members are added at the modified location.
This procedure can be used also in cases where it is necessary to modify the coordinates of a new joint added to the initial design to form the MID, as demonstrated in the numerical examples that follow.

ACKNOWLEDGMENT

The author is indebted to the Alexander von Humboldt Foundation and to the Fund for the Promotion of Research at the Techniom for supporting this work.

REFERENCES

1. Kirsch U.: Optimal topologies of structures, Appl. Mech. Rev., 42 (1989), 223-239.
2. Rozvany, G.I.N., Bendsoe, M. P. and Kirsch, U.: Layout optimization of structures, Appl. Mech. Rev., 48 (1995), 41-119.
3. Barthelemy, J-F.M., and Haftka, R. T.: Recent advances in approximation concepts for optimum structural design. In: Proceedings of NATO/DFG ASI on Optimization of large structural systems. Berchtesgaden, Germany, September 1991
4. Fox, R.L. and Miura, H.: An approximate analysis technique for design calculations. J. AIAA, 9 (1971), 177-179
5. Haftka, R. T., Nachlas, J. A., Watson, L. T., Rizzo, T. and Desai, R.: Two-point constraint approximation in structural optimization, Comp. meth. appl. mech. engrg., 60 (1989), 289-301
6. Kirsch, U.: Structural optimizations, fundamentals and applications Springer-Verlag, Heidelbrg 1993
7. Noor, A. K.: Recent advances and applications of reduction methods, Appl. Mech. Rev., 47 (1994), 125-146
8. Fleury, C.: Efficient approximation concepts using second order Information, Int. J. for Num. Meth. in Engrg., 28 (1989), 2041-2058.
9. Fuchs, M. B.: Linearized homogeneous constraints in structural design, Int. J. Mech. Scien. 22 (1980), 333-400
10. Schmit, L. A. and Farshi, B.: Some approximation concepts for structural synthesis, AIAA J., 11 (1974), 489-494
11. Starnes, J.H. Jr., and Haftka, R.T.: Preliminary design of composite wings for buckling stress and displacement constraints. J. Aircraft 16 (1979), 564-570
12. Fleury, C. and Braibant, V.: Structural optimization: a new dual method using mixed variables. Int. J. Num. Meth. Engrg. 23 (1986), 409-428
13. Svanberg, K.: The method of moving asymptotes - a new method for structural optimization. Int. J. Num. Meth. Engrg. 24 (1987), 359-373
14. Kirsch, U. and Toledano, G.: Approximate reanalysis for modifications of structural geometry, Computers and structures, 16 (1983), 269-279
15. Kirsch, U.: Approximate behavior models for optimum structural design, in New directions in optimum structural design (Eds. E. Atrek, et al), John Wiley &Sons (1984)
16. Hjali, R.M. and Fuchs, M.B.: Generalized approximations of homogeneous constraints in optimal structural design, in Computer aided optimum design of structures (Eds. C.A. Brebbia and S. Hernandez), Springer-Verlag, Berlin , (1989)
17. Kirsch, U.: Reduced basis approximations of structural displacements for optimal design. J. AIAA 29 (1991), 1751-1758
18. Kirsch, U.: Approximate reanalysis methods, Structural optimization: status and promise, Ed. M.P. Kamat, AIAA 1993
19. Kirsch, U.: Approximate reanalysis for topological optimization, Structural optimization, 6 (1993), 143-150.
20. Kirsch, U.: Effective reanalysis of structures, Department of Civil Engineering, Technion, January 1996.
21. Kirsch, U. and Liu, S.: Exact structural reanalysis by a first-order reduced basis approach, Structural Optimization, 10 (1995), 153-158.
22. Kirsch, U. and Liu, S.: Structural reanalysis for general layout modifications, to be publlished, AIAA Journal.

REANALYSIS MODELS FOR TOPOLOGY OPTIMIZATION
PART 2 - APPLICATIONS

U. Kirsch

Israel Institute of Technology, Haifa, Israel

1 CHANGES WITHOUT INCREASING THE NUMBER OF DOF

1.1 Approximate Solution

Nineteen bar truss To illustrate reanalysis for topological and cross-sectional changes consider the tower truss shown in Fig. 1 with nineteen cross-sectional area design variables $X_i (i = 1, ..., 19)$, subjected to a single loading condition of two concentrated loads. The modulus of elasticity is $E = 10\,000$ and the twelve unknowns are the horizontal (to the right) and the vertical (upward) displacements in joints 2, 3, 4, 6, 7, and 8, respectively. Assuming initial cross-sectional areas $\mathbf{X}_0 = \mathbf{1.0}$ and starting with the nineteen-bar topology, the following cases have been solved by considering zero cross sections for the eliminated members:

a.   Elimination of six members to obtain the topology shown in Fig. 2*a*.
b.   Elimination of seven members to obtain the topology shown in Fig. 2*b*.
c.   Elimination of nine members and two joints to obtain the topology shown in Fig. 2*c*.
d.   Elimination of nine members and two joints as in case *c* above and optimization of the remaining members subject to the stress constraints $\sigma \leq 20.0$. The resulting optimal cross sections are given by

$$\mathbf{X}^T = \{2.0, 0, 0.83, 1.33, 0.67, 0.83, 0, 1.33, 0.67, 0, 0.83, 0.1, 0.75, 0,$$

$$0, 0, 0, 0, 0\}$$

The results in Table 1 show the high quality of the approximations achieved by the CA2 for topological and cross-sectional changes in trusses.

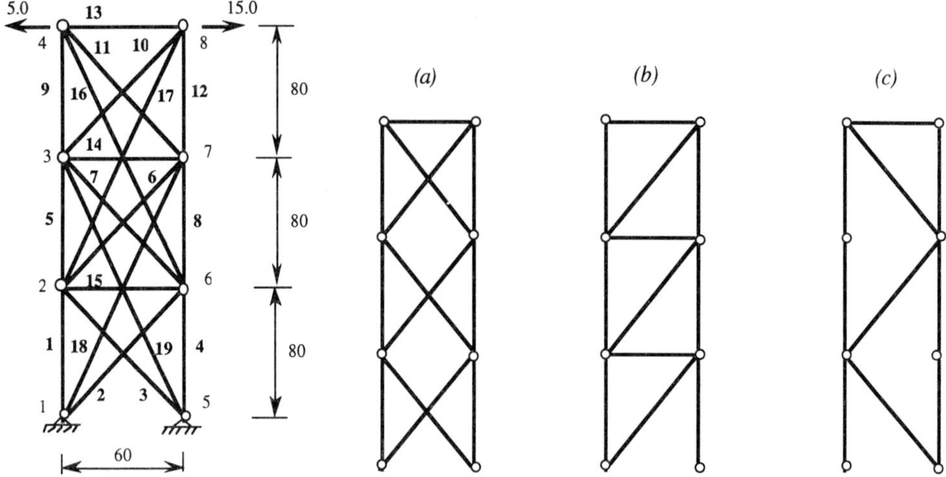

Fig. 1 Nineteen-bar tower truss. **Fig. 2**. Nineteen-bar truss, modified topologies.

Table 1 Displacements, nineteen-bar truss, deletion of members and joints.

| Displ. | Case *a* | | Case *b* | | Case *c* | | Case *d* | |
No	Exact	CA2	Exact	CA2	Exact	CA2	Exact	CA2
1	0.48	0.49	0.76	0.78	0.70	0.71	0.55	0.49
2	0.26	0.27	0.21	0.21	0.32	0.32	0.16	0.16
3	1.58	1.57	2.09	2.07	-	-	-	-
4	0.43	0.43	0.32	0.33	-	-	-	-
5	2.88	2.88	3.62	3.61	3.53	3.53	2.92	2.82
6	0.48	0.48	0.32	0.35	0.53	0.54	0.48	0.52
7	0.50	0.50	0.70	0.72	-	-	-	-
8	-0.27	-0.27	-0.32	-0.32	-	-	-	-
9	1.54	1.54	2.03	2.00	1.98	1.97	1.52	1.40
10	-0.43	-0.42	-0.53	-0.51	-0.43	-0.42	-0.32	-0.30
11	2.93	2.93	3.65	3.64	3.62	3.62	3.04	2.95
12	-0.48	-0.48	-0.64	-0.63	-0.43	-0.42	-0.32	-0.50

Ten bar truss Consider the truss shown in Fig. 3 with ten cross-sectional area variables $X_i (i = 1, ..., 10)$, subjected to a single loading condition of two concentrated loads. The modulus of elasticity is $E = 30000$ and the eight unknowns are the horizontal (to the right) and the vertical (upward) displacements in joints 1, 2, 3 and 4, respectively. The initial cross-sectional areas is unity, that is, the initial design is $X_0 = 1.0$. In order to allow comparisons with exact solutions, a cross-sectional area of 0.001 is assumed for the eliminated members. Assuming the stress constraints $\sigma \leq 25.0$, the resulting optimum is

$$\mathbf{X}_{opt}^T = \{8.0, 0.001, 8.0, 4.0, 0.001, 0.001, 5.66, 5.66, 5.66, 0.001\}$$

Assuming the line between the initial design $\mathbf{X}_0$ and the optimal design $\mathbf{X}_{opt}$

$$\mathbf{X} = \mathbf{X}_0 + \left(\mathbf{X}_{opt} - \mathbf{X}_0\right)\alpha$$

where α is a step size variable, results for $\alpha = 0.5$ and $\alpha = 1.0$ (the optimum) are given in Table 2. It can be observed that at the optimum the cross sections of members 2, 5, 6, 10 are equal to the lower limit ($X = 0.001$), therefore these members and joint 2 are practically eliminated.

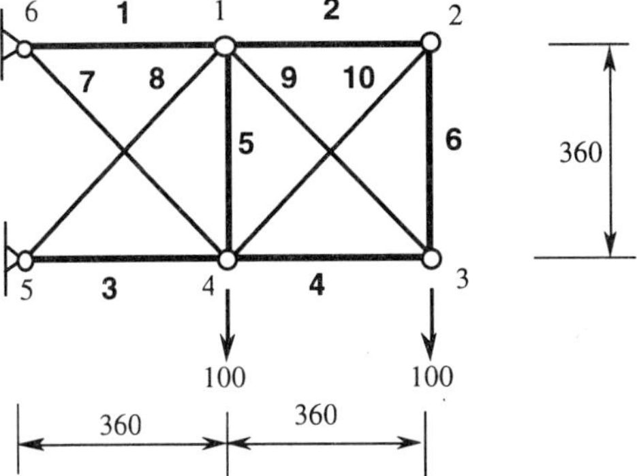

Fig. 3 Ten-bar truss

Table 2 Ten-bar truss, approximations along a line to the optimum.

Case	Method	Displacements							
		1	2	3	4	5	6	7	8
$\alpha=0.5$	Exact	0.52	-1.49	0.77	-3.65	-0.98	-3.90	-0.55	-1.62
	CA2	0.51	-1.46	0.76	-3.63	-0.98	-3.87	-0.56	-1.64
$\alpha=1.0$	Exact	0.30	-0.90	*	*	-0.60	-2.40	-0.30	-0.90
	CA2	0.29	-0.84	*	*	-0.61	-2.34	-0.31	-0.95

* Joint 2 is eliminated

Fifty bar truss To illustrate a near exact solution for structures with a larger number of degrees of freedom, consider the cantilever truss shown in Fig. 4a. The design variables are the fifty cross-sectional areas $X_i (i = 1, ..., 50)$ and the initial cross sections are $\mathbf{X}_0 = \mathbf{1.0}$. The truss is subjected to a single load at the end. The modulus of elasticity is $E = 10\,000$ and the forty unknowns are the X direction (to the right) and the Y direction (upward) displacements in joints 2 through 21, respectively. Eliminating ten diagonal members to obtain the topology shown in Fig. 4b, the resulting displacements at joints 1 - 11 are given in Tables 3. It can be observed that results achieved by the CA1 and CA2 are almost equal to the exact solution.

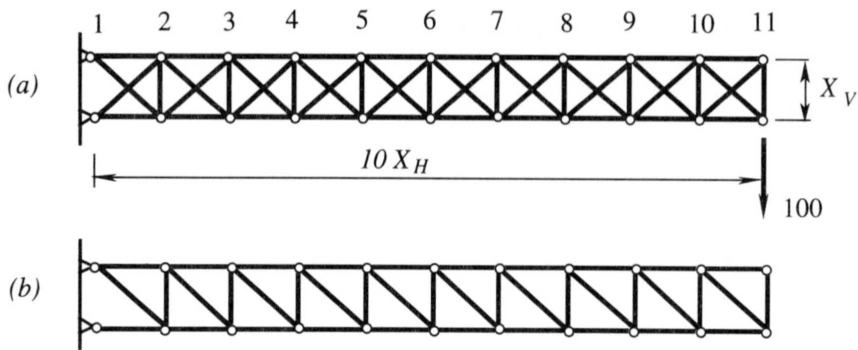

Fig.4 Fifty-bar truss: *a*. initial topology *b*. Final topology.

Table 3 Displacements, fifty-bar truss, deletion of members.

Joint	Direction	Exact	CA2	CA1
2	X	0.0900	0.0900	0.0890
	Y	-0.1383	-0.1384	-0.1429
3	X	0.1700	0.1699	0.1693
	Y	-0.4567	-0.4571	-0.4600
4	X	0.2401	0.2400	0.2394
	Y	-0.9351	-0.9352	-0.9377
5	X	0.3001	0.3000	0.2996
	Y	-1.554	-1.554	-1.555
6	X	0.3501	0.3500	0.3497
	Y	-2.292	-2.292	-2.293
7	X	0.3901	0.3901	0.3899
	Y	-3.131	-3.130	-3.131
8	X	0.4201	0.4201	0.42
	Y	-4.049	-4.049	-4.048
9	X	0.4401	0.4401	0.4401
	Y	-5.028	-5.027	-5.026
10	X	0.4501	0.4501	0.4502
	Y	-6.046	-6.046	-6.044
11	X	0.4501	0.4501	0.4500
	Y	-7.075	-7.074	-7.074

1.2 EXACT SOLUTION

Ten bar truss Consider again the initial topology of the ten-bar truss shown in Fig. 3. Assuming the modified topologies shown in Fig. 5, obtained by elimination of members and joints from the structure, the exact results achieved by Eq. (28) are summarized in Table 4. It is interesting to note that the same results have been achieved by the CA2 (considering only three basis vectors instead of $m+1$, where m is the number of changed members), due to linear dependence of the basis vectors.

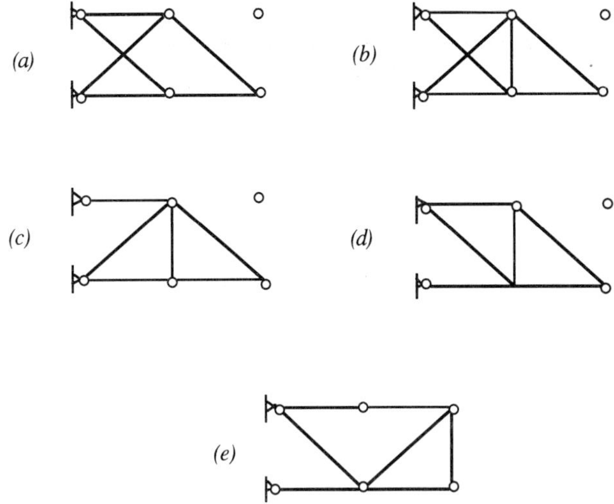

Fig. 5 Ten-bar truss, modified topologies.

Table 4 Ten-bar truss, exact solutions achieved by the CA2.

	Members	Displacements							
Case	eliminated	1	2	3	4	5	6	7	8
Fig. 5*a*	2,5,6,10	2.40	5.79	*	*	-3.60	15.18	-2.40	5.80
Fig. 5*b*	2,6,10	2.40	5.79	*	*	-3.60	15.18	-2.40	5.79
Fig. 5*c*	2,6,7,10	3.60	10.37	*	*	-2.40	19.77	-1.20	11.57
Fig. 5*d*	2,6,8,10	1.20	11.57	*	*	-4.80	20.96	-3.60	10.37
Fig. 5*e*	5,8,9	*	*	2.40	-19.76	-3.60	20.96	-3.60	10.38

* = eliminated joint.

2. GEOMETRICAL AND TOPOLOGICAL MODIFICATIONS

Ten-bar truss To illustrate results for the case of simultaneous elimination of members and joints and modifications in the geometry, consider again the initial ten-bar truss design shown in Fig. 3 Results for three different cases of modified designs due to changes in the structural layout will be demonstrated:

a. Elimination of members 2,6, 10.

b. Changing the geometry by increasing the depth of the truss by 100% . This is a change in six members but, to achieve the exact solution, it can be viewed as a change in twelve members (elimination of six members and addition of six members).

c. Combination of the previous two cases, that is, simultaneous elimination of members 2, 6, 10 and increasing the depth by 100% .

Approximate results achieved by the CA1 and CA2 (considering only two and three basis vectors, respectively) are summarized in Table 5. It can be observed that the results achieved by the CA2 are very close to the exact solution. Good approximations have been achieved also by the CA1.

Table 5 Results, elimination of members and change of geometry

Case	Method	Displacements							
a	Exact	2.40	5.79	-3.60	15.18	-2.40	5.79	*	*
	CA2	2.40	5.79	-3.60	15.18	-2.40	5.79	*	*
	CA1	2.32	5.47	-3.67	14.83	-2.48	6.11	*	*
b	Exact	1.15	3.67	-1.66	7.36	-1.25	4.24	1.34	6.60
	CA2	1.14	3.67	-1.68	7.35	-1.24	4.25	1.34	6.62
	CA1	1.17	3.78	-1.61	7.29	-1.28	4.27	1.26	6.72
c	Exact	1.20	3.95	-1.80	8.80	-1.20	3.96	*	*
	CA2	1.19	3.93	-1.81	8.79	-1.20	3.97	*	*
	CA1	1.25	3.99	-1.89	8.59	-1.25	4.12	*	*

* Joint 4 is eliminated.

Fifty-bar truss Consider again the cantilever truss shown in Fig. 4a . Considering the two geometric variables X_V , X_H and assuming the initial geometry $X_V = X_H = 1.0$, two cases of layout modifications have been solved:

a. Geometrical modifications, where the modified geometry is given by $X_V = 1.2$ (a change of 20% in the depth).

b. Geometrical modifications, where the modified geometry is given by $X_V = 2.0, X_H = 1.9$ (a change of 100% in the depth and 90% in the width).

The stiffness coefficients of thirty members have been changed, therefore exact reanalysis is not efficient. Assuming the CA1, the results are given in Table 6. Comparing the results obtained for the two cases of geometrical modifications it can be seen that better approximations have been achieved in case b, for larger changes in the geometry. The better results in case b are attributed to the fact that the modified geometry is relatively close to a scaled geometry ($X_V = X_H$), for which the CA1 provide the exact solution.

3. ADDITION OF MEMBERS AND JOINTS

3.1 Establishing an MID

To illustrate the solution procedure for the case of addition of members and joints, consider the initial seven-bar truss shown in Fig. 6a. The truss is subjected to a single loading condition of two concentrated loads, the modulus of elasticity is $E = 30\,000$ and the six analysis unknowns are the horizontal (to the right) and the vertical (downward) displacements at joints 1, 2 and 3, respectively. Assuming initial cross-sectional areas $\mathbf{X}_0 = 1.0$, then the initial inverse of the stiffness matrix, the initial load vector, and the initial displacement vector are given by

$$\mathbf{K}_0^{-1} = 1/83.333 \begin{bmatrix} 0.89 & & & & & \\ 0.56 & 2.14 & & \text{Symmetric} & & \\ -0.12 & -0.44 & 1.88 & & & \\ 1.56 & 3.14 & -2.44 & 9.97 & & \\ -0.12 & -0.44 & 0.88 & -1.44 & 0.88 & \\ 0.44 & 1.69 & -0.56 & 2.69 & -0.56 & 2.14 \end{bmatrix} \tag{a}$$

Table 6 Displacements, fifty-bar truss, geometrical modifications

Joint	Direction	Case b		Case c	
		Exact	CA1	Exact	CA1
2	X	0.079	0.089	0.200	0.204
	Y	0.079	0.106	0.240	0.247
3	X	0.150	0.160	0.379	0.384
	Y	0.283	0.347	0.878	0.897
4	X	0.213	0.216	0.537	0.541
	Y	0.599	0.693	1.871	1.901
5	X	0.267	0.260	0.674	0.677
	Y	1.011	1.117	3.175	3.213
6	X	0.312	0.293	0.789	0.790
	Y	1.507	1.602	4.744	4.786
7	X	0.350	0.317	0.884	0.883
	Y	2.072	2.130	6.535	6.576
8	X	0.379	0.335	0.958	0.954
	Y	2.693	2.689	8.503	8.538
9	X	0.400	0.346	1.010	1.005
	Y	3.356	3.268	10.604	10.629
10	X	0.413	0.353	1.042	1.036
	Y	4.046	3.861	12.794	12.805
11	X	0.416	0.354	1.052	1.045
	Y	4.748	4.454	15.024	15.019

$$\mathbf{R}^T = [0,\ 0,\ 0,\ 100,\ 0,\ 100] \tag{b}$$

$$\mathbf{r}_0^T = \{2.40,\ 5.80,\ -3.60,\ 15.19,\ -2.40,\ 5.80\} \tag{c}$$

Adding the new joint 4 and the horizontal and vertical members 8, 9 (connecting this joint with the existing joints 1 and 2, respectively), the resulting MID is shown in Fig. 1b. To introduce the inverse of the modified stiffness matrix [Eqs. (30), (31)], we note that the sub matrix $\mathbf{F}_{00} = \mathbf{K}_0^{-1}$ is already given from analysis of the initial design [Eq. (a)]. In addition, the elements of $\mathbf{F}_{M0}$ are simply rows number one and four of $\mathbf{K}_0^{-1}$. (Similarly, the elements of $\mathbf{F}_{0M}$ are columns number one and four of $\mathbf{K}_0^{-1}$). That is

$$\mathbf{F}_{M0} = \mathbf{F}_{0M}^T = 1/83.333 \begin{bmatrix} 0.89 & 0.56 & -0.12 & 1.56 & -0.12 & 0.44 \\ 1.56 & 3.14 & -2.44 & 9.97 & -1.44 & 2.69 \end{bmatrix} \tag{d}$$

The elements of $\mathbf{F}_{MM}$ can readily be determined by adding the inverse of the stiffness of the new joint (a diagonal 2×2 matrix) to the corresponding elements of $\mathbf{K}_0^{-1}$, namely, the elements 11, 14, 41, 44 in Eq. (a). The resulting sub matrix is

$$\mathbf{F}_{MM} = 1/83.333 \left(\begin{bmatrix} 0.89 & 1.56 \\ 1.56 & 9.97 \end{bmatrix} + \begin{bmatrix} 1.0 & \\ & 1.0 \end{bmatrix} \right) = 1/83.333 \begin{bmatrix} 1.89 & 1.56 \\ 1.56 & 10.97 \end{bmatrix} \tag{e}$$

Finally, the modified initial displacement vector is [Eq. (32)]

$$\mathbf{r}_M^T = \{\mathbf{r}_0^T, \ \Delta\mathbf{r}_0^T\} = \{2.40, \ 5.80, \ -3.60, \ 15.19, \ -2.40, \ 5.80, \ 2.40, \ 15.19\} \qquad (f)$$

in which the displacements of the new joint, $\Delta\mathbf{r}_0$, are the corresponding given displacements

$$r_{M7} = r_1 \qquad r_{M8} = r_4 \qquad (g)$$

3.2 Further topological modifications

Considering the MID of Fig. 6*b* and adding member 10, we obtain the ten-bar truss shown in Fig. 6*c*. Since only a single member has been added, the exact solution can be determined by considering only the two basis vectors $\mathbf{r}_M$, $\mathbf{r}_{1M}$. The vector $\mathbf{r}_M$ is given by Eq. (*f*) whereas the vector $\mathbf{r}_{1M}$ is calculated by Eq. (33)

$$\mathbf{r}_{1M}^T = \{0.19, \ 0.72, \ -1.44, \ 6.94, \ 0.19, \ -0.72, \ -1.44, \ 8.56\} \qquad (h)$$

The coefficients $\mathbf{y}$, calculated by Eq. (18), are $y_0 = 1.0$, $y_1 = -0.2964$ and the resulting exact displacements are [Eq. (28)]

$$\mathbf{r}^T = \{2.34, \ 5.58, \ -3.17, \ 13.13, \ -2.46, \ 6.01, \ 2.83, \ 12.65\} \qquad (i)$$

Other members can be deleted from the MID or added to it and the exact solution is achieved in a similar way. If the number of modified elements is large, then an approximate solution can efficiently be achieved.

4. CONCLUSIONS

Approximations of the structural behavior in terms of the design variables are essential in optimization of large scale structures, where the time consuming analysis must be repeated many times. Local approximations, such as the Taylor series or the binomial series are most efficient, but the quality of the results might be insufficient, particularly in cases of large changes in the design.

In topological optimization the changes in the design are very large, therefore the commonly used series approximations are not suitable. Another basic problem is that the structural model is itself allowed to vary during the design process. Since members are deleted or added during the design process, both the finite element model and the set of design variables change.

The method presented in this study is based on combining the computed terms of a series expansion, used as high quality basis vectors, and coefficients of a reduced basis expression. The latter coefficients can readily be determined by solving a reduced set of analysis equations. The advantage is that the efficiency of local approximations and the improved quality of global approximations are combined to obtain an effective solution procedure.

The method is most suitable for topological optimization. It is shown that high quality approximations can be achieved with a small computational effort for topological changes. The presented method is based on results of a single exact analysis and can be used with a general finite element program. It is suitable for different types of structure, such as trusses, frames, grillages, etc. Calculation of derivatives is not required, and the errors involved in the approximations can readily be evaluated.

Several numerical examples illustrate the effectiveness of the solution process. It is shown that excellent results can be achieved by the presented procedure in various cases of very large changes in the cross sections and in the topology of the structure. In conclusion, the method is a powerful tool to achieve efficient and high quality approximations of the structural behavior in layout optimization problems.

Fig. 6 *a*. Initial design, seven-bar truss. *b*. MID (addition of joint 4 and members 8, 9). *c*. Addition of joint 4 and members 8, 9, 10.

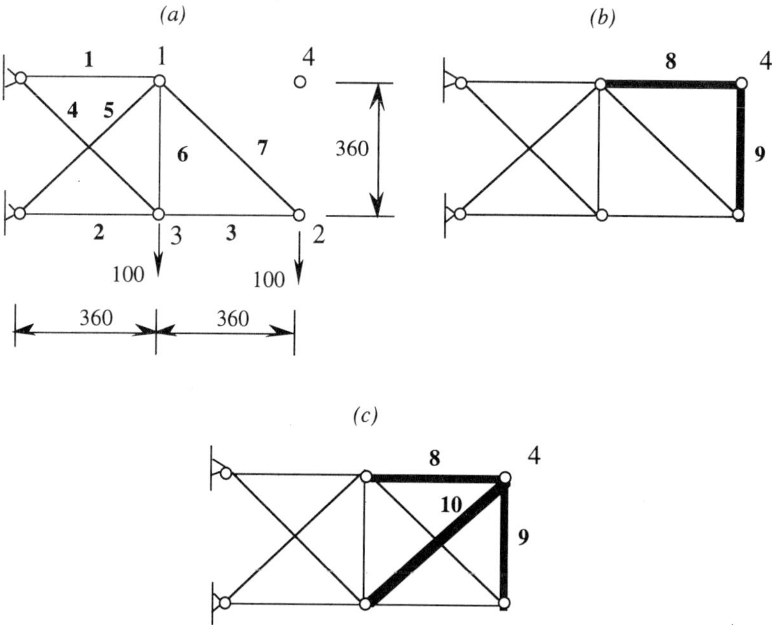

ACKNOWLEDGMENT

The author is indebted to the Alexander von Humboldt Foundation and to the Fund for the Promotion of Research at the Techniom for supporting this work.

TOPOLOGY AND REINFORCEMENT LAYOUT OPTIMIZATION OF DISK, PLATE, AND SHELL STRUCTURES

L.A. Krog and N. Olhoff
Aalborg University, Aalborg, Denmark

Abstract

This chapter deals with topology optimization problems for disks, plates and shells, and with problems of layout optimization of different types of reinforcement of plates and shells. Special emphasis is devoted to the solution of multiple load case stiffness maximization problems and to the solution of eigenfrequency maximization problems.

Two design parametrizations based on the application of layered microstructures of different rank are applied. In both formulations, a parametrization of the material/reinforcement distribution for a disk, plate or shell structure is obtained by modelling the material within each of the elements of a finite element discretized structure as a layered microstructure with a continuously variable density of material/reinforcement. The design variables of the optimization problem are the variables which control the composition of the layered microstructure, and hereby the density of material/reinforcement within each finite element. Furthermore, by allowing for four different configurations of each of the layered microstructures, we obtain formulations for solution of both topology optimization problems for disk, plate and shell structures, and for solution of reinforcement layout optimization problems for rib-stiffened plates and shells and internally stiffened honeycomb and sandwich plates and shells.

Stiffness maximization problems are treated as minimization problems for total elastic energy, and in the case of several independent load cases we either minimize a weighted sum of the total elastic energies, or the maximum total elastic energy from among all the load cases. The optimization problems are solved by means of mathematical programming based on analytical design sensitivity analysis, and examples of solution of maximum stiffness layout problems are presented.

Eigenfrequency maximization problems are considered as maximization problems for the lower bound on a given set of eigenvalues of vibration. The main difficulty associated with solution of such problems is that multiple eigenvalues may exist and that these are non-differentiable with respect to the design parameters. However, it is shown that, despite the lack of usual differentiability properties, such problems may be treated like differentiable optimization problems, if some restrictions are imposed on the vector of design changes at each iteration. In this way, we have developed a new general method for solution of eigenfrequency optimization problems which can handle problems with simple as well as multiple eigenvalues of vibration. This method is particularly attractive since it only requires ordinary methods for design sensitivity analysis and mathematical programming. Several numerical examples pertaining to solution of layout optimization problems with multiple eigenvalues are presented.

1   Introduction

"Topology optimization" is often referred to as "layout optimization" or "generalized shape optimization" in the literature (Rozvany, Bendsøe and Kirch[1]), and these labels will be used interchangeably here. The importance of this type of problem lies in the fact that the choice of the appropriate topology of a structure is generally decisive for the cost-efficiency of the structure. Moreover, usual sizing and shape optimization methods cannot change the structural topology during the solution process, so a solution obtained by one of these methods will have the same topology as that of the initial design. For these reasons, topology or layout optimization methods are most valuable as preprocessing tools for sizing and shape optimization, see Olhoff, Bendsøe & Rasmussen[2].

Depending on the type of structure to be considered, two types of topology optimization exist: for inherently discrete structures, the topology or layout problem consists in determining the optimum number, positions, and mutual connectivity of the structural members, while for continuum structures the shape of the external as well as internal boundaries and the number of inner holes are to be determined.

The reader is referred to, e.g., Rozvany, Bendsøe and Kirch[1] and the part of the present book written by Rozvany for an up-to-date account of the area of layout optimization of discrete structures which has been active for almost a century and been developed largely by Prager and Rozvany.

The present part of this book is devoted to topology optimization of continuum structures. This research area has been extremely active since the publication of the landmark papers by Bendsøe and Kikuchi[3] in 1988 and Bendsøe[4] in 1989.

Bendsøe & Kikuchi[3] treated the topology optimization problem for a plane disk structure as a material distribution problem where the material is assumed to have a perforated microstructure with a continuously variable orientation and density of material. By application of the perforated microstructure the authors obtained a formulation where the material within each of the elements of a finite element discretized structure in a continuous manner is allowed to change from a fully isotropic solid material, over a composite of variable density, to void. Each finite element in the discretized structure has hereby become a potential solid or void subdomain of the structure, and the general design description allows for prediction of the optimum layout of the structure. It should be noted that the topology optimization problem considered in [3] may be also formulated by means of layered materials where a very compliant material plays the role of void, see Bendsøe[4].

Layout optimization problems in the form of determining the optimal layout of stiffener reinforcements on Kirchoff plates were actually considered almost a decade before Bendsøe & Kikuchi[3] solved the topology optimization problem for plane disks, see e.g. Olhoff, Lurie, Cherkaev & Fedorov[5] and Cheng & Olhoff[6]. In these papers, the layout optimization problem was also based on modelling the material as a layered microstructure, i.e., by using a formulation very similar to the formulation later applied by Bendsøe[4]. For the solution of the plate problems the application of layered microstructures appeared to be necessary in order to obtain a well-posed problem, see [5], [6] and [7]. This feature generally also holds true for the plane disk problem considered by Bendsøe & Kikuchi[3] and Bendsøe[4], and was studied in papers by e.g. Kohn & Strang[8],[9].

The purpose of adopting microstructural materials in the formulation of topology optimization problems is not only to obtain a convenient continuous formulation of the material distribution problem. Thus, if the problem had been stated as an integer optimization problem such that either material or no-material would be generated within each of the finite elements of the discretized structure, then the formulation would in general have been ill-posed. Hence, if a series of solutions obtained for sequentially refined finite element discretizations is considered, then this series will, in general, not converge to a fixed limiting design. A key to circumvent this inherent problem is to enlarge the design space such as to include materials with microstructure. This process is termed relaxation and has been studied in various contexts, see e.g. [5], [6], [8], [9], [10], [11], and [12]. Originally it was thought that one must consider the totality of all possible composites in order to obtain a well-posed problem. However, for certain problems (stiffness maximization) it has been shown that only the family of finite rank layered microstructures needs to be considered, see e.g. Avellaneda[10], Lipton[11], and Diaz, Lipton & Soto[12].

The topology optimization problems dealt with in the initial papers by Bendsøe and Kikuchi[3] and Bendsøe[4] concerned stiffness maximization for a single load case, and recent extensions include handling of multiple load cases, Diaz & Bendsøe[13]; bi-material structures, Olhoff, Thomsen & Rasmussen[14]; plate and shell bending problems, Suzuki & Kikuchi[15], Soto & Diaz[16],[17], and Diaz, Lipton & Soto[12]; eigenfrequency optimization, Diaz & Kikuchi[13], Soto & Diaz[18], and Krog & Olhoff[19]; and buckling eigenvalue optimization problems, Folgado, Rodriques & Guedes[20]. Overview of the research activities in layout and topology optimization is given in the recent survey paper by Rozvany, Bendsøe & Kirsch[1] and monograph by Bendsøe[21], and much recent work in the area is also presented in recent proceedings edited by Bendsøe & Mota Soares[22], Pedersen[23] and Olhoff & Rozvany[24].

It is the objective of the work presented in this chapter to develop a methodology for solution of layout optimization problems for plane disk structures and Mindlin plate and shell structures, and especially to consider multiobjective formulations for solution of both multiple load case maximum stiffness layout design problems and maximum eigenfrequency layout design problems. The presentation lends itself on recent work reported in Krog[25].

The present chapter is organized into six subsequent sections and an Appendix.

Section 2 gives a detailed description of the layered microstructures which form the basis for the design parametrization used for solution of layout optimization problems for disks, plates, and shells.

Section 3 gives a presentation of the the structural design criteria used for the definition and solution of the current layout optimization problems and a description of the design sensitivity analysis for these criteria.

Section 4 then presents the different formulations of maximum stiffness and maximum eigenfrequency layout design problems considered, and discusses the solution strategies for solution of these problems.

Section 5 presents examples of solution of layout optimization problems for both statically loaded and freely vibrating disk structures and Mindlin plate structures.

Finally, the work is concluded in Section 6. A reference list and an appendix are found at the end of the chapter.

All the developments reported in this text have been implemented in the <u>O</u>ptimum <u>DES</u>ign <u>SY</u>stem - ODESSY which is beeing developed at Aalborg University, and all the numerical results presented in Section 5 have been obtained using ODESSY.

2 Design Parametrization

This section firstly briefly discusses the need for introduction of microstructures for solution of topology and layout design problems for continuum structures (Sub-section 2.1), and then describes the construction and the parametrization of the layered microstructures used here for optimum topology and layout design of disks, plates and shells (Sub-section 2.2).

Next, in Sub-section 2.3, we derive analytical expressions for the effective stiffness properties of simple layered microstructures of first rank, i.e., microstructures with layers of materials aligned in a single direction only. These simple material models are generally not applied directly for solution of layout optimization problems, but provide basis for the construction of more general microstructures.

In Sub-section 2.4 we then derive analytical expressions for the effective stiffness properties of layered microstructures of any finite rank, i.e., microstructures with layers of material oriented along any finite number of arbitrary directions. As a special case, we establish the stiffness properties of an orthogonal, layered layered microstructure of second rank as this microstructure has been shown to be optimal for single load case stiffness design problems. However, for more general problems more general microstructures are needed, and at the end of this section, we therefore present a so-called moment formulation by which the effective stiffness properties of layered microstructuctures of any finite rank can be fully described in terms of only five design variables.

2.1   The Need for Introduction of Microstructures

As seems to have been originally pointed out by Lurie from a fluid problem in magnetohydrodynamics (see [26] for pertinent references), there need not exist an optimum design or optimal distribution of material properties within some initial formulations of optimization problems for 2-D or 3-D continua (fluids and/or solids). In solid mechanics, the typical problems in this category involve thickness optimization of elastic [5], [7], [6] and elasto-plastic [27], [28] and [29] plates and shells, optimization of non-homogeneous materials, and shape optimization problems where some global or local measure of the solution to a 2-D or 3-D boundary value problem is to be extremized.

The reason for the lack of an optimum solution within an initial or "traditional" formulation (e.g. based on isotropic material behaviour) is that the set of feasible (or admissible) designs, i.e., the design space, is not closed in the appropriate sense, which means that the problem is not well-posed. The remedy for this is to ensure closure of the set of feasible designs via a regularization of the mathematical formulation of the optimization problem.

Mathematical indications of the need for regularization are generation of anisotropy in the design and the impossiblity of satisfying second order necessary conditions for optimality in certain subregions of the structural domain. Numerically, the need manifests itself by lack of convergence or by dependence of the designs on the size of the applied finite element mesh. In particular, it is not possible to obtain a limiting, numerically stable design by consecutively decreasing the mesh size.

Within the area of solid mechanics, the lack of existence of an optimal solution within a discretized formulation of a continuum layout optimization problem was first demonstrated in a paper by Cheng & Olhoff[7] on optimal design of solid elastic plates of variable thickness. The paper clearly illustrated the influence of the mesh size on the optimal layout of integral stiffeners on rectangular and axisymmetric Kirchhoff plates and the need for regularization of the problem. Cheng & Olhoff concluded: *"... we find that a number of local optimal solutions exist and that a possible global optimal plate thickness function does neither exist in the class of smooth functions nor in the class of smooth functions with a finite number of discontinuities. ... The current results indicate clearly that the global optimal design is a plate which, at least in some regions, is equipped with an infinite number of infinitely thin stiffeners. ..."*.

The remedy to ensure existence of an optimal solution to the plate problem was studied in Olhoff, Lurie, Cherkaev & Fedorov[5] and Cheng & Olhoff[6] and was found to be the introduction of anisotropic plate microstructures as admissible designs. Thus, convergence towards a limiting design was achieved for a regularized formulation in which the material in each subdomain of the discretized structure was modelled as a microstructural material which inherently allowed for the infinitely fast variation of thickness that was expected to appear locally in the optimal solution.

Later, the lack of existence of an optimal solution to some initial formulations of other layout optimization problems was discussed by Kohn & Strang[8],[9], and the remedy was again found to be to enlarge (relax) the design space such as to include the optimal solution in a natural way, i.e., to include microstructural materials having the infinitely fast variation of material/no-material expected in the optimal solution.

In their important paper on maximum stiffness topology optimization in plane elasticity Bendsøe & Kikuchi[3] used a material with a periodic perforated microstructure as shown in Figure 1, and in the subsequent paper Bendsøe[4] applied a layered

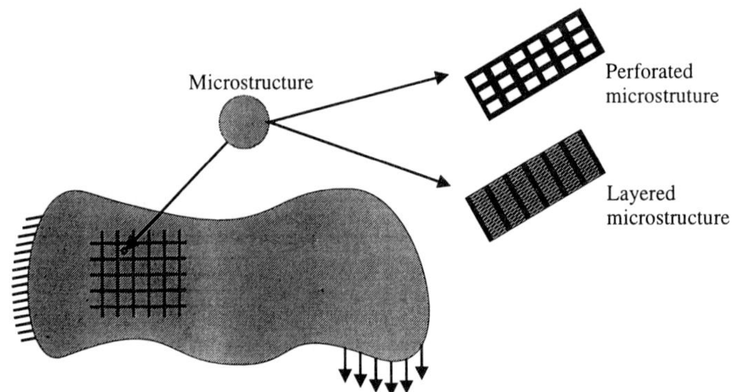

Figure 1: Perforated and layered microstructures for solution of layout optimization problems for plane disks.

medium in which a very compliant material played the role of void. Figure 2 illustrates the perforated and layered microstructures applied in Bendsøe & Kikuchi[3] and Bendsøe[4], respectively. The microstructures depicted in Figure 2 both implies an orthotropic material behaviour.

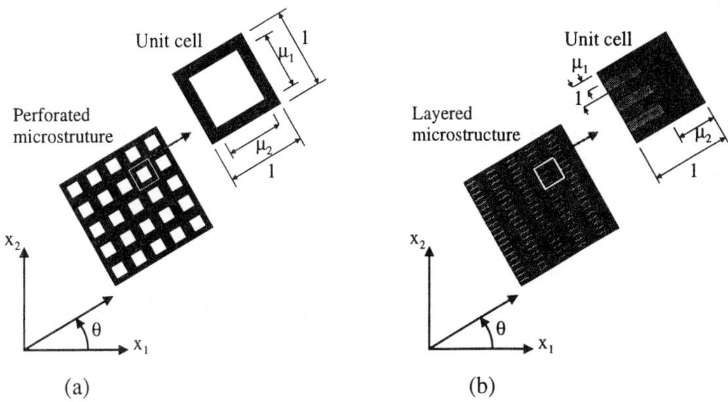

Figure 2: *Microstructures applied for the solution of topology optimization problems for plane disks. (a) Perforated microstructure with rectangular holes in square unit cells. (b) Layered microstructure constructed from two different isotropic materials.*

2.2   Construction of Layered Microstructures

Following an approach described in recent work by Soto & Diaz[30] and Soto[31], we now describe the construction of a set of layered plate microstructures which enables us to solve both topology problems for disk, plate, and shell structures, and a number of reinforcement layout optimization problems for plate and shell structures.   Layered

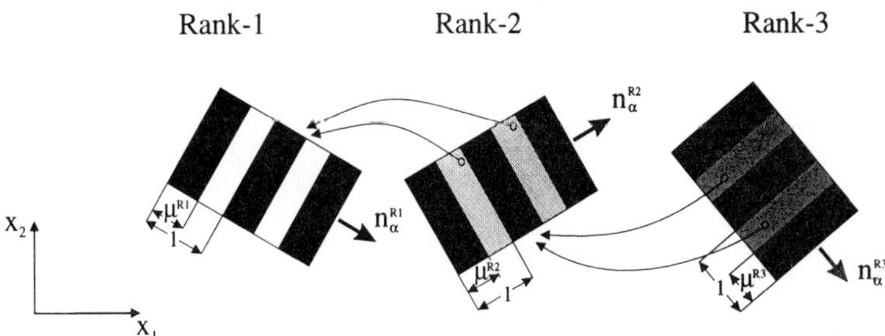

Figure 3: *Construction of first, second, and third-rank microstructures by successive layering along different directions. Black and white domains are occupied by isotropic materials with the stiffness tensors $E^{+}_{\alpha\beta\kappa\gamma}$ and $E^{-}_{\alpha\beta\kappa\gamma}$, respectively, while grey domains represent areas occupied by layered materials composed of the two base materials.*

planar microstructures are constructed through a repetitive process where a new layering of given direction is added to the microstructure a given number of times, see Figure 3. The number of times this has been performed is per definition the *rank* of the layered microstructure. Hence, given two isotropic base materials with the material stiffness tensors $E^{+}_{\alpha\beta\kappa\gamma}$ and $E^{-}_{\alpha\beta\kappa\gamma}$ and taking $E^{+}_{\alpha\beta\kappa\gamma} > E^{-}_{\alpha\beta\kappa\gamma}$, a first-rank planar microstructure with a given density μ^{R1} of the stiffer material[1] is constructed by stacking alternately thin layers of the two isotropic base materials along a given direction characterized by the normal vector n^{R1}_{α}, while a second-rank planar microstructure is build by stacking alternately thin layers of the stiff isotropic material and the first-rank planar microstructure along a new direction characterized by n^{R2}_{α}. This process may be continued to build layered planar microstructures of any finite rank RI. A particular advantage of the layered planar microstructures is that their effective material stiffness properties, in the limit where the width of the layers tends to zero, may be calculated analytically applying either a mathematically based homogenization procedure as used by e.g. Bendsøe[4], or a more physically based smear-out procedure as used by e.g. Olhoff, Lurie, Cherkaev & Fedorov[5] and Thomsen[32]. Hence, considering the first-rank planar microstructure depicted in Figure 3, we may symbolically write

$$E^{R1}_{\alpha\beta\kappa\gamma} = E(\mu^{R1}, n^{R1}_{\alpha}, E^{+}_{\alpha\beta\kappa\gamma}, E^{-}_{\alpha\beta\kappa\gamma}) \tag{1}$$

indicating that the components of the effective material stiffness tensor $E^{R1}_{\alpha\beta\kappa\gamma}$ for the first-rank planar microstructure are given as analytical functions of the density variable μ^{R1}, the layering orientation n^{R1}_{α} and the stiffness tensors for the two isotropic base materials. The simple iterative procedure used for the construction of multi-rank planar microstructures is generally reflected in an equally simple iterative derivation of analytical expressions for the components of the effective material stiffness tensors for multi-rank planar microstructures. The effective material stiffness tensors for the second-rank and third-rank planar microstructures depicted in Figure 3 may for example be obtained analytically by means of Eq(1) by simply substituting the material stiffness tensor $E^{-}_{\alpha\beta\kappa\gamma}$ by, firstly, the effective material stiffness tensor for the first-rank planar microstructure, and secondly by the effective stiffness tensor for the second-rank planar microstructure, i.e.,

$$E^{R2}_{\alpha\beta\kappa\gamma} = E(\mu^{R2}, n^{R2}_{\alpha}, E^{+}_{\alpha\beta\kappa\gamma}, E^{R1}_{\alpha\beta\kappa\gamma})$$

$$E^{R3}_{\alpha\beta\kappa\gamma} = E(\mu^{R3}, n^{R3}_{\alpha}, E^{+}_{\alpha\beta\kappa\gamma}, E^{R2}_{\alpha\beta\kappa\gamma}) \tag{2}$$

Thus, analytical expressions for the components of the effective material stiffness tensor for layered planar microstructures of any finite rank may be obtained through the iterative procedure

$$E^{Ri}_{\alpha\beta\kappa\gamma} = E(\mu^{Ri}, n^{Ri}_{\alpha}, E^{+}_{\alpha\beta\kappa\gamma}, E^{R(i-1)}_{\alpha\beta\kappa\gamma}) \ ; \quad i = 1, \ldots, I \tag{3}$$

The class of layered planar microstructures just described represents a particularly important class of microstructures for the solution of generalized shape optimization problems for planar structures, especially since second-rank planar microstructures with orthogonal layers have been shown to be optimal for solution of single load case

---

[1]This material will be termed "stiff" in the following but has finite stiffness

stiffness design problems, while third-rank planar microstructures with non-orthogonal layers should be used for the solution of multiple load case stiffness design problems, see Avellaneda[10]. For the solution of other elasticity driven layout optimization problems for plane disks parameterized by use of layered planar microstructures, it should be mentioned that no further generality is obtained by application of more than a third-rank planar microstructure with non-orthogonal layers, as the whole range of effective stiffness properties for all finite rank planar microstructures is obtainable by application of third-rank planar microstructures with non-orthogonal layers. This observation follows indirectly from the work by Avellaneda & Milton[33] and Lipton[11].

Layered plate microstructures for solution of layout optimization problems for plate and shell structures may be constructed in a similar way as the layered planar microstructures described above. For the construction of layered plate microstructures we start by considering two isotropic and symmetric plates, a stiff and a compliant one, respectively. From these two isotropic base plates we may construct a layered plate microstructure of any finite rank, applying exactly the same procedure as used for the construction of the planar microstructures, i.e., by a bottom up approach where we start by constructing a first-rank plate microstructure by stacking alternately thin slices of the two base plates along a given direction, and end by building a rank-I plate microstructure by stacking alternately thin slices of the stiff plate and thin slices of a $rank(I - 1)$ plate microstructure along a given direction. The general idea to build a layered plate microstructure by a periodic layering of thin layers of plates with different stiffness properties was in fact used already in the early work by Olhoff, Lurie, Cherkaev & Fedorov[5] and Cheng & Olhoff[7] where a first-rank plate microstructure for the modelling of surface stiffened axisymmetric Kirchoff plates was obtained by stacking alternately thin plate slices of different hights along a single direction only. In both papers the effective bending stiffness properties were derived by a physically based smear-out procedure. However, from these two presentations only the results in [5] were presented in a pure tensor form which, following the approach outlined for the layered planar microstructures, may be used directly for the derivation of the effective bending stiffness tensor for multi-rank plate microstructures.

Using the general ideas outlined above and following an approach described in the recent work by Soto & Diaz[30] and Soto[31] we shall now construct a family of layered plate microstructures that will be used later for solution of different layout optimization problems for disk, plate, and shell structures. It is a particular feature of this approach that layered plate microstructures for solution of layout optimization problems are obtained for different types of plates by simply allowing for different configurations of the compliant parts of the plate. Figure 4 illustrates four different types of first-rank plate microstructures which have been obtained by taking the stiff plate as a solid plate of thickness h_2 made of a stiff isotropic material, while allowing for different configurations of the compliant plate. The four different first-rank plate microstructures depicted in Figure 4 are generally not used directly for solution of layout optimization problems, but represent the basis for the construction of more

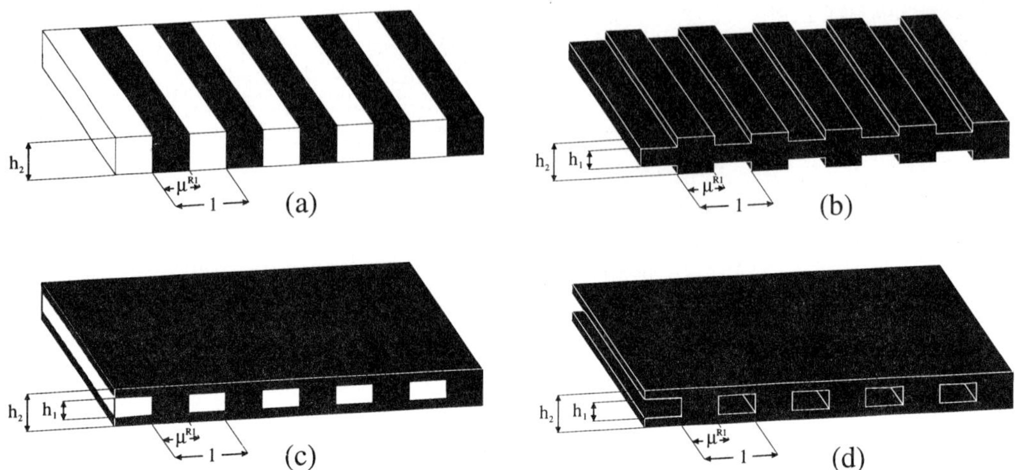

Figure 4: First-rank plate microstructures used for the modelling of: (a) Perforated solid plates. (b) Surface stiffened solid plates. (c) Core stiffened sandwich plates. (d) Core stiffened honeycomb plates. Black and white domains in the microstructures are filled by a stiff and a more compliant isotropic material, respectively.

general multi-rank plate microstructures. The four different types of multi-rank plate microstructures which may be obtained using the different configurations of the first-rank microstructure depicted in Figure 4 shall in general be used for the solution of:

- <u>Configuration a:</u> Generalized shape optimization problems for finding the optimal layout of disk, plate, and shell structures.

- <u>Configuration b:</u> Reinforcement layout optimization problems for finding the optimal layout of surface stiffeners on plate, and shell structures.

- <u>Configuration c:</u> Reinforcement layout optimization problems for finding the optimal layout of core stiffeners in sandwich plate and shell structures.

- <u>Configuration d:</u> Reinforcement layout optimization problems for finding the optimal layout of core stiffeners in honeycomb plate and shell structures.

In the calculation of the effective stiffness properties for the four first-rank plate microstructures shown in Figure 4 it shall generally be avoided to deal with each microstructure independently by simply characterizing the two isotropic base plates used to construct the layered plate microstructures by their stiffness tensors. Realizing that all the first-rank plate microstructures depicted in Figure 4 are constructed symmetrically and adopting a Mindlin plate theory we may characterize the stiff and compliant parts of the plates by their membrane, bending, and transverse stiffness tensors, $A^+_{\alpha\beta\kappa\gamma}$, $D^+_{\alpha\beta\kappa\gamma}$, $S^+_{\alpha\beta}$, and $A^-_{\alpha\beta\kappa\gamma}$, $D^-_{\alpha\beta\kappa\gamma}$, $S^-_{\alpha\beta}$, respectively. Assuming usual plane stress conditions and taking the stiff and compliant isotropic materials used to build the two base plates to have the Young's moduli E^+ and E^- and a common Poisson's ratio ν, we have the following general relationships between the non-zero components of the stiffness tensors for the two base plates

$$A^{(+/-)}_{1111} = A^{(+/-)}_{2222} = A^{(+/-)}_0 \quad ; \quad A^{(+/-)}_{1122} = A^{(+/-)}_{2211} = \nu A^{(+/-)}_0$$

$$A^{(+/-)}_{1212} = A^{(+/-)}_{1221} = A^{(+/-)}_{2121} = A^{(+/-)}_{2121} = \frac{1-v}{2} A^{(+/-)}_0$$

$$D^{(+/-)}_{1111} = D^{(+/-)}_{2222} = D^{(+/-)}_0 \quad ; \quad D^{(+/-)}_{1122} = D^{(+/-)}_{2211} = \nu D^{(+/-)}_0 \tag{4}$$

$$D^{(+/-)}_{1212} = D^{(+/-)}_{1221} = D^{(+/-)}_{2121} = D^{(+/-)}_{2121} = \frac{1-v}{2} D^{(+/-)}_0$$

$$S^{(+/-)}_{11} = S^{(+/-)}_{22} = S^{(+/-)}_0$$

where A^+_0, D^+_0, S^+_0 and A^-_0, D^-_0, S^-_0 are membrane, bending, and transverse shear stiffness constants for the stiff and the compliant plate, respectively. These constants naturally depend on both the sectional geometry of the base plates, and on the stiffness properties of the two isotropic materials used to build the two base plates. The set of stiffness constants for the stiff and compliant plate slice components entering in the different configurations of the first-rank plate microstructures depicted in Figure 4 are given in the following.

Stiff plate slices:

The stiff plate components in all four layered plate microstructures are represented by solid plate slices of thickness (height) h_2 made of an isotropic material with a Young's modulus E^+ and a Poisson's ratio ν. This gives the following set of stiffness constants for the stiff plate slices which appear in all plate configurations:

$$A^+_0 = \frac{h_2 E^+}{1-\nu^2} \quad ; \quad D^+_0 = \frac{h_2^3 E^+}{12(1-\nu^2)} \quad ; \quad S^+_0 = \frac{h_2 E^+}{2c(1+\nu)} \tag{5}$$

It should be noted that the transverse shear stiffness constant has been divided by the so-called Cowper factor $c = 1.2$ in order to account for the parabolic distribution of transverse shear stresses over the thickness of a solid plate section, which Mindlin plate theory generally does not account for.

Compliant plate slices in plate configuration a:

The compliant plate components are here solid plate slices of thickness (height) h_2 made of a compliant isotropic material with a Young's modulus E^- and a Poisson's ratio ν. This yields the following set of stiffness constants for the compliant plate slices used in plate configuration a:

$$A_0^- = \frac{h_2 E^-}{1 - \nu^2} \quad ; \quad D_0^- = \frac{h_2^3 E^-}{12\,(1 - \nu^2)} \quad ; \quad S_0^- = \frac{h_2 E^-}{2c\,(1 + \nu)} \tag{6}$$

Again the transverse shear stiffness constant has been divided by the Cowper factor $c = 1.2$.

Compliant plate slices in plate configuration b:
The compliant plate components are here solid isotropic plate slices of thickness (height) h_1 made of a stiff isotropic material with a Young's modulus E^+ and a Poisson's ratio ν. This yields the following set of stiffness constants for the compliant plate slices used in plate configuration b:

$$A_0^- = \frac{h_1 E^+}{1 - \nu^2} \quad ; \quad D_0^- = \frac{h_1^3 E^+}{12\,(1 - \nu^2)} \quad ; \quad S_0^- = \frac{h_1 E^+}{2c\,(1 + \nu)} \tag{7}$$

Again, the Cowper factor $c = 1.2$ is used.

Compliant plate slices in plate configuration c:
The compliant plate components are here slices of a isotropic sandwich plate with core thickness h_1 and an over-all thickness h_2. The material used for the core is a compliant material with a Young's modulus E^- and a Poisson's ratio ν, while the material used for the face layers is a stiff isotropic material with a Young's modulus E^+ and a Poisson's ratio ν. Expressions for the stiffness of such a plate can be found in standard textbooks dealing with lamination theory. A lamination theory based on a Mindlin plate theory is available in the book by Vinson & Sierakowski[34], from which we obtain the following set of stiffness constants for the compliant plate slices used in plate configuration c.

$$A_0^- = \frac{(h_2 - h_1)E^+ + h_1 E^-}{1 - \nu^2} \quad ; \quad D_0^- = \frac{(h_2^3 - h_1^3)E^+ + h_1^3 E^-}{12\,(1 - \nu^2)} \quad ;$$

$$S_0^- = \frac{5}{4}\left((h_2 - h_1) - \frac{h_2^3 - h_1^3}{3h_2^2}\right)\frac{E^+}{1 + \nu} + \frac{5}{4}\left(h_1 - \frac{h_1^3}{3h_2^2}\right)\frac{E^-}{1 + \nu} \tag{8}$$

Here, the expression for the transverse shear stiffness constant is based on the assumption that the transverse shear stresses are distributed parabolically across the thickness of the composite plate. This assumption is generally only valid for small differences in the Young's moduli E^+ and E^-. Layered plate microstructures based on this first-rank plate microstructure therefore cannot be used for the modelling of honeycomb plates where the material with modulus E^- would play the role as void.

Compliant plate slices in plate configuration d:
The compliant plate components are here hollow isotropic plate slices with an overall thickness h_2 assembled from two thin isotropic cover plates separated at a distance h_1. The material used for the thin isotropic plate slices is a stiff isotropic

material with Young's modulus E^+ and Poisson's ratio ν. This yields the following set of stiffness constants used for the compliant plate slices in configuration d.

$$A_0^- = \frac{(h_2 - h_1)E^+}{1 - \nu^2} \quad ; \quad D_0^- = \frac{(h_2^3 - h_1^3)E^+}{12\,(1 - \nu^2)} \quad ; \quad S_0^- = \frac{(h_2 - h_1)E^+}{2c(1 + \nu)} \tag{9}$$

Here, the transverse shear stiffness constant has been calculated as the sum of the transverse shear stiffness constants for the two thin plates, and should account for a parabolic distribution of the transverse shear stresses across the thickness of each of the thin cover plates. Note that the set of stiffness constants given above are generally only valid when the two thin cover plates are separated at a distance h_1, i.e., when the layered plate microstructure depicted in Figure 4d contains at least a small volume fraction of the stiff plate slices that separate the two thin plates.

The effective membrane, bending, and transverse shear stiffness tensors for the four first-rank Mindlin plate microstructures described above may also be found analytically by means of either a mathematically based homogenization procedure, as used by Soto & Diaz[30] and Soto[31], or a more physically based smear-out procedure, as used by Diaz, Lipton & Soto[12] and Soto[30]. Hence, adopting the notation in Figure 3 we may write

$$A_{\alpha\beta\kappa\gamma}^{R1} = A(\mu^{R1}, n_\alpha^{R1}, A_{\alpha\beta\kappa\gamma}^+(A_0^+), A_{\alpha\beta\kappa\gamma}^-(A_0^-))$$

$$D_{\alpha\beta\kappa\gamma}^{R1} = D(\mu^{R1}, n_\alpha^{R1}, D_{\alpha\beta\kappa\gamma}^+(D_0^+), D_{\alpha\beta\kappa\gamma}^-(D_0^-)) \tag{10}$$

$$S_{\alpha\beta}^{R1} = S(\mu^{R1}, n_\alpha^{R1}, S_{\alpha\beta}^+(S_0^+), S_{\alpha\beta}^-(S_0^-))$$

which symbolically states that the components of the effective membrane, bending, and transverse shear stiffness tensors for the first-rank plate microstructures depicted in Figure 4 are given as analytical functions of the density variable μ^{R1}, the layering orientation n_α^{R1}, and the set of stiffness constants for the two base plates. The derivation of analytical expressions for the effective stiffness properties for multi-rank plate microstructures may generally be performed applying the iterative procedure described for the layered planar microstructures. Such an approach was also taken in Soto & Diaz[30], Diaz, Lipton & Soto[12], and Soto[12]. The derivation of effective stiffness properties for first-rank plate microstructures shall be considered in more detail in Sub-section 2.3.1, and the calculation of effective stiffness properties of multi-rank plate microstructures in Sub-section 2.4 and Sub-section 2.4.2.

2.3   First-Rank Mindlin Plate Microstructures

2.3.1   Smear-Out Technique for Mindlin Plates

We shall here describe how the effective stiffness tensors for microstructurally layered Mindlin plates can be derived analytically by means of a simple averaging technique which basically determines the effective stiffness tensors for the Mindlin plate from constitutive relationships between average force/strain tensors and average moment/curvature tensors in the microstructure, obtained by use of interface conditions.

Such so-called *smear-out* techniques have previously been used by e.g. Olhoff, Lurie, Cherkaev & Fedorov[5] for the derivation of effective bending stiffness tensors for microstructurally layered Kirchhoff plates, by Thomsen[32] for the derivation of effective material stiffness tensors for first-rank planar disk microstructures, and recently also by Soto & Diaz[30] and Soto[31] for the derivation of effective bending and transverse shear stiffness tensors for microstructurally layered Mindlin plates. The smear-out technique considered here actually is identical to the smear-out technique developed by Soto & Diaz[30] and Soto[31]. However, for the presentation of the basic ideas behind this technique we shall mainly follow the work by Olhoff, Lurie, Cherkaev & Fedorov[5], while the final derivation of expressions for the effective stiffness tensors naturally follows Soto & Diaz[30] and Soto[31]. For the derivation in the following we shall adopt a Mindlin plate theory which encounters both membrane, bending, and transverse shear action.

Following the method used in [5] we start by considering a small rectangular plate element Ω, as depicted in Figure 5. The size of this element, which consists of a finite number of parallel plate slices taken from a stiff and a compliant isotropic plate, respectively, is assumed to be small in comparison with the dimensions of the entire plate structure, but at the same time considered to be large in comparison with its underlying first-rank microstructure. The orientation of the element in the plate mid-plane coordinate system $x_1 x_2$, see Figure 5, is given by the unit vectors n_α^{R1} and t_α^{R1} which are perpendicular to and parallel with the layers, respectively, i.e., $n_\alpha^{R1} n_\alpha^{R1} = 1$, $t_\alpha^{R1} t_\alpha^{R1} = 1$, and $n_\alpha^{R1} t_\alpha^{R1} = 0$. The stress-strain field within the element is assumed to be homogeneous from a macroscopic point of view, while at the microscroscale, i.e., at the level of subdomains Ω^+ and Ω^-, the stress-strain field is assumed to be piece-wise homogenous. In accordance with the notation used in Sub-section 2.2 we shall denote the membrane, bending, and transverse shear stiffness tensors for the stiff and compliant plate slices, located in subdomains Ω^+ and Ω^-, by $A_{\alpha\beta\kappa\gamma}^+$, $D_{\alpha\beta\kappa\gamma}^+$, $S_{\alpha\beta}^+$ and $A_{\alpha\beta\kappa\gamma}^-$, $D_{\alpha\beta\kappa\gamma}^-$, $S_{\alpha\beta}^-$, respectively. Furthermore, we shall assume that the non-zero components of these tensors are given in the form shown in Eq(4), such that the expressions derived in the following directly apply to the calculation of the effective stiffness tensors for the four first-rank Mindlin plate microstructures depicted in Figure 4. Following the general ideas of the averaging (smear-out) technique described in Olhoff, Lurie, Cherkaev, & Fedorov[5], we determine the effective stiffness properties of our first-rank microstructure shown in Figure 5, as the effective membrane, bending, and transverse shear stiffness tensors $A_{\alpha\beta\kappa\gamma}^{R1}$, $D_{\alpha\beta\kappa\gamma}^{R1}$ and $S_{\alpha\beta}^{R1}$ in the following constitutive relationships for the small rectangular plate domain Ω,

$$N_{\alpha\beta}^{avr} = A_{\alpha\beta\kappa\gamma}^{R1} \varepsilon_{\kappa\gamma}^{avr}$$

$$M_{\alpha\beta}^{avr} = D_{\alpha\beta\kappa\gamma}^{R1} \kappa_{\kappa\gamma}^{avr} \tag{11}$$

$$Q_\alpha^{avr} = S_{\alpha\beta}^{R1} \gamma_\beta^{avr}$$

Here, the Greek indices refer to the axes of a global coordinate system $x_1 x_2$ embedded in the mid-plane of the plate, see Figure 5, and the tensors $\varepsilon_{\alpha\beta}^{avr}$, $\kappa_{\alpha\beta}^{avr}$, γ_α^{avr} and $N_{\alpha\beta}^{avr}$, $M_{\alpha\beta}^{avr}$,

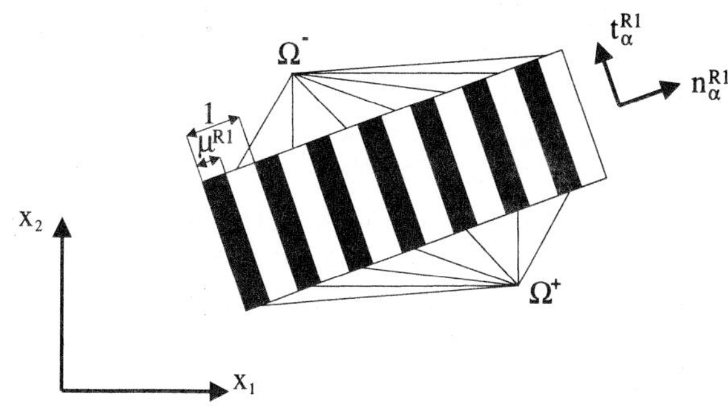

Figure 5: Plate element with an underlying first-rank microstructure.

Q_α^{avr} are direct averages of the membrane strain, bending curvature, and transverse shear strain tensors, and of the membrane force, bending moment, and transverse shear force tensors for the small plate element Ω. Defining homogeneous membrane strain, curvature, and transverse shear strain tensors $\varepsilon_{\alpha\beta}^+$, $\kappa_{\alpha\beta}^+$, γ_α^+ and $\varepsilon_{\alpha\beta}^-$, $\kappa_{\alpha\beta}^-$, γ_α^-, and homogeneous membrane force, bending moment, and transverse shear force tensors $N_{\alpha\beta}^+$, $M_{\alpha\beta}^+$, Q_α^+ and $N_{\alpha\beta}^-$, $M_{\alpha\beta}^-$, Q_α^- within subdomains Ω^+ and Ω^-, respectively, we obtain the following constitutive relations for the subdomains Ω^+ and Ω^-,

$$N_{\alpha\beta}^+ = A_{\alpha\beta\kappa\gamma}^+ \varepsilon_{\kappa\gamma}^+ \quad ; \quad N_{\alpha\beta}^- = A_{\alpha\beta\kappa\gamma}^- \varepsilon_{\kappa\gamma}^-$$

$$M_{\alpha\beta}^+ = D_{\alpha\beta\kappa\gamma}^+ \kappa_{\kappa\gamma}^+ \quad ; \quad M_{\alpha\beta}^- = D_{\alpha\beta\kappa\gamma}^- \kappa_{\kappa\gamma}^- \tag{12}$$

$$Q_\alpha^+ = S_{\alpha\beta}^+ \gamma_\beta^+ \quad ; \quad Q_\alpha^- = S_{\alpha\beta}^- \gamma_\beta^-$$

and the following expressions for the direct averages of the membrane strain, curvature, and transverse shear strain tensors and of the membrane force, bending moment, and transverse shear force tensors within the domain Ω,

$$\varepsilon_{\alpha\beta}^{avr} = \mu^{R1}\varepsilon_{\alpha\beta}^+ + (1-\mu^{R1})\varepsilon_{\alpha\beta}^- \quad ; \quad N_{\alpha\beta}^{avr} = \mu^{R1}N_{\alpha\beta}^+ + (1-\mu^{R1})N_{\alpha\beta}^-$$

$$\kappa_{\alpha\beta}^{avr} = \mu^{R1}\kappa_{\alpha\beta}^+ + (1-\mu^{R1})\kappa_{\alpha\beta}^- \quad ; \quad M_{\alpha\beta}^{avr} = \mu^{R1}M_{\alpha\beta}^+ + (1-\mu^{R1})M_{\alpha\beta}^- \tag{13}$$

$$\gamma_\alpha^{avr} = \mu^{R1}\gamma_\alpha^+ + (1-\mu^{R1})\gamma_\alpha^- \quad ; \quad Q_\alpha^{avr} = \mu^{R1}Q_\alpha^+ + (1-\mu^{R1})Q_\alpha^-$$

The problem is now to establish analytical expressions for the effective membrane, bending, and transverse shear stiffness tensors, $A_{\alpha\beta\kappa\gamma}^{R1}$, $D_{\alpha\beta\kappa\gamma}^{R1}$ and $S_{\alpha\beta}^{R1}$, as functions of the density variable μ^{R1}, and the layering orientation n_α^{R1}, see Figure 5. Such analytical expressions are derived in the Appendix by the use of Eqs(11)-(13) and two sets of interface conditions that express the mechanics of the microstructure.

Firstly, from considerations of static equilibrium across the interfaces between adjacent subdomains Ω^+ and Ω^- we can write the following set of continuity/discontinuity conditions for the components of the membrane force, bending moment, and transverse shear force tensors acting parallel and orthogonal to the interface,

$$(N^+_{\alpha\beta} - N^-_{\alpha\beta})n^{R1}_\alpha n^{R1}_\beta = 0 \ ; \quad (N^+_{\alpha\beta} - N^-_{\alpha\beta})t^{R1}_\alpha t^{R1}_\beta \neq 0 \ ; \quad (N^+_{\alpha\beta} - N^-_{\alpha\beta})n^{R1}_\alpha t^{R1}_\beta = 0$$

$$(M^+_{\alpha\beta} - M^-_{\alpha\beta})n^{R1}_\alpha n^{R1}_\beta = 0 \ ; \quad (M^+_{\alpha\beta} - M^-_{\alpha\beta})t^{R1}_\alpha t^{R1}_\beta \neq 0 \ ; \quad (M^+_{\alpha\beta} - M^-_{\alpha\beta})n^{R1}_\alpha t^{R1}_\beta = 0 \tag{14}$$

$$(Q^+_\alpha - Q^-_\alpha)n^{R1}_\alpha = 0 \ ; \quad (Q^+_\alpha - Q^-_\alpha)t^{R1}_\alpha \neq 0$$

Similarly, by geometric compatibility considerations we may set up the following continuity/discontinuity conditions for the components of the membrane strain, curvature, and transverse shear strain tensors acting parallel and orthogonal to the interface:

$$(\varepsilon^+_{\alpha\beta} - \varepsilon^-_{\alpha\beta})n^{R1}_\alpha n^{R1}_\beta \neq 0 \ ; \quad (\varepsilon^+_{\alpha\beta} - \varepsilon^-_{\alpha\beta})t^{R1}_\alpha t^{R1}_\beta = 0 \ ; \quad (\varepsilon^+_{\alpha\beta} - \varepsilon^-_{\alpha\beta})n^{R1}_\alpha t^{R1}_\beta \neq 0$$

$$(\kappa^+_{\alpha\beta} - \kappa^-_{\alpha\beta})n^{R1}_\alpha n^{R1}_\beta \neq 0 \ ; \quad (\kappa^+_{\alpha\beta} - \kappa^-_{\alpha\beta})t^{R1}_\alpha t^{R1}_\beta = 0 \ ; \quad (\kappa^+_{\alpha\beta} - \kappa^-_{\alpha\beta})n^{R1}_\alpha t^{R1}_\beta \neq 0 \tag{15}$$

$$(\gamma^+_\alpha - \gamma^-_\alpha)n^{R1}_\alpha \neq 0 \ ; \quad (\gamma^+_\alpha - \gamma^-_\alpha)t^{R1}_\alpha = 0$$

However, following Soto & Diaz[30] and Soto[31], we shall state the last set of interface conditions, Eq(15), in the more compact form,

$$\varepsilon^+_{\alpha\beta} - \varepsilon^-_{\alpha\beta} = \beta_1 n^{R1}_\alpha n^{R1}_\beta + \beta_2(t^{R1}_\alpha n^{R1}_\beta + n^{R1}_\alpha t^{R1}_\beta)$$

$$\kappa^+_{\alpha\beta} - \kappa^-_{\alpha\beta} = \beta_3 n^{R1}_\alpha n^{R1}_\beta + \beta_4(t^{R1}_\alpha n^{R1}_\beta + n^{R1}_\alpha t^{R1}_\beta) \tag{16}$$

$$\gamma^+_\alpha - \gamma^-_\alpha = \beta_5 n^{R1}_\alpha$$

where the scalars $\beta_1, \ldots, \beta_5$ designate the jumps of those components of the membrane strains, curvature, and transverse shear strain tensors that exhibit discontinuity across the interfaces between Ω^+ and Ω^- subdomains.

Eqs(11)-(14) and Eq(16) together with Eqs(4)-(9) constitute the set of equations needed for the derivation of the effective membrane, bending, and transverse shear stiffness tensors. The remaining part of the derivation, which is performed in the Appendix, consists of simple but cumbersome algebraic manipulations of these equations. The derivations in the Appendix yield the following general set of analytical expressions for the effective membrane, bending, and transverse stiffness tensors for the different plate microstructures depicted in Figure 4,

$$A^{R1}_{\alpha\beta\kappa\gamma} = A^+_{\alpha\beta\kappa\gamma} - (1 - \mu^{R1})\left(\left(A^+_{\alpha\beta\kappa\gamma} - A^-_{\alpha\beta\kappa\gamma}\right)^{-1} - \frac{\mu^{R1}}{A^+_0}\Lambda^{A,R1}_{\alpha\beta\kappa\gamma}\right)^{-1}$$

$$D^{R1}_{\alpha\beta\kappa\gamma} = D^+_{\alpha\beta\kappa\gamma} - (1 - \mu^{R1})\left(\left(D^+_{\alpha\beta\kappa\gamma} - D^-_{\alpha\beta\kappa\gamma}\right)^{-1} - \frac{\mu^{R1}}{D^+_0}\Lambda^{D,R1}_{\alpha\beta\kappa\gamma}\right)^{-1} \tag{17}$$

$$S^{R1}_{\alpha\beta} = S^+_{\alpha\beta} - (1 - \mu^{R1})\left(\left(S^+_{\alpha\beta} - S^-_{\alpha\beta}\right)^{-1} - \frac{\mu^{R1}}{S^+_0}\Lambda^{S,R1}_{\alpha\beta}\right)^{-1}$$

with specific expressions for stiffness entries defined in Eqs(5)-(9). The tensors $\Lambda^{A,R1}_{\alpha\beta\kappa\gamma}$,

$\Lambda^{D,R1}_{\alpha\beta\kappa\gamma}$, and $\Lambda^{S,R1}_{\alpha\beta}$ which contain information about layer orientation, are defined as

$$
\Lambda^{A,R1}_{\alpha\beta\kappa\gamma} = \Lambda^{D,R1}_{\alpha\beta\kappa\gamma} = n^{R1}_\alpha n^{R1}_\beta n^{R1}_\kappa n^{R1}_\gamma + \frac{1}{2(1-\nu)}\Big(t^{R1}_\alpha n^{R1}_\beta n^{R1}_\kappa t^{R1}_\gamma + n^{R1}_\alpha t^{R1}_\beta n^{R1}_\kappa t^{R1}_\gamma
$$
$$
+ t^{R1}_\alpha n^{R1}_\beta t^{R1}_\kappa n^{R1}_\gamma + n^{R1}_\alpha t^{R1}_\beta t^{R1}_\kappa n^{R1}_\gamma \Big) \tag{18}
$$

$$
\Lambda^{S,R1}_{\alpha\beta} = n^{R1}_\alpha n^{R1}_\beta
$$

It is worth mentioning that the isotropy of the stiffness tensors $A^+_{\alpha\beta\kappa\gamma}$, $D^+_{\alpha\beta\kappa\gamma}$, and $S^+_{\alpha\beta}$ for the material in the subdomains Ω^+ has been used in the derivation of the expressions in Eq(17), while no such restrictions have been imposed on the stiffness tensors $A^-_{\alpha\beta\kappa\gamma}$, $D^-_{\alpha\beta\kappa\gamma}$, and $S^-_{\alpha\beta}$ for the material in the subdomains Ω^-. This important feature makes these equations applicable for calculation of effective stiffness properties of multi-rank microstructures.

2.3.2 Matrix Form of Effective Stiffness Tensor Expressions

Following the ideas outlined in Lipton[11], Diaz, Lipton & Soto[12] and Soto[31] we in the present section introduce a simple transformation which allows us to restate the tensor expressions given in Eq(17) in a convenient matrix form. The basic idea behind this transformation is to express the membrane strain and curvature tensors for the Mindlin plate in the following convenient basis of second order tensors

$$
\xi^1 = \frac{1}{\sqrt{2}}\begin{bmatrix} 1 & 0 \\ 0 & -1 \end{bmatrix}, \quad \xi^2 = \frac{1}{\sqrt{2}}\begin{bmatrix} 0 & 1 \\ 1 & 0 \end{bmatrix}, \quad \xi^3 = \frac{1}{\sqrt{2}}\begin{bmatrix} 1 & 0 \\ 0 & 1 \end{bmatrix} \tag{19}
$$

such that

$$
\varepsilon_{\alpha\beta} = c^i \xi^i_{\alpha\beta} = \varepsilon_{\kappa\gamma}\xi^i_{\kappa\gamma}\xi^i_{\alpha\beta} = \frac{\varepsilon_{11}-\varepsilon_{22}}{\sqrt{2}}\xi^1_{\alpha\beta} + \frac{2\varepsilon_{12}}{\sqrt{2}}\xi^2_{\alpha\beta} + \frac{\varepsilon_{11}+\varepsilon_{22}}{\sqrt{2}}\xi^3_{\alpha\beta}
$$
$$
\kappa_{\alpha\beta} = c^i \xi^i_{\alpha\beta} = \kappa_{\kappa\gamma}\xi^i_{\kappa\gamma}\xi^i_{\alpha\beta} = \frac{\kappa_{11}-\kappa_{22}}{\sqrt{2}}\xi^1_{\alpha\beta} + \frac{2\kappa_{12}}{\sqrt{2}}\xi^2_{\alpha\beta} + \frac{\kappa_{11}+\kappa_{22}}{\sqrt{2}}\xi^3_{\alpha\beta}
$$
$$\tag{20}$$

These expressions for the membrane strain and curvature tensors allow us to express the strain energy density u in a Mindlin plate in the following simple way

$$
\begin{aligned}
u &= \tfrac{1}{2}\varepsilon_{\alpha\beta}A_{\alpha\beta\kappa\gamma}\varepsilon_{\kappa\gamma} + \tfrac{1}{2}\kappa_{\alpha\beta}D_{\alpha\beta\kappa\gamma}\kappa_{\kappa\gamma} + \tfrac{1}{2}\gamma_\alpha S_{\alpha\beta}\gamma_\beta \\[4pt]
&= \tfrac{1}{2}(\varepsilon_{\psi\eta}\xi^i_{\psi\eta}\xi^i_{\alpha\beta})A_{\alpha\beta\kappa\gamma}(\varepsilon_{\zeta\nu}\xi^j_{\zeta\nu}\xi^j_{\kappa\gamma}) + \tfrac{1}{2}(\kappa_{\psi\eta}\xi^i_{\psi\eta}\xi^i_{\alpha\beta})D_{\alpha\beta\kappa\gamma}(\kappa_{\zeta\nu}\xi^j_{\zeta\nu}\xi^j_{\kappa\gamma}) + \tfrac{1}{2}\gamma_\alpha S_{\alpha\beta}\gamma_\beta \\[4pt]
&= \tfrac{1}{2}(\varepsilon_{\psi\eta}\xi^i_{\psi\eta})(\xi^i_{\alpha\beta}A_{\alpha\beta\kappa\gamma}\xi^j_{\kappa\gamma})(\varepsilon_{\zeta\nu}\xi^j_{\zeta\nu}) + \tfrac{1}{2}(\kappa_{\psi\eta}\xi^i_{\psi\eta})(\xi^i_{\alpha\beta}D_{\alpha\beta\kappa\gamma}\xi^j_{\kappa\gamma})(\kappa_{\zeta\nu}\xi^j_{\zeta\nu}) + \tfrac{1}{2}\gamma_\alpha S_{\alpha\beta}\gamma_\beta \\[4pt]
&= \tfrac{1}{2}\{\varepsilon_i\}[A_{ij}]\{\varepsilon_j\} + \tfrac{1}{2}\{\kappa_i\}[D_{ij}]\{\kappa_j\} + \tfrac{1}{2}\{\gamma_\alpha\}[S_{\alpha\beta}]\{\gamma_\beta\} \\[4pt]
&= \tfrac{1}{2}\varepsilon^T A\varepsilon + \tfrac{1}{2}\kappa^T D\kappa + \tfrac{1}{2}\gamma^T S\gamma
\end{aligned} \tag{21}
$$

where it should be noted how the transverse shear strain and stiffness tensors, γ_α and $S_{\alpha\beta}$, have been replaced by their vector and matrix intrepretations γ and S, how the second-order membrane strain and curvature tensors $\varepsilon_{\alpha\beta}$ and $\kappa_{\alpha\beta}$ have been transformed into vectors $\boldsymbol{\varepsilon}$ and $\boldsymbol{\kappa}$ with components ε_i and κ_i given by Eq(20), and how the fourth-order membrane and bending stiffness tensors $A_{\alpha\beta\kappa\gamma}$ and $D_{\alpha\beta\kappa\gamma}$ have been transformed into matrices $\boldsymbol{A}$ and $\boldsymbol{D}$ with components A_{ij} and D_{ij}. The components of the membrane, bending, and transverse shear stiffness matrices, defined in Eq(21), may in general be written as

$$\boldsymbol{A} = [A_{ij}] = [\xi^i_{\alpha\beta}\xi^j_{\kappa\gamma}A_{\alpha\beta\kappa\gamma}] = \begin{bmatrix} \frac{1}{2}(A_{1111} + A_{2222}) - A_{1122} & A_{1112} - A_{2221} & \frac{1}{2}(A_{1111} - A_{2222}) \\ & 2A_{1212} & A_{1112} + A_{2221} \\ sym & & \frac{1}{2}(A_{1111} + A_{2222}) + A_{1122} \end{bmatrix}$$

$$\boldsymbol{D} = [D_{ij}] = [\xi^i_{\alpha\beta}\xi^j_{\kappa\gamma}D_{\alpha\beta\kappa\gamma}] = \begin{bmatrix} \frac{1}{2}(D_{1111} + D_{2222}) - D_{1122} & D_{1112} - D_{2221} & \frac{1}{2}(D_{1111} - D_{2222}) \\ & 2D_{1212} & D_{1112} + D_{2221} \\ sym & & \frac{1}{2}(D_{1111} + D_{2222}) + D_{1122} \end{bmatrix}$$

$$\boldsymbol{S} = [S_{\alpha\beta}] = \begin{bmatrix} S_{11} & S_{12} \\ sym & S_{22} \end{bmatrix}$$

$$\tag{22}$$

Here it should be emphasized that the components A_{ij} and D_{ij} of the membrane and bending stiffness matrices in general do not correspond to any of the components $A_{\alpha\beta\kappa\gamma}$ and $D_{\alpha\beta\kappa\gamma}$ of the original fourth-order membrane and bending stiffness tensors. However, the components of these tensors may easily be expressed in terms of the components of the membrane and bending stiffness matrices. Thus, by inverting the first two systems of linear equations in Eqs(22) we obtain:

$$\begin{bmatrix} A_{1111} & A_{1122} & A_{1112} \\ & A_{2222} & A_{2221} \\ sym & & A_{1212} \end{bmatrix} = \begin{bmatrix} \frac{1}{2}(A_{11} + A_{33}) + A_{13} & -\frac{1}{2}(A_{11} - A_{33}) & \frac{1}{2}(A_{12} + A_{23}) \\ & \frac{1}{2}(A_{11} + A_{33}) - A_{13} & -\frac{1}{2}(A_{12} - A_{23}) \\ sym & & \frac{1}{2}A_{22} \end{bmatrix}$$

$$\begin{bmatrix} D_{1111} & D_{1122} & D_{1112} \\ & D_{2222} & D_{2221} \\ sym & & D_{1212} \end{bmatrix} = \begin{bmatrix} \frac{1}{2}(D_{11} + D_{33}) + D_{13} & -\frac{1}{2}(D_{11} - D_{33}) & \frac{1}{2}(D_{12} + D_{23}) \\ & \frac{1}{2}(D_{11} + D_{33}) - D_{13} & -\frac{1}{2}(D_{12} - D_{23}) \\ sym & & \frac{1}{2}D_{22} \end{bmatrix}$$

$$\tag{23}$$

With the set of transformations in Eq(22) at hand, we are now able to transform the tensor equations for the effective membrane, bending, and transverse shear stiffness tensors for first-rank Mindlin plate microstructures derived in Sub-section 2.3.1 into simple matrix equations, and also, by the use of the "inverse transformations" given in Eq(23), to retain the components of the original stiffness tensors.

Thus, using the transformations in Eq(22) directly to the tensor equations for the effective membrane, bending, and transverse shear stiffness tensors, $A^{R1}_{\alpha\beta\kappa\gamma}$, $D^{R1}_{\alpha\beta\kappa\gamma}$, and $S^{R1}_{\alpha\beta}$, given in Eq(17) and making use of the following relation

$$\xi^i_{\alpha\beta}\xi^j_{\kappa\gamma}\left(A_{\alpha\beta\kappa\gamma}\right)^{-1} = \left(\xi^i_{\alpha\beta}\xi^j_{\kappa\gamma}A_{\alpha\beta\kappa\gamma}\right)^{-1} \tag{24}$$

which can be proven to hold for fourth order tensors with all major and minor symmetries, we may derive the following set of matrix equations

$$\boldsymbol{A}^{R1} = [A^{R1}_{ij}] = [\xi^i_{\alpha\beta}\xi^j_{\kappa\gamma}A^{R1}_{\alpha\beta\kappa\gamma}] = \boldsymbol{A}^+ - (1-\mu^{R1})\left(\left(\boldsymbol{A}^+ - \boldsymbol{A}^-\right)^{-1} - \frac{\mu^{R1}}{A^+_0}\boldsymbol{\Lambda}^{A,R1}\right)^{-1}$$

$$\boldsymbol{D}^{R1} = [D^{R1}_{ij}] = [\xi^i_{\alpha\beta}\xi^j_{\kappa\gamma}D^{R1}_{\alpha\beta\kappa\gamma}] = \boldsymbol{D}^+ - (1-\mu^{R1})\left(\left(\boldsymbol{D}^+ - \boldsymbol{D}^-\right)^{-1} - \frac{\mu^{R1}}{D^+_0}\boldsymbol{\Lambda}^{D,R1}\right)^{-1} \tag{25}$$

$$\boldsymbol{S}^{R1} = [S^{R1}_{\alpha\beta}] = \boldsymbol{S}^+ - (1-\mu^{R1})\left(\left(\boldsymbol{S}^+ - \boldsymbol{S}^-\right)^{-1} - \frac{\mu^{R1}}{S^+_0}\boldsymbol{\Lambda}^{S,R1}\right)^{-1}$$

Here $\boldsymbol{A}^+$, $\boldsymbol{A}^-$, $\boldsymbol{\Lambda}^{A,R1}$ and $\boldsymbol{D}^+$, $\boldsymbol{D}^-$, $\boldsymbol{\Lambda}^{D,R1}$ are (3x3) matrices obtained by using the transformations given in Eq(22) directly to the fourth order tensors $A^+_{\alpha\beta\kappa\gamma}$, $A^-_{\alpha\beta\kappa\gamma}$, $\Lambda^{A,R1}_{\alpha\beta\kappa\gamma}$ and $D^+_{\alpha\beta\kappa\gamma}$, $D^-_{\alpha\beta\kappa\gamma}$, $\Lambda^{D,R1}_{\alpha\beta\kappa\gamma}$, respectively, while $\boldsymbol{S}^+$, $\boldsymbol{S}^-$ and $\boldsymbol{\Lambda}^{S,R1}$ are (2x2) matrices obtained as simple matrix interpretations of the second order tensors $S^+_{\alpha\beta}$, $S^-_{\alpha\beta}$, and $\Lambda^{S,R1}_{\alpha\beta}$. Adopting the notation introduced in Eq(4), and taking advantage of the isotropy of the tensors $A^+_{\alpha\beta\kappa\gamma}$, $D^+_{\alpha\beta\kappa\gamma}$, $S^+_{\alpha\beta}$ and $A^-_{\alpha\beta\kappa\gamma}$, $D^-_{\alpha\beta\kappa\gamma}$, $S^-_{\alpha\beta}$ we easily derive the following set of expressions for the matrices $\boldsymbol{A}^+$, $\boldsymbol{D}^+$, $\boldsymbol{S}^+$ and $\boldsymbol{A}^-$, $\boldsymbol{D}^-$, $\boldsymbol{S}^-$,

$$\boldsymbol{A}^{(+/-)} = [A^{(+/-)}_{ij}] = [\xi^i_{\alpha\beta}\xi^j_{\kappa\gamma}A^{(+/-)}_{\alpha\beta\kappa\gamma}] = A^{(+/-)}_0 \begin{bmatrix} 1-\nu & 0 & 0 \\ 0 & 1-\nu & 0 \\ 0 & 0 & 1+\nu \end{bmatrix}$$

$$\boldsymbol{D}^{(+/-)} = [D^{(+/-)}_{ij}] = [\xi^i_{\alpha\beta}\xi^j_{\kappa\gamma}D^{(+/-)}_{\alpha\beta\kappa\gamma}] = D^{(+/-)}_0 \begin{bmatrix} 1-\nu & 0 & 0 \\ 0 & 1-\nu & 0 \\ 0 & 0 & 1+\nu \end{bmatrix} \tag{26}$$

$$\boldsymbol{S}^{(+/-)} = [S^{(+/-)}_{ij}] = S^{(+/-)}_0 \begin{bmatrix} 1 & 0 \\ 0 & 1 \end{bmatrix}$$

Finally, taking the normal and tangential vectors, n^{R1}_α and t^{R1}_α, in the expressions for the tensors $\Lambda^{A,R1}_{\alpha\beta\kappa\gamma}$, $\Lambda^{D,R1}_{\alpha\beta\kappa\gamma}$, and $\Lambda^{S,R1}_{\alpha\beta}$ given in Eq(18), as $n^{R1}_\alpha = \{cos(\theta^{R1}), sin(\theta^{R1})\}^T$ and $t^{R1}_\alpha = \{-sin(\theta^{R1}), cos(\theta^{R1})\}^T$, we obtain the following set of expressions for the matrices $\boldsymbol{\Lambda}^{A,R1}$, $\boldsymbol{\Lambda}^{D,R1}$, and $\boldsymbol{\Lambda}^{S,R1}$,

$$\boldsymbol{\Lambda}^{A/D,R1} = [\xi^i_{\alpha\beta}\xi^j_{\kappa\gamma}\Lambda^{A/D,R1}_{\alpha\beta\kappa\gamma}] = \begin{bmatrix} \dfrac{3-\nu-(1+\nu)cos(4\theta^{R1})}{4(1-\nu)} & -\dfrac{(1+\nu)sin(4\theta^{R1})}{4(1-\nu)} & \dfrac{cos(2\theta^{R1})}{2} \\[2ex] & \dfrac{3-\nu+(1+\nu)cos(4\theta^{R1})}{4(1-\nu)} & \dfrac{sin(2\theta^{R1})}{2} \\[2ex] sym & & \dfrac{1}{2} \end{bmatrix}$$

$$\boldsymbol{\Lambda}^{S,R1} = \Lambda^{R1}_{\alpha\beta} = \begin{bmatrix} \dfrac{1+cos(2\theta^{R1})}{2} & \dfrac{sin(2\theta^{R1})}{2} \\[2ex] sym & \dfrac{1-cos(2\theta^{R1})}{2} \end{bmatrix}$$

$$(27)$$

The matrix expressions given in Eqs(25)-(27) together with the inverse transformations given in Eq(23) form a simple set of matrix equations which easily admits derivation of analytical expressions for the components of the effective membrane, bending, and transverse shear stiffness tensors for the first-rank Mindlin plate microstructures described in Section 2.2. This set of matrix expressions shall therefore also be used as the basis for the derivation of effective stiffness properties of the more general multi-rank Mindlin plate microstructures considered in the remainder of this section.

2.4   Multi-Rank Mindlin Plate Microstructures

We shall now establish a set of simple analytical expressions for the effective membrane, bending, and transverse shear stiffness matrices for layered Mindlin plate microstructures of any finite rank via the procedure described in Sub-section 2.2. To avoid repetitive details, only the derivations for the effective membrane stiffness matrix will be presented here. Expressions for the effective bending and transverse shear stiffness matrices are readily obtained following the same procedure, and expressions for these matrices are given at the end of this sub-section.

Following the procedure for derivation of effective stiffness properties of multi-rank microstructures outlined in Sub-section 2.2 and making use of the expression for the effective membrane stiffness matrix derived in Eq(26), we get the following expression which can be used to determine effective membrane stiffness matrices for Mindlin plate microstructures of any finite rank

$$\boldsymbol{A}^{Ri} = \boldsymbol{A}^+ - (1-\mu^{Ri})\left[\left(\boldsymbol{A}^+ - \boldsymbol{A}^{R(i-1)}\right)^{-1} - \frac{\mu^{Ri}}{A^+_0}\boldsymbol{\Lambda}^{A,Ri}\right]^{-1} \ , \quad i = 1,\dots,I \qquad (28)$$

Let us now show how analytical expressions for the effective membrane stiffness matrices for first-rank, second-rank, and third-rank microstructures are established using Eq(28). The procedure always starts with the derivation of an expression for the effective membrane stiffness matrix for a first-rank microstructure, i.e.,

$$A^{R1} = A^+ - (1 - \mu^{R1}) \left[\left(A^+ - A^-\right)^{-1} - \frac{\mu^{R1}}{A_0^+} \Lambda^{A,R1} \right]^{-1} \tag{29}$$

Next, inserting Eq(29) in Eq(28) with $Ri = R2$, we establish an expression for the effective membrane stiffness matrix for a second-rank microstructure,

$$\begin{aligned} A^{R2} &= A^+ - (1 - \mu^{R2}) \left[\left(A^+ - A^{R1}\right)^{-1} - \frac{\mu^{R2}}{A_0^+} \Lambda^{A,R2} \right]^{-1} \\[2mm] &= A^+ - (1 - \mu^{R1})(1 - \mu^{R2}) \left[\left(A^+ - A^-\right)^{-1} - \frac{\mu^{R1}\Lambda^{A,R1} + (1 - \mu^{R1})\mu^{R2}\Lambda^{A,R2}}{A_0^+} \right]^{-1} \end{aligned} \tag{30}$$

and inserting Eq(30) in Eq(28) with $Ri = R3$, we obtain the following expression for the effective membrane stiffness matrix for a third-rank microstructure

$$\begin{aligned} A^{R3} &= A^+ - (1 - \mu^{R3}) \left(\left(A^+ - A^{R2}\right)^{-1} - \frac{\mu^{R3}}{A_0^+} \Lambda^{A,R3} \right)^{-1} \\[2mm] &= A^+ - (1 - \mu^{R1})(1 - \mu^{R2})(1 - \mu^{R3}) \left[\left(A^+ - A^-\right)^{-1} \right. \\[2mm] &\quad \left. - \frac{\mu^{R1}\Lambda^{A,R1} + (1 - \mu^{R1})\mu^{R2}\Lambda^{A,R2} + (1 - \mu^{R1})(1 - \mu^{R2})\mu^{R3}\Lambda^{A,R3}}{A_0^+} \right]^{-1} \end{aligned} \tag{31}$$

Performing these calculations recursively we may establish expressions for the effective membrane stiffness matrices for a layered Mindlin plate microstructure of any finite rank. However, the resulting expressions will very rapidly grow large and become difficult to handle. In the following we shall therefore introduce a variable substitution which allows us to restate the expressions given in Eqs(29)-(31) in a more compact form. To this end we start by realizing that the density of stiff plate slices ϱ^{R1} in a first-rank, ϱ^{R2} in a second-rank, and ϱ^{R3} in a third-rank Mindlin plate microstructure are given by

$$\varrho^{R1} = \mu^{R1} = p_1 \varrho^{R1}$$

$$\varrho^{R2} = \mu^{R2} + (1 - \mu^{R2})\varrho^{R1} = \mu^{R1} + (1 - \varrho^{R1})\mu^{R2} = (p_1 + p_2)\varrho^{R2} \tag{32}$$

$$\varrho^{R3} = \mu^{R3} + (1 - \mu^{R3})\varrho^{R2} = \mu^{R1} + (1 - \varrho^{R1})\mu^{R2} + (1 - \varrho^{R2})\mu^{R3} = (p_1 + p_2 + p_3)\varrho^{R3}$$

with the factors p_i, for a rank-I microstructure, defined as,

$$p_i = \frac{(1 - \varrho^{R(i-1)})\mu^{Ri}}{\varrho^{RI}} \quad , \quad i = 1, \dots, I \tag{33}$$

$$\sum_{i=1}^{I} p_i = 1 \quad ; \quad p_i \geq 0 \quad , \quad i = 1, \dots, I \tag{34}$$

Using the expressions in Eq(32) we derive the following set of expressions for the density of compliant plate slices in the first-, second-, and third-rank Mindlin plate microstructure,

$$1 - \varrho^{R1} = (1 - \mu^{R1})$$

$$1 - \varrho^{R2} = (1 - \mu^{R1})(1 - \mu^{R2}) \tag{35}$$

$$1 - \varrho^{R3} = (1 - \mu^{R1})(1 - \mu^{R2})(1 - \mu^{R3})$$

Finally, introducing the expressions in Eq(32)-Eq(35) into the expressions for the effective membrane stiffness matrices for the first-, second-, and third-rank Mindlin plate microstructures given in Eqs(29)-(31), we obtain

$$\boldsymbol{A}^{R1} = \boldsymbol{A}^{+} - (1 - \varrho^{R1})\left((\boldsymbol{A}^{+} - \boldsymbol{A}^{-})^{-1} - \frac{\varrho^{R1}}{A_0^{+}} \sum_{i=1}^{1} p_i \boldsymbol{\Lambda}^{A,Ri} \right)^{-1}$$

$$\boldsymbol{A}^{R2} = \boldsymbol{A}^{+} - (1 - \varrho^{R2})\left((\boldsymbol{A}^{+} - \boldsymbol{A}^{-})^{-1} - \frac{\varrho^{R2}}{A_0^{+}} \sum_{i=1}^{2} p_i \boldsymbol{\Lambda}^{A,Ri} \right)^{-1} \tag{36}$$

$$\boldsymbol{A}^{R3} = \boldsymbol{A}^{+} - (1 - \varrho^{R3})\left((\boldsymbol{A}^{+} - \boldsymbol{A}^{-})^{-1} - \frac{\varrho^{R3}}{A_0^{+}} \sum_{i=1}^{3} p_i \boldsymbol{\Lambda}^{A,Ri} \right)^{-1}$$

Thus, by substituting the original density variables μ^{Ri}, $i = 1, \ldots, I$, by a new set of variables ϱ^{RI} and p_i, $i = 1, \ldots, I$, we have obtained three very simple and almost identical expressions. Continuing the process used to establish the expressions in Eq(36), we see that for a general rank-I microstructure the effective membrane stiffness matrix can be written as

$$\boldsymbol{A}^{RI} = \boldsymbol{A}^{+} - (1 - \varrho^{RI})\left((\boldsymbol{A}^{+} - \boldsymbol{A}^{-})^{-1} - \frac{\varrho^{RI}}{A_0^{+}} \sum_{i=1}^{I} p_i \boldsymbol{\Lambda}^{A,Ri} \right)^{-1} \tag{37}$$

A general proof of the validity of this equation can be found in Soto[31], see also Lipton[35]. Using a similar procedure as used to establish Eq(37) we may derive the following expressions for the effective bending and transverse shear stiffness matrices, see [12] and [31],

$$\boldsymbol{D}^{RI} = \boldsymbol{D}^{+} - (1 - \varrho^{RI})\left((\boldsymbol{D}^{+} - \boldsymbol{D}^{-})^{-1} - \frac{\varrho^{RI}}{D_0^{+}} \sum_{i=1}^{I} p_i \boldsymbol{\Lambda}^{D,Ri} \right)^{-1} \tag{38}$$

$$\boldsymbol{S}^{RI} = \boldsymbol{S}^{+} - (1 - \varrho^{RI})\left((\boldsymbol{S}^{+} - \boldsymbol{S}^{-})^{-1} - \frac{\varrho^{RI}}{S_0^{+}} \sum_{i=1}^{I} p_i \boldsymbol{\Lambda}^{S,Ri} \right)^{-1} \tag{39}$$

From Eqs(37)-(39) one now has the effective membrane bending and transverse shear stiffness matrices for any finite-rank Mindlin plate microstructure given in terms of $(2I + 1)$ parameters, namely the total density ϱ^{R1} of stiff plate slices in the microstructure, the I new density variables p_i, and the I layer directions in the microstructure given by the angles of rotation θ^{Ri}, see Eq(23).

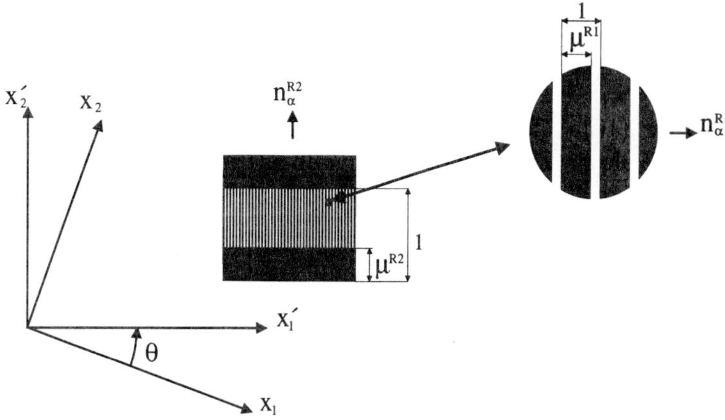

Figure 6: Second-rank microstructure with orthogonal layers.

For solution of layout optimization of plane disks it is well-known that the optimal microstructures for single and multiple load case maximum stiffness layout design problems is a second-rank microstructure with orthogonal layers and a third-rank microstructure with non-orthogonal layers, respectively, while a third-rank microstructure with non-orthogonal layers should be used for solution of both single and multiple load case maximum stiffness layout design problems for Mindlin plates and shells. Thus, for the treatment of maximum stiffness layout design problems we generally only need to consider Eqs(37)-(39) for $I = 2$ and $I = 3$. Also, it was shown in Diaz, Lipton & Soto[12] and Soto[31] that all effective stiffness properties for the finite-rank Mindlin plate microstructures are obtained by microstructures with at most three non-orthogonal layers. For the treatment of other elasticity driven problems than stiffness design problems it should therefore be recognized that no further generality is obtained by using layered microstructures with more than three non-orthogonal layers.

2.4.1 Second-Rank Mindlin Plate Microstructures

Second-rank microstructures with orthogonal layers play an important role in the design parametrization of layout optimization problems for planar disks, since second-rank microstructures with orthogonal layers have been shown to be optimal for the solution of single load case stiffness design problems for planar disks. In this sub-section we therefore consider a simple second-rank Mindlin plate microstructure with orthogonal layers for which the effective stiffness properties may be given in terms of three variables only, namely two density variables μ^{R1} and μ^{R2} which control the densities of stiff plate slices along two mutually orthogonal directions, and an angle of rotation θ which controls the overall rotation of the microstructure relative to a fixed global coordinate system x_1x_2, see Figure 6.   Analytical expressions for the non-zero components of the effective membrane, bending, and transverse shear stiffness tensors for this second-rank Mindlin plate microstructure are in the following established relative to a fixed material coordinate system $x_1' x_2'$ with axes parallel to the two orthogonal layering directions in the microstructure, see Figure 6. The derived expressions will therefore only be functions of the two density variables μ^{R1} and μ^{R2}. However, the second-rank Mindlin plate microstructure will, in general, be orthotropic and the overall angle of rotation θ of the microstructure naturally enters as a third design variable via usual transformation formulas for rotation of anisotropic materials, see for example the book by Vinson & Sierakovski[34].

Now, let us start by establishing the equations needed in order to determine the effective membrane, bending, and transverse shear stiffness matrices for a second-rank Mindlin plate microstructure with orthogonal layers. From the derivations in Subsection 2.4 we have the following expression for the effective membrane stiffness matrix for a general second-rank Mindlin plate microstructure with non-orthogonal layers.

$$\boldsymbol{A}^{R2} = \boldsymbol{A}^{+} - (1 - \mu^{R1})(1 - \mu^{R2}) \left[(\boldsymbol{A}^{+} - \boldsymbol{A}^{-})^{-1} - \frac{\mu^{R1}\boldsymbol{\Lambda}^{A,R1} + (1 - \mu^{R1})\mu^{R2}\boldsymbol{\Lambda}^{A,R2}}{A_0^{+}} \right]^{-1} \quad (40)$$

Following the procedure used for the derivation of this expression, see Eqs(29)-(30), we may easily derive the corresponding expressions for the effective bending and transverse shear stiffness matrices

$$\boldsymbol{D}^{R2} = \boldsymbol{D}^{+} - (1 - \mu^{R1})(1 - \mu^{R2}) \left[(\boldsymbol{D}^{+} - \boldsymbol{D}^{-})^{-1} - \frac{\mu^{R1}\boldsymbol{\Lambda}^{D,R1} + (1 - \mu^{R1})\mu^{R2}\boldsymbol{\Lambda}^{D,R2}}{D_0^{+}} \right]^{-1} \quad (41)$$

$$\boldsymbol{S}^{R2} = \boldsymbol{S}^{+} - (1 - \mu^{R1})(1 - \mu^{R2}) \left[(\boldsymbol{S}^{+} - \boldsymbol{S}^{-})^{-1} - \frac{\mu^{R1}\boldsymbol{\Lambda}^{S,R1} + (1 - \mu^{R1})\mu^{R2}\boldsymbol{\Lambda}^{S,R2}}{S_0^{+}} \right]^{-1} \quad (42)$$

The first two sets of matrices used in Eqs(40)-(42) are the membrane, bending, and transverse shear stiffness matrices for the stiff and the compliant plate slices in the Mindlin plate microstructure, $\boldsymbol{A}^{+}$, $\boldsymbol{D}^{+}$, $\boldsymbol{S}^{+}$ and $\boldsymbol{A}^{-}$, $\boldsymbol{D}^{-}$, $\boldsymbol{S}^{-}$, which are generally defined as in Eq(26). The other two sets of matrices used in Eqs(40)-(42) are the geometrical matrices for the first and second layering in the second-rank Mindlin plate microstructure, $\boldsymbol{\Lambda}^{A,R1}$, $\boldsymbol{\Lambda}^{D,R1}$, $\boldsymbol{\Lambda}^{S,R1}$ and $\boldsymbol{\Lambda}^{A,R2}$, $\boldsymbol{\Lambda}^{D,R2}$, $\boldsymbol{\Lambda}^{S,R2}$. The geometrical matrices, which depend on the angles of rotation θ^{R1} and θ^{R2} of the normal vectors n_α^{R1} and n_α^{R1} relative to the layers in the microstructure, see Figure 6, are generally defined as in Eq(27). Thus, taking the normal vectors n_α^{R1} and n_α^{R2} as orthogonal vectors characterized by angles of rotation $\theta^{R1} = 0°$ and $\theta^{R2} = 90°$ we obtain the following set of geometric matrices for a second-rank Mindlin plate microstructure with orthogonal layers,

$$
\boldsymbol{\Lambda}^{A/D,R1} = \begin{bmatrix} \dfrac{1}{2} & 0 & \dfrac{1}{2} \\ 0 & \dfrac{1}{1-\nu} & 0 \\ \dfrac{1}{2} & 0 & \dfrac{1}{2} \end{bmatrix} \quad ; \quad
\boldsymbol{\Lambda}^{A/D,R2} = \begin{bmatrix} \dfrac{1}{2} & 0 & -\dfrac{1}{2} \\ 0 & \dfrac{1}{1-\nu} & 0 \\ -\dfrac{1}{2} & 0 & \dfrac{1}{2} \end{bmatrix} \tag{43}
$$

$$
\boldsymbol{\Lambda}^{S,R1} = \begin{bmatrix} 1 & 0 \\ 0 & 0 \end{bmatrix} \quad ; \quad
\boldsymbol{\Lambda}^{S,R2} = \begin{bmatrix} 0 & 0 \\ 0 & 1 \end{bmatrix}
$$

With the definitions in Eq(26) and Eq(43), and the expressions in Eqs(40)-(42) we are now able to derive analytical expressions for the effective membrane, bending, and transverse shear stiffness matrices for a second-rank Mindlin plate microstructure with orthogonal layers. However, it should be recalled that the components of the membrane and bending stiffness matrices $\boldsymbol{A}^{R2}$ and $\boldsymbol{D}^{R2}$ do not correspond to the components of the effective membrane and bending stiffness tensors $A^{R2}_{\alpha\beta\kappa\gamma}$ and $D^{R2}_{\alpha\beta\kappa\gamma}$. To obtain the latter we must first calculate the effective membrane and bending stiffness matrices and then apply the "inverse transformations" given in Eq(23). Performing these derivations using *Mathematica*, we have established the set of expressions given below for the non-zero components of the effective membrane, bending, and transverse shear stiffness tensors for a second-rank Mindlin plate microstructure with orthogonal layers. Note that we have here replaced the usual symbols for the density variables μ^{R1} and μ^{R2} by the short-hand symbols μ_1 and μ_2, in order to enhance the readability of the expressions.

$$
A^{R2}_{1111} = \frac{A^{R2}_0 + A^{R2}_1 \left(\mu_2 - \mu_1\mu_2 - \mu_2^2 + 2\mu_1\mu_2^2 - \mu_1^2\mu_2^2\right)}{A^{R2}_2 + A^{R2}_3} \quad ; \quad
A^{R2}_{2222} = \frac{A^{R2}_0 + A^{R2}_1 \left(\mu_1 - \mu_1^2\right)}{A^{R2}_2 + A^{R2}_3}
$$

$$
A^{R2}_{1122} = \frac{\nu A^{R2}_0 + \nu A^{R2}_1 \left(\mu_1\mu_2 - \mu_1^2\mu_2\right)}{A^{R2}_2 + A^{R2}_3} \quad ; \quad
A^{R2}_{1212} = \frac{(1-\nu)A_0^- A_0^+}{2A^{R2}_4}
$$

$$
D^{R2}_{1111} = \frac{D^{R2}_0 + D^{R2}_1 \left(\mu_2 - \mu_1\mu_2 - \mu_2^2 + 2\mu_1\mu_2^2 - \mu_1^2\mu_2^2\right)}{D^{R2}_2 + D^{R2}_3} \quad ; \quad
D^{R2}_{2222} = \frac{D^{R2}_0 + D^{R2}_1 \left(\mu_1 - \mu_1^2\right)}{D^{R2}_2 + D^{R2}_3}
$$

$$
D^{R2}_{1122} = \frac{\nu D^{R2}_0 + \nu D^{R2}_1 \left(\mu_1\mu_2 - \mu_1^2\mu_2\right)}{D^{R2}_2 + D^{R2}_3} \quad ; \quad
D^{R2}_{1212} = \frac{(1-\nu)D_0^- D_0^+}{2D^{R2}_4} \tag{44}
$$

$$
S_{11} = S_0^+ - \frac{S_0^+ \left(S_0^+ - S_0^-\right)\left(1 - \mu_1 - \mu_2 + \mu_1\mu_2\right)}{(1-\mu_1)S_0^+ + \mu_1 S_0^-}
$$

$$
S_{22} = S_0^+ - \frac{S_0^+ \left(S_0^+ - S_0^-\right)\left(1 - \mu_1 - \mu_2 + \mu_1\mu_2\right)}{S_0^+ - \left(S_0^+ - S_0^-\right)\left(\mu_2 - \mu_1\mu_2\right)}
$$

with

$$A_0^{R2} = A_0^- (A_0^+)^2 \quad ; \quad A_1^{R2} = \left(1 - \nu^2\right) \left(A_0^- - A_0^+\right)^2 A_0^+$$

$$A_2^{R2} = (A_0^+)^2 + A_0^+ \left(A_0^- - A_0^+\right)(\mu_1 + \mu_2 - \mu_1\mu_2) \quad ; \quad A_3^{R2} = \left(1 - \nu^2\right)\left(A_0^- - A_0^+\right)^2 \left(\mu_1\mu_2 - \mu_1^2\mu_2\right)$$

$$A_4^{R2} = A_0^+ + \left(A_0^- - A_0^+\right)(\mu_1 + \mu_2 - \mu_1\mu_2)$$

$$D_0^{R2} = D_0^- (D_0^+)^2 \quad ; \quad D_1^{R2} = \left(1 - \nu^2\right)\left(D_0^- - D_0^+\right)^2 D_0^+$$

$$D_2^{R2} = (D_0^+)^2 + D_0^+ \left(D_0^- - D_0^+\right)(\mu_1 + \mu_2 - \mu_1\mu_2) \quad ; \quad D_3^{R2} = \left(1 - \nu^2\right)\left(D_0^- - D_0^+\right)^2 \left(\mu_1\mu_2 - \mu_1^2\mu_2\right)$$

$$D_4^{R2} = D_0^+ + \left(D_0^- - D_0^+\right)(\mu_1 + \mu_2 - \mu_1\mu_2)$$

$$\tag{45}$$

For the solution of layout optimization problems for disk, plate, and shell structures we may now apply a design parametrization where the material in each domain of a finite element discretized structure is modelled as an orthotropic material with the stiffness properties of the second-rank Mindlin plate microstructure considered above. This would, in general, give us a nice finite parameter, continuous formulation of the different types of layout optimization problems for disk, plate, and shell structures that corresponds to the different configurations of Mindlin plate microstructures discussed in Sub-section 2.2, and the structures can easily be analyzed by means of flat orthotropic disk, plate, or shell finite elements. We shall omit here to discuss the type of behavioural objective and constraint functions which may be considered using such a design parametrization, and only present the following general formulation for the layout optimization problem

$$
\begin{aligned}
&Objective: && \min_{\mu_{1e},\mu_{2e},\theta_e} \text{ or } \max \; \left[\, f(\mu_{1e}, \mu_{2e}, \theta_e)\,\right]\,, && e = 1, \ldots, N_e \\[2mm]
&Subject\ to: && g_j(\mu_{1e}, \mu_{2e}, \theta_e) \leq 0\,, && e = 1, \ldots, N_e\,, \;\; j = 1, \ldots, p \\[2mm]
& && h_j(\mu_{1e}, \mu_{2e}, \theta_e) = 0\,, && e = 1, \ldots, N_e\,, \;\; j = 1, \ldots, q \\[2mm]
& && 0 \leq \mu_{1e} \leq 1\,, && e = 1, \ldots, N_e \\[2mm]
& && 0 \leq \mu_{2e} \leq 1\,, && e = 1, \ldots, N_e
\end{aligned}
\tag{46}
$$

Here f, g_j, and h_j are objective and constraint functions for the optimization, and N_e denotes the total number of finite elements in the discretized structure. It is important to realize that from an optimization point of view this problem is inherently a very large scale problem; the number of design variables equals three times the number of finite elements in the discretized structure. This naturally imposes some restrictions on the type of objective and constraint functions that may be considered, and the problem generally requires solution by means of special optimization procedures in order to deal with the many design variables in an efficient way. These issues will be further discussed in Sections 3 and 4.

2.4.2   The Moment Formulation

The second-rank Mindlin plate microstructure considered in the preceding sub-section is generally only optimal for solution of single load case stiffness design problems for planar disks. For solution of layout design problems for Mindlin plates and multiple load case stiffness design problems it would be needed to apply a more general third-rank Mindlin plate microstructure with non-orthogonal layers, and for the solution of other types of design problems, where the optimal microstructure usually is unknown, one should use as general a microstructure as possible. In both situations a design parametrization using a third-rank Mindlin plate microstructure seems reasonable; the full range of effective stiffness properties for all finite-rank Mindlin plate microstructures are described by only six independent variables, and the effective stiffness properties may be calculated analytically using the expressions derived in Section 2.4. Note however, that a design parametrization of stiffnesses directly in terms of layering directions is associated with certain problems, especially since local optimal solutions to design problems are well-known to exist with respect to directions of anisotropy, see e.g. Pedersen[36],[37],[38].

In the following we shall therefore consider a special restatement of the expressions for the effective stiffness properties of finite-rank Mindlin plate microstructures in terms of so-called moment variables. Moment formulations for description of effective stiffness properties of layered microstructures, as introduced by Francfort & Murat[39] and applied by Avellaneda & Milton[33] in studies on bounds of the elastic stiffness tensors for layered planar microstructures of arbitrary rank, have in relation to layout optimization problems been used by e.g. Lipton[11], Diaz, Lipton & Soto[12] and Soto[31] for the description of effective stiffness properties of microstructurally layered Kirchhoff and Mindlin plates, and by Lipton & Diaz[40] for description of effective stiffness properties of three dimensional microstructures. The moment formulation for microstructurally layered Mindlin plates considered here follows the work by Diaz, Lipton & Soto[12] and Soto[31], and it is attractive for the following reasons:

- Firstly, as shown in [12] and [31], the moment formulation gives a full description of the effective stiffness properties for all finite-rank Mindlin plate microstructures using only five design variables, see also [11] for Kirchhoff plates.

- Secondly, for a fixed strain field, the strain energy density has been found to be concave in the set of variables used to describe the anisotropy of the microstructure, see [12] and [11]. Thus, for the solution of stiffness design problems, the possibility of convergence to local extrema is eliminated in the computation of the optimal anisotropy.

- Thirdly, for the solution of other types of problems than stiffness design problems, it is worth mentioning that all application of periodic functions of layering orientations generally is avoided in the expressions for the effective stiffness properties, so difficulties with local extrema with respect to layer orientations therefore can be expected to be reduced for other types of problems as well.

Now, in order to describe the applied moment formulation for layered Mindlin plate microstructures we start by considering the set of expressions for the effective membrane, bending, and transverse shear stiffness matrices for a general rank-I Mindlin plate microstructure given in Eqs(37)-(39),

$$\boldsymbol{A}^{RI} = \boldsymbol{A}^+ - (1 - \varrho^{RI}) \left((\boldsymbol{A}^+ - \boldsymbol{A}^-)^{-1} - \frac{\varrho^{RI}}{A_0^+} \sum_{i=1}^{I} p_i \boldsymbol{\Lambda}^{A,Ri} \right)^{-1}$$

$$\boldsymbol{D}^{RI} = \boldsymbol{D}^+ - (1 - \varrho^{RI}) \left((\boldsymbol{D}^+ - \boldsymbol{D}^-)^{-1} - \frac{\varrho^{RI}}{D_0^+} \sum_{i=1}^{I} p_i \boldsymbol{\Lambda}^{D,Ri} \right)^{-1} \qquad (47)$$

$$\boldsymbol{S}^{RI} = \boldsymbol{S}^+ - (1 - \varrho^{RI}) \left((\boldsymbol{S}^+ - \boldsymbol{S}^-)^{-1} - \frac{\varrho^{RI}}{S_0^+} \sum_{i=1}^{I} p_i \boldsymbol{\Lambda}^{S,Ri} \right)^{-1}$$

It is evident from the foregoing that Eq(27) is invariant if $R1$ is replaced by Ri, so we have the following definitions for the set of matrices $\boldsymbol{\Lambda}^{A,Ri}$, $\boldsymbol{\Lambda}^{D,Ri}$, and $\boldsymbol{\Lambda}^{S,Ri}$,

$$\boldsymbol{\Lambda}^{A/D,Ri} = \begin{bmatrix} \dfrac{3 - \nu - (1+\nu)cos(4\theta^{Ri})}{4(1-\nu)} & -\dfrac{(1+\nu)sin(4\theta^{Ri})}{4(1-\nu)} & \dfrac{cos(2\theta^{Ri})}{2} \\[3mm] & \dfrac{3 - \nu + (1+\nu)cos(4\theta^{Ri})}{4(1-\nu)} & \dfrac{sin(2\theta^{Ri})}{2} \\[3mm] sym & & \dfrac{1}{2} \end{bmatrix} \qquad (48)$$

$$\boldsymbol{\Lambda}^{S,Ri} = \begin{bmatrix} \dfrac{1 + cos(2\theta^{Ri})}{2} & \dfrac{sin(2\theta^{Ri})}{2} \\[3mm] sym & \dfrac{1 - cos(2\theta^{Ri})}{2} \end{bmatrix}$$

We now perform a variable substitution in which we condense all information on the layering densities and orientations, given by the design variables p_i and θ^{Ri}, $i = 1, \ldots, I$, into four new variables. For this purpose we define four so-called moment variables

$$m_1 = \sum_{i=1}^{I} p_i cos(2\theta^{Ri}) \,, \qquad m_2 = \sum_{i=1}^{I} p_i sin(2\theta^{Ri})$$

$$\qquad (49)$$

$$m_3 = \sum_{i=1}^{I} p_i cos(4\theta^{Ri}) \,, \qquad m_4 = \sum_{i=1}^{I} p_i sin(4\theta^{Ri})$$

which, by application of the condition $\sum_{i=1}^{I} p_i = 1$ from Eq(34) and the expressions in

Eq(48), allows us to define the following set of matrices

$$M^{A/D} = \sum_{i=1}^{I} p_i \Lambda^{A/D,Ri} = \begin{bmatrix} \dfrac{3 - \nu - (1+\nu)m_3}{4(1-\nu)} & -\dfrac{(1+\nu)m_4}{4(1-\nu)} & \dfrac{m_1}{2} \\[2ex] & \dfrac{3 - \nu + (1+\nu)m_3}{4(1-\nu)} & \dfrac{m_2}{2} \\[2ex] sym & & \dfrac{1}{2} \end{bmatrix} \tag{50}$$

$$M^S = \sum_{i=1}^{I} p_i \Lambda^{S,Ri} = \begin{bmatrix} \dfrac{1+m_1}{2} & \dfrac{m_2}{2} \\[2ex] sym & \dfrac{1-m_1}{2} \end{bmatrix}$$

Using these matrices, we may rewrite the expressions for the effective membrane, bending and transverse shear stiffness matrices given in Eq(47) in the following convenient forms

$$A^{RI} = A^+ - (1 - \varrho^{RI})\left((A^+ - A^-)^{-1} - \frac{\varrho^{RI}}{A_0^+} M^A \right)^{-1}$$

$$D^{RI} = D^+ - (1 - \varrho^{RI})\left((D^+ - D^-)^{-1} - \frac{\varrho^{RI}}{D_0^+} M^D \right)^{-1} \tag{51}$$

$$S^{RI} = S^+ - (1 - \varrho^{RI})\left((S^+ - S^-)^{-1} - \frac{\varrho^{RI}}{S_0^+} M^S \right)^{-1}$$

Hence, by the variable substitution in Eq(49) we have obtained a set of very simple expressions which gives the full range of effective stiffness properties of all finite-rank Mindlin plate microstructures using only five design variables, namely the density ϱ^{RI} of stiff plate slices in the microstructure and the four moment variables $m_1 - m_4$ defined in Eq(49). From the definitions in Eq(49) it should be clear that the moment variables, in general, must satisfy the simple side constraints

$$-1 \leq m_i \leq 1 \ , \quad i = 1, \ldots, 4 \tag{52}$$

However, the moment variables $m_1 - m_4$ are obviously not mutually independent due to the trigonometric relations between the functions applied to define the moments, and we should therefore impose some additional constraints which restrict the combinations of the moment variables to some feasible domain. This set of additional constraints on the moment variables was first established in Avellaneda & Milton[33] by use of the solution to the trigonometric moment problem as treated in Krein & Nudelman[41]. The trigonometric moment problem treated by Krein & Nudelman reads: *"We wish to find necessary and sufficient conditions for a prescribed set of complex numbers γ_k, $k = 0, \ldots, K$, to admit a representation:*

$$\int_0^{2\pi} \sigma(t) e^{-ikt} dt = \gamma_k \ , \quad k = 0, \ldots, K \tag{53}$$

where $\sigma(t)$, $0 \leq t \leq 2\pi$, is a distribution.", and it was found in [41] to be a necessary and sufficient condition for such a representation that the equation shown below is fulfilled for arbitrary complex and complex conjugate numbers x_p and $\overline{x}_q$, with $\gamma_{-k} = \overline{\gamma}_k$.

$$\sum_{p=0}^{K}\sum_{q=0}^{K} \gamma_{p-q} x_p \overline{x}_q \geq 0 \tag{54}$$

At a first glance it may be difficult to see how this may help us to establish the set of constraints which identifies the feasible combinations of the moment variables. However, if we consider the definitions of the moment variables given in Eq(49) in the limit where we range over all possible layer combinations as well as layer directions, we get

$$m_1 = \int_0^{2\pi} p(\theta)cos(2\theta)d\theta \ , \qquad m_2 = \int_0^{2\pi} p(\theta)sin(2\theta)d\theta$$

$$m_3 = \int_0^{2\pi} p(\theta)cos(4\theta)d\theta \ , \qquad m_4 = \int_0^{2\pi} p(\theta)sin(4\theta)d\theta \tag{55}$$

and realizing that

$$\int_0^{2\pi} p(\theta)d\theta = 1 \ ; \quad p(\theta) = p(\theta + \pi) \tag{56}$$

we may, inspired by Eq(53), define the following set of complex numbers

$$\gamma_0 = \int_0^{2\pi} p(\theta)e^{-i(0\theta)}d\theta = \int_0^{2\pi} p(\theta)cos(0\theta)d\theta - i\int_0^{2\pi} p(\theta)sin(0\theta)d\theta = 1$$

$$\gamma_1 = \int_0^{2\pi} p(\theta)e^{-i(1\theta)}d\theta = \int_0^{2\pi} p(\theta)cos(1\theta)d\theta - i\int_0^{2\pi} p(\theta)sin(1\theta)d\theta = 0$$

$$\gamma_2 = \int_0^{2\pi} p(\theta)e^{-i(2\theta)}d\theta = \int_0^{2\pi} p(\theta)cos(2\theta)d\theta - i\int_0^{2\pi} p(\theta)sin(2\theta)d\theta = m_1 - im_2 \tag{57}$$

$$\gamma_3 = \int_0^{2\pi} p(\theta)e^{-i(3\theta)}d\theta = \int_0^{2\pi} p(\theta)cos(3\theta)d\theta - i\int_0^{2\pi} p(\theta)sin(3\theta)d\theta = 0$$

$$\gamma_4 = \int_0^{2\pi} p(\theta)e^{-i(4\theta)}d\theta = \int_0^{2\pi} p(\theta)cos(4\theta)d\theta - i\int_0^{2\pi} p(\theta)sin(4\theta)d\theta = m_3 - im_4$$

The relation to the trigonometric moment problem in Eq(53) is now obvious, and from the condition in Eq(54) we get the following expression to be fulfilled for all complex vectors $\{x_0, x_1, x_2, x_3, x_4\}^T$

$$\begin{Bmatrix} x_0 \\ x_1 \\ x_2 \\ x_3 \\ x_4 \end{Bmatrix}^T \begin{bmatrix} 1 & 0 & m_1 - im_2 & 0 & m_3 - im_4 \\ 0 & 1 & 0 & m_1 - im_2 & 0 \\ m_1 + im_2 & 0 & 1 & 0 & m_1 - im_2 \\ 0 & m_1 + im_2 & 0 & 1 & 0 \\ m_3 + im_4 & 0 & m_1 + im_2 & 0 & 1 \end{bmatrix} \begin{Bmatrix} \overline{x}_0 \\ \overline{x}_1 \\ \overline{x}_2 \\ \overline{x}_3 \\ \overline{x}_4 \end{Bmatrix} \geq 0 \tag{58}$$

Hence, the complex matrix should be positive definite. Expressing Eq(58) using determinants, i.e., requiring that the determinants of all submatrices should be positive, we arrive at the following two constraint equations which the feasible combinations of the moment variables must necessarily satisfy:

$$g_1(m_1, m_2) = m_1^2 + m_2^2 \leq 1$$

$$g_2(m_1, m_2, m_3, m_4) = 2m_1^2(1 - m_3) + 2m_2^2(1 + m_3) + (m_3^2 + m_4^2) - 4m_1 m_2 m_4 \leq 1$$

$$(59)$$

For the solution of layout optimization problems for disk, plate, and shell structures we may now consider a design parametrization where the material in each subdomain of a finite element discretized structure is modelled as an anisotropic material with stiffness properties given by Eq(51). Adopting such a design parametrization we may state a general layout optimization problem as follows

$$Objective: \quad \min_{\varrho_e^{RI},\, m_{1e},\, m_{2e},\, m_{3e},\, m_{4e}} or\ max \quad [\, f(\varrho_e^{RI}, m_{1e}, m_{2e}, m_{3e}, m_{4e}) \,] \,, \quad e = 1, \ldots, N_e$$

$$Subject\ to: \quad g_j(\varrho_e^{RI}, m_{1e}, m_{2e}, m_{3e}, m_{4e}) \leq 0 \,, \qquad e = 1, \ldots, N_e \,, \quad j = 1, \ldots, p$$

$$h_j(\varrho_e^{RI}, m_{1e}, m_{2e}, m_{3e}, m_{4e}) = 0 \,, \qquad e = 1, \ldots, N_e \,, \quad j = 1, \ldots, q$$

$$g_{1e}(m_{1e}, m_{2e}) \leq 1 \,, \qquad e = 1, \ldots, N_e$$

$$g_{2e}(m_{1e}, m_{2e}, m_{3e}, m_{4e}) \leq 1 \,, \qquad e = 1, \ldots, N_e$$

$$0 \leq \varrho_e^{RI} \leq 1 \,, \qquad e = 1, \ldots, N_e$$

$$-1 \leq m_{ie} \leq 1 \,, \qquad i = 1, \ldots, 4 \,, \quad e = 1, \ldots, N_e$$

$$(60)$$

where f, g_j, and h_j are the objective and constraint functions for the optimization, g_{1e} and g_{2e} are local element constraints which must be satisfied by the set of moment variables associated with each finite element, and N_e denotes the total number of finite elements in the discretized structure. Comparing with the layout optimization problem formulated in Sub-section 2.4.1, we see that the optimization problem just stated, in general is much larger both in the number of design variables and in the number of constraint equations. It should be noted that the solution to the optimization problem just stated directly yields the optimum material densities, and hence the global distribution of material over the structure. In addition, the solution contains the optimum values of the moment variables. These , however , do not give us the optimal layer densities and layer directions directly. A method for finding the local microstructure that corresponds to a given moment combination can be found in e.g. Lipton[11], Diaz, Lipton, & Soto[12], Soto[31].

3   Structural Design Criteria

This section presents the structural design criterion or objective functions which are used in the definition and solution of single- and multicriterion layout optimization problems in the following sections, namely integral structural stiffness (inverse of the total elastic energy in the structure) and eigenfrequencies of structural vibrations. An account of the design sensitivity analysis of these criterion functions is also given. Since multiple eigenvalues very often occur as a result of optimization, the eigenvalue sensitivity analysis is not only presented for simple eigenvalues, but also carried out for the much more complex case of multiple eigenvalues.

3.1　Structural Stiffness

In design for optimal stiffness of statically loaded linearly elastic structures we generally seek to minimize the displacements caused by a set of loads acting on the structure. In the case of a single point load we usually minimize the magnitude of the displacement at the location of the applied load. In the case of a distributed loading we generally need some integral measure for the structural stiffness which tends to control the size of the displacements at all points acted upon by the distributed loading. An applicable measure for the structural stiffness against a set of distributed surface tractions and body forces is therefore given by the so-called compliance L, defined as the work W done by the applied loads at the equilibrium state of the structure, i.e.,

$$L(\boldsymbol{u}^*(\boldsymbol{x})) = W(\boldsymbol{u}^*(\boldsymbol{x})) = \int_\Omega (\boldsymbol{u}^*(\boldsymbol{x}))^T \boldsymbol{F}_v(\boldsymbol{x})dV + \int_\Gamma (\boldsymbol{u}^*(\boldsymbol{x}))^T \boldsymbol{F}_s(\boldsymbol{x})dS \tag{61}$$

Here $\boldsymbol{x}$ represents an arbitrary vector of spatial coordinates within the structural domain Ω, whose surface is denoted by Γ. $\boldsymbol{F}_v(\boldsymbol{x})$ and $\boldsymbol{F}_s(\boldsymbol{x})$ are volume forces and surface traction vector functions, and $\boldsymbol{u}^*(\boldsymbol{x})$ is the displacement vector function for the equilibrium state of the structure. Hence, by minimizing the compliance of the structure we should minimize the displacements, and thus maximize the structural stiffness. Note that the evaluation of the compliance, as defined in Eq(61), requires that the static equilibrium problem for the structure is solved in advance. Let us therefore consider the derivation of the basic finite element equations needed for a static analysis. The starting point for these derivations is the expression for the total potential energy Π of a linear elastic structure, defined as the elastic strain energy U minus the work W done by the applied body forces and surface tractions, i.e.,

$$\Pi = U - W = \int_\Omega \varepsilon^T(\boldsymbol{x})\boldsymbol{E}(\boldsymbol{x})\varepsilon(\boldsymbol{x})dV - \int_\Omega u^T(\boldsymbol{x})\boldsymbol{F}_v(\boldsymbol{x})dV - \int_\Gamma u^T(\boldsymbol{x})\boldsymbol{F}_s(\boldsymbol{x})dS \tag{62}$$

Here $\boldsymbol{\varepsilon}(\boldsymbol{x})$ is the strain vector function, $\boldsymbol{u}(\boldsymbol{x})$ is an admissible displacement vector function, and $\boldsymbol{E}(\boldsymbol{x})$ is the material stiffness matrix which is assumed to be variable over the domain Ω of the structure. Next, we discretize the structure into N_e finite elements, and we select an admissible displacement field described in a picewise fashion where the displacements within any element are interpolated from the nodal degrees of freedom of that element. Denoting the vector of nodal degrees of freedom and the interpolation function matrix for an element by $\boldsymbol{d}_e$ and $\boldsymbol{I}_e(\boldsymbol{x})$, respectively, the element displacement field vector $\boldsymbol{u}_e(\boldsymbol{x})$ is obtained as

$$\boldsymbol{u}_e(\boldsymbol{x}) = \boldsymbol{I}_e(\boldsymbol{x})\boldsymbol{d}_e \tag{63}$$

and the corresponding element strain field vector $\varepsilon_e(\boldsymbol{x})$ is readily determined by differentiation. Thus, taking ∂ as a differential operator matrix, we get

$$\varepsilon_e(\boldsymbol{x}) = \partial \boldsymbol{u}_e(\boldsymbol{x}) \quad which\ yields \quad \varepsilon_e(\boldsymbol{x}) = \boldsymbol{B}_e(\boldsymbol{x})\boldsymbol{d}_e \ , \quad where \quad \boldsymbol{B}_e(\boldsymbol{x}) = \partial \boldsymbol{I}_e(\boldsymbol{x}) \tag{64}$$

Substitution of the expressions for $\boldsymbol{u}_e(\boldsymbol{x})$ and $\varepsilon_e(\boldsymbol{x})$ into Eq(62) yields

$$\Pi = \sum_{e=1}^{N_e} \boldsymbol{d}_e^T \boldsymbol{k}_e \boldsymbol{d}_e - \sum_{e=1}^{N_e} \boldsymbol{d}_e^T \boldsymbol{r}_e \tag{65}$$

where $\boldsymbol{k}_e$ and $\boldsymbol{r}_e$, defined below, are element stiffness matrices and nodal load vectors, respectively.

$$\boldsymbol{k}_e = \int_{\Omega_e} \boldsymbol{B}_e^T(\boldsymbol{x})\boldsymbol{E}_e(\boldsymbol{x})\boldsymbol{B}_e(\boldsymbol{x})dV \tag{66}$$

$$\boldsymbol{r}_e = \int_{\Omega_e} \boldsymbol{I}_e^T(\boldsymbol{x})\boldsymbol{F}_v(\boldsymbol{x})dV + \int_{\Gamma_e} \boldsymbol{I}_e^T(\boldsymbol{x})\boldsymbol{F}_s(\boldsymbol{x})dS \tag{67}$$

Finally, by performing a usual assembly of local element vectors and matrices into global ones, i.e.,

$$\boldsymbol{D} \stackrel{A}{=} \sum_{e=1}^{N_e} \boldsymbol{d}_e \ ; \quad \boldsymbol{R} \stackrel{A}{=} \sum_{e=1}^{N_e} \boldsymbol{r}_e \ ; \quad \boldsymbol{K} \stackrel{A}{=} \sum_{e=1}^{N_e} \boldsymbol{k}_e \tag{68}$$

we get the following expression for the total potential energy Π,

$$\Pi = U - W = \tfrac{1}{2}\boldsymbol{D}^T \boldsymbol{K} \boldsymbol{D} - \boldsymbol{D}^T \boldsymbol{R} \tag{69}$$

and by use of the principle of minimum total potential energy, requiring stationarity of Π with respect to all degrees of freedom in the global displacement vector $\boldsymbol{D}$, we get the well-known expression for the static equilibrium of a linearly elastic structure

$$\frac{\partial \Pi}{\partial \boldsymbol{D}} = 0 \quad which\ yields \quad \boldsymbol{K}\boldsymbol{D} = \boldsymbol{R} \tag{70}$$

Let us now return to the stiffness design problem considered in the beginning of this section and restate the expression for the compliance of a structure given in Eq(61) using finite element notation. Thus

$$L(\boldsymbol{D}) = W(\boldsymbol{D}) = \boldsymbol{D}^T \boldsymbol{R} \tag{71}$$

where $\boldsymbol{D}$ is the global displacement vector for the equilibrium state. Denoting the vector of generalized design variables by $\boldsymbol{a} = \{a_1, \ldots, a_I\}$, we may now state the minimum compliance/maximum stiffness design problem as

$$Objective: \quad \min_{\boldsymbol{a}} [\ L(\boldsymbol{D})\ ]$$

$$\tag{72}$$

$$Subject\ to: \quad \boldsymbol{K}\boldsymbol{D} = \boldsymbol{R}$$

Two alternative formulations of this problem are easily obtained by the use of Clayperon's work theorem which gives us the following relations between the total potential energy, the total elastic strain energy, and the work performed by the applied loads at the equilibrium state of the structure,

$$\Pi = U - W = -\frac{1}{2}W \quad \Leftrightarrow \quad W = 2U \tag{73}$$

From the relations in Eq(73) and the identity between the compliance and the work done by the applied loads at equilibrium, see Eq(71), we firstly obtain a formulation of the stiffness design problem stated as a minimization problem for the elastic strain energy, i.e.,

$$Objective: \quad \min_{a} \left[\, U(\boldsymbol{D}) \,\right]$$

$$\tag{74}$$

$$Subject\ to: \quad \boldsymbol{KD} = \boldsymbol{R}$$

and secondly, by applying the principle of minimum total potential energy to enforce equilibrium, we get the following formulation of the stiffness design problem stated as a max-min problem for the total potential energy.

$$Objective: \quad \max_{a}\ \min_{\boldsymbol{D}}\ \left[\, \Pi(\boldsymbol{D}) \,\right] \tag{75}$$

Hence, Eq(72), Eq(74) and Eq(75) actually give us three alternative formulations of the stiffness design problem, and thus we are free to choose which criterion to use for the structural stiffness. However, in the remainder of this thesis we shall apply the elastic strain energy as the only measure for the structural stiffness. Note that the total elastic strain energy for a finite element discretized linearly elastic structure may be calculated in a number of different ways, i.e.,

$$U = \frac{1}{2}\boldsymbol{D}^T\boldsymbol{KD} = \sum_{e=1}^{N_e}\frac{1}{2}\boldsymbol{d}_e^T\boldsymbol{k}_e\boldsymbol{d}_e = \sum_{e=1}^{N_e}\frac{1}{2}\int_{\Omega_e}\boldsymbol{\varepsilon}_e^T(\boldsymbol{x})\boldsymbol{E}_e(\boldsymbol{x})\boldsymbol{\varepsilon}_e(\boldsymbol{x})dV = \sum_{e=1}^{N_e}\frac{1}{2}\int_{\Omega_e}\boldsymbol{\sigma}_e^T(\boldsymbol{x})\boldsymbol{C}_e(\boldsymbol{x})\boldsymbol{\sigma}_e(\boldsymbol{x})dV$$

$$\tag{76}$$

where $\boldsymbol{\sigma}_e(\boldsymbol{x})$ is the element stress field vector, and $\boldsymbol{C}_e(\boldsymbol{x}) = \boldsymbol{E}_e^{-1}(\boldsymbol{x})$ is the element material compliance matrix. Also, it should be noted that the last equality, in general, only holds due to the identity between the strain energy and its complementary energy in linear elasticity.

3.1.1 Design Sensitivities of the Elastic Strain Energy

We shall here consider the derivation of a set of finite element based expressions for the design sensitivities of the total elastic strain energy in statically loaded, linearly elastic structures with respect to a set of generalized design variables a_i, $i = 1,\dots,I$. Using a direct approach we may derive expressions for the design sensitivities by partial differentiation of any of the expressions for the elastic strain energy given in Eq(76) with respect to the design variables. Thus, taking the first expression in Eq(76) as a basis for the design sensitivity analysis, we may derive the following expression for the design sensitivities of the elastic strain energy,

$$\frac{\partial U}{\partial a_i} = \frac{1}{2}\left(\frac{\partial \boldsymbol{D}^T}{\partial a_i}\boldsymbol{KD} + \boldsymbol{D}^T\frac{\partial \boldsymbol{K}}{\partial a_i}\boldsymbol{D} + \boldsymbol{D}^T\boldsymbol{K}\frac{\partial \boldsymbol{D}}{\partial a_i}\right), \quad i = 1,\dots,I \tag{77}$$

and taking advantage of the symmetry of the stiffness matrix, whereby

$$\frac{\partial \boldsymbol{D}^T}{\partial a_i}(\boldsymbol{K}\boldsymbol{D}) = (\boldsymbol{K}\boldsymbol{D})^T \frac{\partial \boldsymbol{D}}{\partial a_i} = \boldsymbol{D}^T \boldsymbol{K}^T \frac{\partial \boldsymbol{D}}{\partial a_i} = \boldsymbol{D}^T \boldsymbol{K} \frac{\partial \boldsymbol{D}}{\partial a_i} \tag{78}$$

we get

$$\frac{\partial U}{\partial a_i} = \frac{1}{2}\boldsymbol{D}^T \frac{\partial \boldsymbol{K}}{\partial a_i}\boldsymbol{D} + \boldsymbol{D}^T \boldsymbol{K} \frac{\partial \boldsymbol{D}}{\partial a_i} \quad , \quad i = 1\ldots, I \tag{79}$$

This expression for the design sensitivities of the elastic strain energy may be further simplified if the applied static loads are independent of design. In this case differentiation of the static equilibrium equation in Eq(70) yields

$$\frac{\partial \boldsymbol{K}}{\partial a_i}\boldsymbol{D} + \boldsymbol{K}\frac{\partial \boldsymbol{D}}{\partial a_i} = \frac{\partial \boldsymbol{R}}{\partial a_i} = 0$$
$$\Updownarrow \tag{80}$$
$$\boldsymbol{K}\frac{\partial \boldsymbol{D}}{\partial a_i} = -\frac{\partial \boldsymbol{K}}{\partial a_i}\boldsymbol{D}$$

Using this relation, the design sensitivity expression in Eq(79) simplifies to

$$\frac{\partial U}{\partial a_i} = -\left(\frac{\partial U}{\partial a_i}\right)_{fixed\ disp} = -\frac{1}{2}\boldsymbol{D}^T \frac{\partial \boldsymbol{K}}{\partial a_i}\boldsymbol{D} = -\frac{1}{2}\sum_{e=1}^{N_e} \boldsymbol{d}_e^T \frac{\partial \boldsymbol{k}_e}{\partial a_i}\boldsymbol{d}_e \quad , \quad i = 1,\ldots, I \tag{81}$$

Hence, assuming *design independent loads*, the design sensitivities of the elastic strain energy in a statically loaded, linearly elastic structure is obtained as the negative of the gradients of the elastic strain energy calculated for a *fixed displacement field.* This rather surprising result was first presented in an early paper by Masur[42] (see also Pedersen[36]), while the derivation above is similar to one in Olhoff, Thomsen & Rasmussen[14], see also Thomsen[32].

A strain based formulation of the design sensitivity expression for the elastic strain energy derived above is readily achieved by substituting the expression for the element stiffness matrix given in Eq(66) into Eq(81), and making use of the finite element strain-displacement relation given in Eq(64). Thus

$$\frac{\partial U}{\partial a_i} = -\frac{1}{2}\sum_{e=1}^{N_e} \boldsymbol{d}_e^T \frac{\partial \boldsymbol{k}_e}{\partial a_i}\boldsymbol{d}_e$$

$$= -\frac{1}{2}\sum_{e=1}^{N_e} \boldsymbol{d}_e^T \frac{\partial}{\partial a_i}\left(\int_{\Omega_e} \boldsymbol{B}_e^T(\boldsymbol{x})\boldsymbol{E}_e(\boldsymbol{x})\boldsymbol{B}_e(\boldsymbol{x})dV\right)\boldsymbol{d}_e \tag{82}$$

$$= -\frac{1}{2}\sum_{e=1}^{N_e} \frac{\partial}{\partial a_i}\left(\int_{\Omega_e} \boldsymbol{\varepsilon}_e^T(\boldsymbol{x})\boldsymbol{E}_e(\boldsymbol{x})\boldsymbol{\varepsilon}_e(\boldsymbol{x})dV\right)_{fixed\ disp} \quad , \quad i = 1,\ldots, I$$

Note that this expression generally only holds for a fixed displacement field. However, if we furthermore assume that the design variables do not change the finite element mesh and recall that the finite element strain-displacement matrix $\boldsymbol{B}_e$, see Eq(64), is found by differentiation of the element shape function matrix $\boldsymbol{I}_e$ which solely depends on the type and the shape of the finite element, we obtain the following design sensitivity expression

$$\frac{\partial U}{\partial a_i} = -\left(\frac{\partial U}{\partial a_i}\right)_{\substack{fixed\ strains \\ fixed\ mesh}} = -\frac{1}{2}\sum_{e=1}^{N_e}\frac{\partial}{\partial a_i}\left(\int_{\Omega_e}\varepsilon_e^T(\boldsymbol{x})\boldsymbol{E}_e(\boldsymbol{x})\varepsilon_e(\boldsymbol{x})dV\right)_{\substack{fixed\ strains \\ fixed\ mesh}} \quad , \quad i=1,\dots,I$$

$$(83)$$

The design sensitivities of the elastic energy in a statically loaded, linearly elastic structure may accordingly also be calculated for a *fixed strain field*.

Finally, considering linear elasticity, in which we have correspondence between the strain energy density and the complementary energy density, the above expression is rewritten as

$$\frac{\partial U}{\partial a_i} = -\frac{1}{2}\sum_{e=1}^{N_e}\frac{\partial}{\partial a_i}\left(\int_{\Omega_e}\boldsymbol{\sigma}_e^T(\boldsymbol{x})\boldsymbol{C}_e(\boldsymbol{x})\boldsymbol{\sigma}_e(\boldsymbol{x})dV\right)_{\substack{fixed\ strains \\ fixed\ mesh}}$$

$$= -\frac{1}{2}\sum_{e=1}^{N_e}\int_{\Omega_e}\left(\boldsymbol{\sigma}_e^T(\boldsymbol{x})\frac{\partial\boldsymbol{C}_e(\boldsymbol{x})}{\partial a_i}\boldsymbol{\sigma}_e + 2\boldsymbol{\sigma}_e^T(\boldsymbol{x})\boldsymbol{C}_e(\boldsymbol{x})\frac{\partial\boldsymbol{\sigma}_e(\boldsymbol{x})}{\partial a_i}\right)_{\substack{fixed\ strains \\ fixed\ mesh}} dV \quad , \quad i=1,\dots,I$$

$$(84)$$

and considering the element strain-stress relation of the form,

$$\varepsilon_e(\boldsymbol{x}) = \boldsymbol{C}_e(\boldsymbol{x})\boldsymbol{\sigma}_e(\boldsymbol{x}) \tag{85}$$

and differentiating this relation with respect to a design variable a_i, assuming fixed strains, we get the following relation

$$\frac{\partial\varepsilon(\boldsymbol{x})}{\partial a_i} = \frac{\partial\boldsymbol{C}_e(\boldsymbol{x})}{\partial a_i}\boldsymbol{\sigma}_e(\boldsymbol{x}) + \boldsymbol{C}_e(\boldsymbol{x})\frac{\partial\boldsymbol{\sigma}_e(\boldsymbol{x})}{\partial a_i} = 0$$

$$\Updownarrow \tag{86}$$

$$\boldsymbol{C}_e(\boldsymbol{x})\frac{\partial\boldsymbol{\sigma}_e(\boldsymbol{x})}{\partial a_i} = -\frac{\partial\boldsymbol{C}_e(\boldsymbol{x})}{\partial a_i}\boldsymbol{\sigma}_e(\boldsymbol{x})$$

by which Eq(84) simplifies to

$$\frac{\partial U}{\partial a_i} = \left(\frac{\partial U}{\partial a_i}\right)_{\substack{fixed\ stresses \\ fixed\ mesh}} = \frac{1}{2}\sum_{e=1}^{N_e}\frac{\partial}{\partial a_i}\left(\int_{\Omega_e}\boldsymbol{\sigma}_e^T(\boldsymbol{x})\boldsymbol{C}_e(\boldsymbol{x})\boldsymbol{\sigma}_e(\boldsymbol{x})dV\right)_{\substack{fixed\ stresses \\ fixed\ mesh}} \quad , \quad i=1,\dots,I$$

$$(87)$$

Hence, the design sensitivities of the elastic energy in a statically loaded, linear elastic structure may also be calculated for a *fixed stress field*. It should be mentioned that the strain and stress based design sensitivity expressions, which have been derived here applying the finite element formulation, originally were derived by Pedersen[37],[38] using a continuum formulation.

3.1.2 Design Sensitivities of the Elastic Strain Energy in Layout Optimization

The design sensitivity expressions for the total elastic strain energy presented in the preceding sub-section were derived without reference to the actual type of design parametrization or type of structure. In the present sub-section we therefore give a short description of the pertinent way of calculating the design sensitivities of the total elastic strain energy with respect to the design variables of the layout optimization problems outlined in Sub-sections 2.4.1 and 2.4.2. As earlier we denote the set of generalized design variables by a_i, $i = 1, \ldots, I$. However, in the following it shall be utilized that each of the design variables a_i in the layout optimization problem, in general, affects the material properties in a single finite element only. Thus, taking the expression in Eq(81) as the basis for the sensitity analysis, we obtain

$$\frac{\partial U}{\partial a_{e_i}} = -\frac{1}{2}\sum_{i=1}^{N_e} d_e^T \frac{\partial k_e}{\partial a_e} d_e = -\frac{1}{2} d_e^T \frac{\partial k_e}{\partial a_i} d_e \ , \quad i = 1, \ldots, I \ , \quad e = 1, \ldots, N_e \tag{88}$$

where a_i is assumed to be a design variable associated with element number e only. Let us now introduce the expression for the local element stiffness matrix Eq(66) into this expression, i.e.,

$$\frac{\partial U}{\partial a_i} = -\frac{1}{2} d_e^T \int_{\Omega_e} B_e^T \frac{\partial E_e}{\partial a_i} B_e \, dV_e \, d_e \ , \quad i = 1, \ldots, I_e \ , \quad e = 1, \ldots, N_e \tag{89}$$

Here Ω_e and B_e are the element domain and the element strain-displacement matrix, both of which remain unchanged since the design variables a_i do not change the finite element mesh, while E_e in the layout optimization problem for disks, plates, and shells described in Sub-sections 2.4.1 and 2.4.2 are defined by the effective material stiffness matrices E_e^A, E_e^D and E_e^S for the membrane, bending, and transverse shear state, respectively,

$$Plane \ Disk : \qquad E_e(a) = E_e^A(a)$$

$$Mindlin \ Plate : \qquad E_e(a) = \begin{bmatrix} E_e^A(a) & 0 & 0 \\ 0 & E_e^D(a) & 0 \\ 0 & 0 & E_e^S(a) \end{bmatrix} \tag{90}$$

The components of these matrices are in general given as analytical functions of the design variables a_i, see Sub-section 2.4.1 and 2.4.1, and the derivatives of the constitutive material stiffness matrix in Eq(89) with respect to the design variables may therefore be easily calculated. Now, if one has direct access to the finite element code, the derivatives of the element stiffness matrix in Eq(89) can be simply obtained by calling the routine which sets up the element stiffness matrix with the derivatives of the material stiffnesses properties instead of the real stiffness properties, whereupon the design sensitivities are determined by evaluating the expression in Eq(88). This way of calculating the design sensitivities for the total elastic strain energy in a structure turns out to be very efficient and it is always consistent with the numerical integration scheme applied for the integration of the element stiffness matrix. The sensitivity analysis will therefore, within numerical accuracy, result in exact design sensitivities.

3.2   Eigenfrequencies of Vibration

Eigenfrequencies or eigenvalues (squared angular frequencies) of free undamped vibrations often appear in the objective function or in the constraint equations of a structural design optimization problem, and we shall therefore give an account of the analysis and the design sensitivity analysis of eigenvalues in the following. As is well known, eigenvalues of free undamped vibrations of linearly elastic, finite element discretized structures are determined by solution of a real, symmetric eigenvalue problem of the form

$$(\boldsymbol{K} - \lambda_n \boldsymbol{M})\boldsymbol{\Phi}_n = 0 \ , \quad n = 1, \ldots, N_{dof} \tag{91}$$

where $\boldsymbol{K}$ and $\boldsymbol{M}$ represent the global stiffness and mass matrices for the discretized structure, respectively, λ_n is the eigenvalue (of order n) of free vibrations, $\boldsymbol{\Phi}_n$ the corresponding eigenvector, and N_{dof} denotes the total number of degrees of freedom of the finite element model and thus the dimension of the eigenvalue problem. The global vectors and matrices in Eq(91) are, as always, formed by assembling local element vectors and matrices, i.e.

$$\boldsymbol{\Phi}_n \overset{A}{=} \sum_{e=1}^{N_e} \phi_{ne} \ ; \quad \boldsymbol{K} \overset{A}{=} \sum_{e=1}^{N_e} \boldsymbol{k}_e \ , \quad \boldsymbol{M} \overset{A}{=} \sum_{e=1}^{N_e} \boldsymbol{m}_e \tag{92}$$

Here N_e is the total number of finite elements in the discretized structure, ϕ_{ne} is an element displacement vector for the n'th eigenmode, and $\boldsymbol{k}_e$ and $\boldsymbol{m}_e$, defined below, are the element stiffness and mass matrices, respectively,

$$\boldsymbol{k}_e = \int_{\Omega_e} \boldsymbol{B}_e^T(\boldsymbol{x})\boldsymbol{E}_e(\boldsymbol{x})\boldsymbol{B}_e(\boldsymbol{x})dV \ ; \quad \boldsymbol{m}_e = \int_{\Omega_e} \varrho_e \boldsymbol{I}_e^T(\boldsymbol{x})\boldsymbol{I}_e(\boldsymbol{x})dV \tag{93}$$

As in the preceeding section, the matrices $\boldsymbol{B}_e(\boldsymbol{x})$, $\boldsymbol{E}_e(\boldsymbol{x})$ and $\boldsymbol{I}_e(\boldsymbol{x})$ are the element strain/displacement, material stiffness, and displacement interpolation matrices, while Ω_e and ϱ_e represent the domain of the element and the mass density of the element material, respectively. Now, the eigenvalue problem in Eq(91) possesses N_{dof} eigenvalues λ_n and corresponding eigenvectors $\boldsymbol{\Phi}_n$, and the eigenvalues, which are all real, can be ordered by magnitude as follows:

$$0 < \lambda_1 \leq \lambda_2 \leq \ldots \leq \lambda_j \leq \ldots \leq \lambda_{N_{dof}} \tag{94}$$

In the following it shall be assumed that the eigenvectors have been $\boldsymbol{M}$-*orthonormalized*, whereby the eigenvectors are also $\boldsymbol{K} - orthogonal$, and we have the following relations where δ_{ij} denotes Kronecker's delta.

$$\boldsymbol{\Phi}_i^T \boldsymbol{M} \boldsymbol{\Phi}_j = \delta_{ij} \ ; \quad \boldsymbol{\Phi}_i^T \boldsymbol{K} \boldsymbol{\Phi}_j = \lambda_j \delta_{ij} \ , \quad i, j = 1, \ldots, N_{dof} \tag{95}$$

3.2.1   Sensitivity Analysis of Simple Eigenvalues

For simple (distinct) eigenvalues $\lambda_{j-1} < \lambda_j < \lambda_{j+1}$ the design sensitivities with respect to a set of generalized design variables a_i, $i = 1, \ldots, I$, are obtained by partial differentiation of the eigenvalue problem in Eq(91), with respect to the design variables

$$\frac{\partial K}{\partial a_i}\boldsymbol{\Phi}_j + (K - \lambda_j M)\frac{\partial \boldsymbol{\Phi}_j}{\partial a_i} = \frac{\partial \lambda_j}{\partial a_i}M\boldsymbol{\Phi}_j + \lambda_j \frac{\partial M}{\partial a_i}\boldsymbol{\Phi}_j, \quad i = 1,\ldots,I \tag{96}$$

By premultiplying this equation by $\boldsymbol{\Phi}_j^T$ and making use of Eq(91), we obtain the following well-known expression for the design sensitivities of a simple eigenvalue, see e.g. Courant & Hilbert[43] and Wittrick[44],

$$\frac{\partial \lambda_j}{\partial a_i} = \frac{\boldsymbol{\Phi}_j^T\left[\dfrac{\partial K}{\partial a_i} - \lambda_j \dfrac{\partial M}{\partial a_i}\right]\boldsymbol{\Phi}_j}{\boldsymbol{\Phi}_j^T M \boldsymbol{\Phi}_j}, \quad i = 1,\ldots,I \tag{97}$$

where the denominator equals one, due to the $M - orthonormalization$ of the eigenvectors.

For the application of usual gradient based methods of optimization it is essential that all design criteria used in the formulation of the optimization problem admit linearization in the design variables. This is the case for simple eigenvalues, as they are differentiable with respect to design. Thus, if all design variables a_i are changed simultaneously then the increment of a simple eigenvalue λ_j is given by the scalar product

$$\Delta \lambda_j = \boldsymbol{\nabla}^T \lambda_j \boldsymbol{\Delta a} \tag{98}$$

where $\boldsymbol{\Delta a} = \{\Delta a_1, \Delta a_2, \ldots, \Delta a_I\}$ is the vector of changes of the design variables a_i, and $\boldsymbol{\nabla}\lambda_j$ is the vector of gradients of the eigenvalue λ_j with respect to the design variables a_i,

$$\boldsymbol{\nabla}\lambda_j = \left\{\boldsymbol{\Phi}_j^T\left[\frac{\partial K}{\partial a_1} - \lambda_j \frac{\partial M}{\partial a_1}\right]\boldsymbol{\Phi}_j, \ldots\ldots, \boldsymbol{\Phi}_s^T\left[\frac{\partial K}{\partial a_I} - \lambda_j \frac{\partial M}{\partial a_I}\right]\boldsymbol{\Phi}_j\right\}^T \tag{99}$$

3.2.2 Sensitivity analysis of multiple eigenvalues

Multiple (or repeated) eigenvalues are not uncommon for structures with symmetry and/or high degree of kinematic constraint. If, in particular, such structures are subjected to optimization with a large number of design variables, it is the rule rather than the exception that the very optimization process drives the structure towards possession of not only a higher number of multiple eigenvalues, but also increases the multiplicity of these. Thus the resulting optimum design may be associated with several clusters of multiple eigenvalues. This must be accounted for in the very formulation of the optimization problem as well as in the solution procedure because multiple eigenvalues generally are not differentiable with respect to design variables. The non-differentiability implies that the analysis of the optimization problem becomes more cumbersome, and that standard methods of solution cannot be applied without modification. The well-known fact that the eigenvalues associated with multiple eigenvalues are not unique, is the cause for the non-differentiability of the eigenvalues. However, work by Courant & Hilbert[43] and Lancaster[45] has provided a basis for calculation of design sensitivities of multiple eigenvalues.

With a view to present a structural design sensitivity analysis with clusters of multiple eigenvalues in mind, we consider the eigenvalues within a subspectrum of interest, that would normally include the fundamental eigenvalue. The eigenvalues of this subspectrum are numbered as

$$\widetilde{\lambda}_m = \lambda_j \ , \quad j = r_m, \ldots, R_m \ , \quad m = 1, \ldots, M \tag{100}$$

where M denotes the total number of simple eigenvalues and clusters of multiple eigenvalues in the subspectrum. Hence, $\widetilde{\lambda}_m$ is either the value of a simple eigenvalue $(r_m = R_m)$ or the common value of a set of multiple eigenvalues where index values r_m and R_m refer to the first and the last member of the multiple eigenvalue.

We now introduce linear combinations of eigenvectors

$$\widetilde{\boldsymbol{\Phi}}_j = \sum_{k=r_m}^{R_m} \beta_{jk} \boldsymbol{\Phi}_k \ , \quad j = r_m, \ldots, R_m \ , \quad m = 1, \ldots, M \tag{101}$$

which always, independently of the numerical values of the scalar coefficients β_{jk}, will satisfy the state equation for the free vibration problem. By use of the notation introduced in Eqs(100)-(101), the state equation in Eq(91) can be written in the alternative form

$$(\boldsymbol{K} - \lambda_j \boldsymbol{M})\widetilde{\boldsymbol{\Phi}}_j = 0 \ , \quad j = r_m, \ldots, R_m \ , \quad m = 1, \ldots, M \tag{102}$$

In the sequel we follow work by Seyranian, Lund & Olhoff[46] and Lund[47] and apply a perturbation technique to establish a design sensitivity expression in the form of a subeigenvalue problem, which by solution yields the design sensitivities of a set of multiple eigenvalues associated with a design change along a given direction in the design space. For this purpose, consider a design perturbation of the form

$$\boldsymbol{a} + \boldsymbol{\Delta a} = \boldsymbol{a} + \varepsilon e \tag{103}$$

where $\boldsymbol{a} = \{a_1, a_2, \ldots, a_I\}^T$ is the vector of current values of the design variables, $\boldsymbol{\Delta a} = \{\Delta a_1, \Delta a_2, \ldots, \Delta a_I\}^T$ is the vector of design perturbations, $e = \{e_1, e_2, \ldots, e_I\}^T$ is a unit vector defining the direction of perturbation in the design space, and ε is a small positive scalar which gives the magnitude of the perturbation in this direction. Due to the design perturbation in Eq(103) both the stiffness and mass matrix will, in general, be incremented and a linear expansion gives

$$\boldsymbol{K}(\boldsymbol{a} + \boldsymbol{\Delta a}) = \boldsymbol{K}(\boldsymbol{a}) + \varepsilon \sum_{i=1}^{I} \frac{\partial \boldsymbol{K}}{\partial a_i} e_i \ ; \quad \boldsymbol{M}(\boldsymbol{a} + \boldsymbol{\Delta a}) = \boldsymbol{M}(\boldsymbol{a}) + \varepsilon \sum_{i=1}^{I} \frac{\partial \boldsymbol{M}}{\partial a_i} e_i \tag{104}$$

Also the eigenvalues and eigenmodes will be perturbed, and an expansion of a set of multiple eigenvalues and corresponding eigenvectors, λ_j and $\widetilde{\boldsymbol{\Phi}}_j$, $j = r_m, \ldots, R_m$, for the perturbed design can be written as

$$\left. \begin{array}{l} \lambda_j(\boldsymbol{a} + \varepsilon e) = \widetilde{\lambda}_m + \varepsilon \mu_j(\boldsymbol{a}, e) \\[2ex] \widetilde{\boldsymbol{\Phi}}_j(\boldsymbol{a} + \varepsilon e) = \widetilde{\boldsymbol{\Phi}}_j + \varepsilon \nu_j(\boldsymbol{a}, e) \end{array} \right\} \ , \quad j = r_m, \ldots, R_m \tag{105}$$

where μ_j and $\boldsymbol{\nu}_j$ are unknown directional derivatives of the eigenvalue $\tilde{\lambda}_j$ and corresponding eigenmodes $\tilde{\boldsymbol{\Phi}}_j$, respectively. Note that the eigenvectors associated with the multiple eigenvalues, in order to cater for the non-uniqueness of the eigenvectors, have been written as a linear combination of the original eigenvectors in Eq(101). Let us now substitute the expressions given in Eqs(104)-(105) into the state equation for the unperturbed design Eq(102) to get a linear approximation to the state equation for the perturbed design,

$$\left(\left(\boldsymbol{K} + \varepsilon \sum_{i=1}^{I} \frac{\partial \boldsymbol{K}}{\partial a_i} e_i\right) - \left(\tilde{\lambda}_m + \varepsilon \mu_j\right)\left(\boldsymbol{M} + \varepsilon \sum_{i=1}^{I} \frac{\partial \boldsymbol{M}}{\partial a_i} e_i\right)\right)\left(\tilde{\boldsymbol{\Phi}}_j + \varepsilon \boldsymbol{\nu}_j\right) = 0 \quad , \quad j = r_m, \ldots, R_m \tag{106}$$

By disregarding higher order terms in ε and making use of Eq(102), Eq(106) simplifies to

$$\left(\sum_{i=1}^{I} \frac{\partial \boldsymbol{K}}{\partial a_i} e_i - \tilde{\lambda}_m \sum_{i=1}^{I} \frac{\partial \boldsymbol{M}}{\partial a_i} e_i - \mu_j \boldsymbol{M}\right)\tilde{\boldsymbol{\Phi}}_j + (\boldsymbol{K} - \tilde{\lambda}_m \boldsymbol{M})\boldsymbol{\nu}_j = 0 \quad , \quad j = r_m, \ldots, R_m \tag{107}$$

Applying this equation we shall now establish an expression which can be used to determine directional derivatives of multiple eigenvalues. We start by premultiplying Eq(107) by the eigenvectors $\boldsymbol{\Phi}_s^T$, $s = r_m, \ldots, R_m$, and obtain

$$\boldsymbol{\Phi}_s^T \left(\sum_{i=1}^{I} \frac{\partial \boldsymbol{K}}{\partial a_i} e_i - \tilde{\lambda}_m \sum_{i=1}^{I} \frac{\partial \boldsymbol{M}}{\partial a_i} e_i - \mu_j \boldsymbol{M}\right)\tilde{\boldsymbol{\Phi}}_j = 0 \quad , \quad s, j = r_m, \ldots, R_m \tag{108}$$

Here all terms of the form $\boldsymbol{\Phi}_s^T(\boldsymbol{K} - \tilde{\lambda}_m \boldsymbol{M})\boldsymbol{\nu}_j = \boldsymbol{\nu}_j^T(\boldsymbol{K} - \tilde{\lambda}_m \boldsymbol{M})\boldsymbol{\Phi}_s$ have dropped out due to the original state equation in Eq(91). Next, by substituting the expression for $\tilde{\boldsymbol{\Phi}}_j$ given in Eq(101) into Eq(108), and by making use of the $\boldsymbol{M} - orthonormalization$ of the eigenvectors we obtain the following system of linear algebraic equations for the unknown coefficients β_{jk} in the expansion $\tilde{\boldsymbol{\Phi}} - \sum_{k=r_m}^{R_m} \beta_{jk}\boldsymbol{\Phi}_k$ in Eq(101),

$$\sum_{k=r_m}^{R_m} \beta_{jk}\left(\boldsymbol{\Phi}_s^T\left(\sum_{i=1}^{I} \frac{\partial \boldsymbol{K}}{\partial a_i} e_i - \tilde{\lambda}_m \sum_{i=1}^{I} \frac{\partial \boldsymbol{M}}{\partial a_i} e_i\right)\boldsymbol{\Phi}_k - \mu_j \delta_{sk}\right) = 0 \quad , \quad s, j = r_m, \ldots, R_m \tag{109}$$

A non-trivial solution to this system of equations only exists when the determinant of the coefficient matrix for the system of equations vanishes, i.e., a necessary condition for a non-trivial solution is expressed by the characteristic equation

$$det\left[\boldsymbol{\Phi}_s^T\left(\sum_{i=1}^{I} \frac{\partial \boldsymbol{K}}{\partial a_i} e_i - \tilde{\lambda}_m \sum_{i=1}^{I} \frac{\partial \boldsymbol{M}}{\partial a_i} e_i\right)\boldsymbol{\Phi}_k - \mu \delta_{sk}\right] = 0 \quad , \quad s, k = r_m, \ldots, R_m \tag{110}$$

Introducing I-*dimensional* gradient vectors $\boldsymbol{f}_{sk}$

$$\mathbf{f}_{sk} = \left\{\boldsymbol{\Phi}_s^T\left[\frac{\partial \boldsymbol{K}}{\partial a_1} - \tilde{\lambda}_m \frac{\partial \boldsymbol{M}}{\partial a_1}\right]\boldsymbol{\Phi}_k, \ldots \ldots, \boldsymbol{\Phi}_s^T\left[\frac{\partial \boldsymbol{K}}{\partial a_I} - \tilde{\lambda}_m \frac{\partial \boldsymbol{M}}{\partial a_I}\right]\boldsymbol{\Phi}_k\right\}^T \tag{111}$$

the expression in Eq(110) takes the simple form

$$det\left[\mathbf{f}_{sk}^T e - \mu\delta_{sk}\right] = 0 \quad , \quad s, k = r_m, \ldots, R_m \tag{112}$$

This equation defines a subeigenvalue problem which, by solution, yields the directional derivatives of the multiple eigenvalue $\tilde{\lambda}_m$ associated with a simultaneous change of all design variables in the direction e in the design space. Thus, knowing the eigenvectors $\boldsymbol{\Phi}_j$, $j = r_m, \ldots, R_m$, and the derivatives of the K and M matrices, we can compute the gradient vectors $\boldsymbol{f}_{sk}$ from Eq(111) and obtain the directional derivatives μ corresponding to any perturbation direction e in the design space, by forming and solving the subeigenvalue problem in Eq(112). In this form Eq(112) was originally established by Bratus & Seyranian[48] and Seyranian[49]. Work on design sensitivity analysis for multiple eigenvalues can also be found in e.g. Haug & Rousselet[50], Masur[51],[52] and Haug, Choi & Komkov[53].

The subeigenvalue problem in Eq(112) may also be written in terms of absolute design changes and eigenvalue increments. Realizing from the foregoing that $\boldsymbol{\Delta}a = \varepsilon e$ and $\Delta\lambda_j = \varepsilon\mu_j$, we obtain the following equation which constitutes an algebraic subeigenvalue problem for increments $\Delta\lambda$ of the members λ_j, $j = r_m, \ldots, R_m$ of a multiple eigenvalue in terms of actual increments Δa_i of the design variables a_i, $i = 1, \ldots, I$.

$$det\left[\mathbf{f}_{sk}^T \boldsymbol{\Delta}a - \Delta\lambda\delta_{sk}\right] = 0 \quad , \quad s, k = r_m, \ldots, R_m \tag{113}$$

Note that the eigenvalue increments $\Delta\lambda$ in Eq(113) are themselves eigenvalues (like the directional derivatives μ in Eq(112)), and that they may be numbered as $\Delta\lambda_j$, $j = r_m, \ldots, R_m$. However, such a numbering will be arbitrary in general; due to the form of Eq(113) with $\boldsymbol{f}_{sk}$ defined by Eq(111), none of the increments $\Delta\lambda_j$ can be set into a one-to-one correspondence with any of the eigenmodes $\boldsymbol{\Phi}_j$, $j = r_m, \ldots, R_m$. Also, note that the increments $\Delta\lambda_j$, $j = r_m, \ldots, R_m$, of the multiple eigenvalue generally are non-linear functions of the design change $\boldsymbol{\Delta}a$. Thus, multiple eigenvalues do not admit a usual linearization as did the simple eigenvalues, and standard gradient based methods for optimization are therefore not directly applicable for the solution of optimization problems with multiple eigenvalues. However, if the vector $\boldsymbol{\Delta}a$ of design changes is chosen such that all off-diagonal terms in Eq(112) vanish,

$$\boldsymbol{f}_{sk}^T \boldsymbol{\Delta}a = 0 \quad , \quad s \neq k \ , \quad s, k = r_m, \ldots, R_m \tag{114}$$

then the linear increments $\Delta\lambda$ of the multiple eigenvalue become

$$\Delta\lambda_j = \boldsymbol{f}_{jj}^T \boldsymbol{\Delta}a \quad , \quad j = r_m, \ldots, R_m \tag{115}$$

where

$$\mathbf{f}_{jj} = \left\{ \boldsymbol{\Phi}_j^T \left[\frac{\partial K}{\partial a_1} - \lambda_j \frac{\partial M}{\partial a_1}\right] \boldsymbol{\Phi}_j, \ldots \ldots, \boldsymbol{\Phi}_j^T \left[\frac{\partial K}{\partial a_I} - \lambda_j \frac{\partial M}{\partial a_I}\right] \boldsymbol{\Phi}_j \right\}^T \tag{116}$$

By comparing Eqs(98)-(99) with Eqs(115)-(116), we now see that the mathematical expression for the gradient vector $\boldsymbol{f}_{jj}$ of a multiple eigenvalue is identical with the usual expression for the gradient vector $\boldsymbol{\nabla}\lambda_j$ of a simple eigenvalue, and thus we may write $\boldsymbol{f}_{jj} \equiv \boldsymbol{\nabla}\lambda_j$. We shall apply this simple idea in Section 4 and develop a simple formulation which implies that optimization problems with both simple and multiple eigenvalues can be solved using ordinary gradient based methods of optimization.

3.2.3   Design Sensitivities of Eigenvalues in Layout Optimization

In the preceding section we established a set of general expressions for the calculation of design sensitivities of both simple and multiple eigenvalues of free vibration with respect to arbitrary design variables. Based on these expressions we shall now briefly describe the calculation of design sensitivities of both simple and multiple eigenvalues of free vibration with respect to the design variables of the layout optimization problems described in Sub-sections 2.4.1 and 2.4.2.

Now, considering the eigenvalues within a subspectrum of interest, we shall again apply the notation in Eq(100), and refer to a set of all in all M simple and multiple eigenvalues as follows,

$$\widetilde{\lambda}_m = \lambda_j \quad , \quad j = r_m, \ldots, R_m \ , \quad m = 1, \ldots, M \tag{117}$$

Thus, $\widetilde{\lambda}_m$ is either the value of a simple eigenvalue ($r_m = R_m$) or the value of a multiple eigenvalue where the index values r_m and R_m refer to the first and the last member.

Applying the design sensitivity expressions derived in Sub-section 3.2.2, we may generally write the set of gradient vectors which needs to be calculated as follows

$$\boldsymbol{f}_{sk} = \left\{ \boldsymbol{\Phi}_s^T \left[\frac{\partial \boldsymbol{K}}{\partial a_1} - \widetilde{\lambda}_m \frac{\partial \boldsymbol{M}}{\partial a_1} \right] \boldsymbol{\Phi}_k, \ldots, \boldsymbol{\Phi}_s^T \left[\frac{\partial \boldsymbol{K}}{\partial a_I} - \widetilde{\lambda}_m \frac{\partial \boldsymbol{M}}{\partial a_I} \right] \boldsymbol{\Phi}_k \right\}^T , \quad s, k = r_m, \ldots, R_m \ , \quad m = 1, \ldots, M \tag{118}$$

Note that this expression, which is identical to the expression in Eq(111), inherently contains the expressions for the gradient vectors associated with both simple and multiple eigenvalues.

Now, considering the design sensitivity analysis with respect to the design variables of the layout optimization problems described in Section 2 and taking advantage of the localized influence of these variables, we obtain the following expression for the components of the gradient vectors $\boldsymbol{f}_{sk}$

$$\boldsymbol{\Phi}_s^T \left[\frac{\partial \boldsymbol{K}}{\partial a_i} - \lambda_j \frac{\partial \boldsymbol{M}}{\partial a_i} \right] \boldsymbol{\Phi}_k = \phi_{se}^T \left[\frac{\partial \boldsymbol{k}_e}{\partial a_i} - \widetilde{\lambda}_m \frac{\partial \boldsymbol{m}_e}{\partial a_i} \right] \phi_{ke} \ , \quad s, k = r_m, \ldots, R_m \ , \quad m = 1, \ldots, M \tag{119}$$

where a_i is assumed to be a design variable that is solely associated with finite element number e. Hence, the components of the gradient vectors $\boldsymbol{f}_{sk}$ may be actually calculated by a simple vector-matrix-vector multiplication between the precalculated element displacement vector and the derivatives of the element stiffness and mass matrices, i.e.,

$$\frac{\partial \boldsymbol{k}_e}{\partial a_i} = \int_{\Omega_e} \boldsymbol{B}_e^T \frac{\partial \boldsymbol{E}_e}{\partial a_i} \boldsymbol{B}_e d\Omega \ ; \quad \frac{\partial \boldsymbol{m}_e}{\partial a_i} = \int_{\Omega_e} \frac{\partial \varrho_e}{\partial a_i} \boldsymbol{I}_e^T \boldsymbol{I}_e d\Omega \tag{120}$$

The derivatives of these matrices are, like in the global stiffness problem, calculated by calling the routines which set up the local finite element stiffness and mass matrices and simply using as entries the derivatives of the (material) stiffness matrix and mass density for the applied microstructural material. These derivatives are easily obtained analytically and applying Eq(119) for the calculation of the components of the gradient vectors $\boldsymbol{f}_{sk}$, we should again, within numerical accuracy, obtain fully correct design sensitivities.

4   Optimum Topology and Layout Design Problems

Both the problem of finding the optimum topology of a disk, plate, or shell structure with possible interior holes and the problem of finding the optimum layout of a stiffener reinforcement for an existing disk, plate or shell structure, consists of finding the optimum spatial distribution of a given isotropic material within an admissible design domain $\Omega \in \Re^2$. In Section 2, two different design parametrizations based on layered microstructures with continously variable layer densities and layer orientations were considered. Firstly, a design parametrization for solution of the topology and reinforcement layout design problems was obtained by modelling the material within each of the elements of the finite element discretized structure as a material with an orthogonal layered microstructure of second rank. In this case the layout design problem is defined through the layer densities μ_{1e}, μ_{2e} and material orientations θ_e associated with the layered material within each finite element. These variables in turn define both the local geometry of the layered microstructure and the global distribution of material/reinforcement throughout the structure. A similar design parametrization, also applicable for solution of both the topology and reinforcement layout optimization problems, was described in Sub-section 2.4.2. Here, the design parametrization was based on a more general layered microstructure of arbitrary rank, and the layout design problem became defined in terms of element density and moment variables $\varrho_e, m_{1e}, \ldots, m_{4e}$, which, respectively, define the distribution of material/reinforcement throughout the structure and the composition of the layered microstucture within each finite element.

As noted in Section 3, it is important to realize that optimization problems formulated by adopting a design parametrization as described above, usually will result in very large scale optimization problems. As an example, a finite element model which is resonable for analysis may easily consist of, say, one thousand finite elements. Thus, by adopting a design parametrization based on the moment formulation described in Sub-section 2.4.2 we will obtain an optimization problem with five thousand design variables and two thousand constraint equations, and that is even before any of the structural design criteria have been included in the problem formulation. Optimization problems of this size are by no means trivial.

For the solution of optimum topology and layout problems in ODESSY it has been chosen mainly to base the optimization on analytical design sensitivity analysis and mathematical programming, and to develop a solution technique which is considered to be general enough to be used for the solution of topology and layout optimization problems with new types of behavioural or non-behavioural criteria as objectives and constraints.

In the following, we in Sub-section 4.1 describe several formulations and solution strategies implemented in ODESSY for the solution of single and multiple load case maximum stiffness layout design problems, and in Sub-section 4.2 describe an important new formulation and solution strategy implemented for solution of optimum eigenfrequency layout design problems that involve simple as well as multiple eigenfrequencies of vibration.

4.1 Stiffness Layout Design Problems

The majority of the layout optimization problems for disk, plate, and shell structures which has been treated in the literature up to now, are single load case maximum stiffness layout design problems, see e.g. Bendsøe & Kikuchi[3], Bendsøe[4], and Olhoff, Thomsen & Rasmussen[14] (plane disks); Suzuki & Kikuchi[15] (Kirchhoff plates and shells); Soto & Diaz[17],[16] (Mindlin plates and shells). An exhaustive description of the material distribution approach to stiffness layout optimization of continuum structures is available in Bendsøe[21]. In the rather few papers that deal with maximum stiffness layout design problems for disk, plate, and shell structures subjected to several independent sets of loads, the authors have, up to now, always avoided to deal directly with the genuine multiobjective problem, and rather considered a simpler substitute problem using a scalarized version of the objective function defined as a weighted average of the original objectives. This appoach was for example taken in Diaz & Bendsøe[13] where stiffness maximization of planar disk structures subjected to several independent loading cases was performed by minimizing a weighted sum of the compliances associated with each load case. Also Diaz, Lipton & Soto[12] and Soto[31] used this approach for multiple loading stiffness maximization problems for Mindlin plate and shell structures. Hence, up to now no attemt has been made to solve the directly relevant "worst case" type multiobjective problem of maximizing the minimum stiffnesses of a structure against several independent sets of loads. In ODESSY, we have implemented all three problem formulations, i.e., single load case as well as the weighted sum and the max-min formulation of the multiple load case problem, and it is the aim of this section to describe the different formulations and to outline the solution strategies implemented in order to solve the problems.

4.1.1 Single Load Case Problem

Single load case maximum stiffness layout design problems are stated in ODESSY as optimization problems for minimum total elastic energy U under static equilibrium and under a constraint for the available volume of material $\overline{V}$ for the structure, and may be solved by means of both a design parametrization where the material is modelled by a second-rank layered microstructure, see Sub-section 2.4.1, and a parametrization where the material is modelled by a layered microstructure of arbitrary rank (moment formulation), see Sub-section 2.4.2. By denoting the vector of design variables for the layout design problem by $\boldsymbol{a} = \{a_1, \ldots, a_I\}^T$, and the admissible set of design variables by $\boldsymbol{a}_{ad}$, the single load case maximum stiffness (minimum total elastic energy) layout design problem takes the following form

$$\begin{aligned} Objective: \quad &\underset{\boldsymbol{a} \in \boldsymbol{a}_{ad}}{minimize} \; [\; U(\boldsymbol{a}) \;] \\[1em] Subject\;to: \quad &\boldsymbol{K}(\boldsymbol{a})\boldsymbol{D}(\boldsymbol{a}) = \boldsymbol{R} \\[1em] &V(\boldsymbol{a}) \le \overline{V} \end{aligned} \tag{121}$$

Here, $\boldsymbol{K}(\boldsymbol{a})$ is the finite element stiffness matrix, $\boldsymbol{R}$ the finite element load vector, $\boldsymbol{D}(\boldsymbol{a})$ the displacement vector which is determined by solution of the static equlibrium equation, and $V(\boldsymbol{a})$ is the volume of material of the structure. The optimization problem in Eq(121) is generally solved iteratively by means of a method of mathematical programming and design sensitivity analysis. In each loop of redesign the equilibrium equation is solved as a first step, then the objective and constraint functions and their design sensitivities are evaluated, whereupon the optimization problem is stated in a linearized form which is then solved for improved values of the design variables by means of a method of mathematical programming. The linearized form of the problem in Eq(121) that is solved in ODESSY may be written as

$$\begin{aligned} Objective: \quad &\underset{\boldsymbol{\Delta} a}{minimize} \; \Big[\; U(\boldsymbol{a}) + \boldsymbol{\nabla}^T U(\boldsymbol{a})\boldsymbol{\Delta a} \;\Big] \\[1em] Subject\;to: \quad &V(\boldsymbol{a}) + \boldsymbol{\nabla}^T V(\boldsymbol{a})\boldsymbol{\Delta a} = \overline{V} \\[1em] &\boldsymbol{a} + \boldsymbol{\Delta a} \in \boldsymbol{a}_{ad} \end{aligned} \tag{122}$$

Here, $\boldsymbol{\Delta a}$ is a vector of improving design variable updates, $U(\boldsymbol{a})$ the total elastic energy for the current design, $V(\boldsymbol{a})$ the current volume of material of the structure, while $\boldsymbol{\nabla} U = \{\partial U/\partial a_1, \ldots, \partial U/\partial a_I\}^T$ is the vector of gradients of the total elastic energy, and $\boldsymbol{\nabla} V = \{\partial V/\partial a_1, \ldots, \partial V/\partial a_I\}^T$ the vector of gradients of the volume of material in the structure. The constraint $\boldsymbol{a} + \boldsymbol{\Delta a} \in \boldsymbol{a}_{ad}$ is to be understood to cover not only the trivial upper and lower bound constraints on the design variables, but for example also the set of non-trivial constraints which each set of moment variables $m_{1e}, \ldots, m_{4e}$ must satisfy when the moment formulation is used.

In a layout design problem parameterized by modelling the material within each of the elements as a second-rank layered microstructure, we use as design variables the layer densities μ_{1e}, μ_{2e} and the angle of rotation θ_e of the second-rank Mindlin plate microstructure described in Sub-section 2.4.1. In this formulation the finite element stiffness matrix $\boldsymbol{K}$ is defined via expressions for the effective stiffness properties of the second-rank Mindlin plate microstructure and standard rotation formulas for orthotropic materials, while the volume V of structural material is found by the summation

$$V = \sum_{e=1}^{N_e} v_e \tag{123}$$

where N_e denotes the number of finite elements and v_e is the volume of material in an element. For the different Mindlin plate microstructure configurations described in Sub-section 2.2, the volume v_e of material in an element is given by:

$$\begin{aligned}
Configuration\ a \quad &: \quad v_e = \varrho_e h_{2e} A_e \\
Configuration\ b \quad &: \quad v_e = (h_{1e} + \varrho_e (h_{2e} - h_{1e})) A_e \\
Configuration\ c,\ d \quad &: \quad v_e = ((h_{2e} - h_{1e}) + \varrho_e h_{1e}) A_e
\end{aligned} \tag{124}$$

Here A_e is element area, h_{1e} and h_{2e} are element plate thicknesses defined in Figure 4, while ϱ_e denotes the bulk density of layers in the second-rank microstructure and is given by

$$\varrho_e = \mu_{2e} + (1 - \mu_{2e})\mu_{1e} \tag{125}$$

An inherent difficulty associated with the solution of layout design problems parameterized by means of second-rank layered microstructures is related to the problem of finding the optimum orientations of the microstructures, see e.g. Pedersen[36],[37],[38]. It is well-known that local extrema of the total elastic energy exist with respect to orientation of an orthotropic material and usual gradient based methods of optimization therefore generally fail in determining optimum orientations of the material. Maximum stiffness layout design problems parameterized using second-rank layered microstructures are then normally solved by means of an iterative two level approach of redesign where in each loop we initially determine optimal orientations of the orthotropic material by means of an optimality criterion approach, whereupon the layer densities are improved applying an ordinary mathematical programming method as described above. This hierarchical redesign procedure has been also implemented in ODESSY.

In maximum stiffness layout design problems where the material is modelled by a layered microstructure of arbitrary rank, we use as design variables the bulk density of layers ϱ_e and the set of moment variables $m_{1e}, \ldots, m_{4e}$ associated with each of the finite elements in the discretized structure. It is a remarkable and most desirable feature of the parametrization in terms of moment variables that in a fixed stress field, the total elastic energy is convex in the set of moment variables, see Lipton[11] and Diaz, Lipton & Soto[12]. This feature rules out the possibility of local minima with respect to the local anisotropy of the layered microstructure, and the optimization problem may therefore be solved directly by applying a mathematical programming approach as described earlier. However, for numerical reasons it may still be advantageous to consider a hierarchical redesign procedure where we for example initially determine improved values of the moment variables $m_{1e}, \ldots, m_{4e}$ and hereafter new values of the density variables ϱ_e in a second level of optimization. Such a hierarchical redesign procedure which allows us to split an optimization problem into several smaller sub-problems has been also implemented in ODESSY for the solution of layout design problems by mathematical programming.

4.1.2   Multiple Load Case Problems - Weighted Sum Formulation

Two different formulations for solution of multiple load case maximum stiffness layout design problems for statically loaded structures have been implemented in ODESSY. A particularly simple formulation of the multiple load case design problem, which may be solved by means of both a parametrization based on a second-rank layered microstructure, see Sub-section 2.4.1, and a parametrization where the material is modelled by a layered microstructure of arbitrary rank (moment formulation), see Sub-section 2.4.2, is obtained by scalarizing the objective function as a weighted sum of the total elastic energies associated with each of the independent load cases. The multiple load case layout design problem is in this setting stated as a single objective minimization problem subject to static equlibrium for each load case and a constraint on the available volume of material $\overline{V}$ for the structure. Thus, by denoting the vector of design variables for the layout design problem as $a = \{a_1, \ldots, a_I\}^T$ and the admissible set of design variables by a_{ad}, the scalarized multiple load case problem takes the following form

$$
\begin{aligned}
Objective: \quad &\underset{a \,\in\, a_{ad}}{minimize} \quad \left[\sum_{j=1}^{J} w_j U_j(a) \right] \\[2mm]
Subject\ to: \quad &K(a)D_j(a) = R_j \ , \qquad\qquad j = 1, \ldots, J \\[2mm]
&V(a) \leq \overline{V}
\end{aligned}
\tag{126}
$$

where J is the number of independent load cases for the structure, and w_j and U_j are the weighting factor and the total elastic energy for the $j'th$ load case, respectively, $K(a)$ is the finite element stiffness matrix, and R_j and D_j are a design independent load vector and the corresponding displacement vector for the $j'th$ load case, while $V(a)$ is the total volume of material of the structure. The optimization problem in Eq(126) may, in a direct approach like for the single load case problem in Eq(121), be solved by means of a method of mathematical programming and design sensitivity analysis. The linearized form of the problem in Eq(126) which is solved in ODESSY, is

$$
\begin{aligned}
Objective: \quad &\underset{\Delta a}{minimize} \quad \left[\sum_{j=1}^{J} w_j U_j(a) + \sum_{j=1}^{J} w_j \nabla U_j^T(a)\Delta a \right] \\[2mm]
Subject\ to: \quad &V(a) + \nabla^T V(a)\Delta a \leq \overline{V} \\[2mm]
&a + \Delta a \in a_{ad}
\end{aligned}
\tag{127}
$$

Here, Δa is a vector of design variable updates, $\nabla U_j = \{\partial U_j/\partial a_1, \ldots, \partial U_j/\partial a_I\}^T$ the vector of gradients of the total elastic energy U_j for the $j'th$ load case and $\nabla V = \{\partial V/\partial a_1, \ldots, \partial V/\partial a_I\}^T$ the vector of gradients of the volume of material V in the structure. The constraint $a + \Delta a \in a_{ad}$ has the same interpretation as in the previous section.

Solution of the scalarized multiple load case stiffness maximization problem in Eq(126), when based on parametrizations where the material is modelled by second-rank layered microstructures, is associated with exactly the same difficulty as the solution of the corresponding single load case stiffness layout design problem considered in Section 4.1.1. Thus, local minima of the objective function (weighted sum of total elastic energies) exist with respect to orientation of the orthotropic material, and usual gradient based optimization methods normally fail in determining the optimum orientations of the material. The present optimization problem is therefore also usually solved iteratively by a two-level redesign procedure where we in each loop initally determine the optimum orientations by means of an optimality criterion based method followed by improvement of the layer density variables by means of analytical design sensitivity analysis and mathematical programming.

The optimization problem in Eq(126) may also be solved using a design parametrization involving layered microstructures of arbitrary rank (moment formulation), and again this design parametrization rules out the posibility of local minima of the objective function with respect to the local anisotropy of the layered microstructure, which is described by the set of moment variables $m_{1e}, \ldots, m_{4e}$. Thus, the optimization problem may be solved applying exactly the same solution procedures as used for the solution of the single load case problem.

4.1.3 Muliple Load Case Problems - Min-Max Formulation

The multiple load case maximum stiffness layout design problem will now be stated directly as a minimization problem for the maximum total elastic energy from among the different load cases under static equilibrim, and subject to a constraint on the available volume of structural material. Again, this problem may be parametrized on the basis of using either second-rank layered microstructures or layered microstructures of arbitrary rank (moment formulation), see Sub-sections 2.4.1 and 2.4.2. The multiple load case maximum stiffness layout design problem is now considered in the form

$$
\begin{aligned}
Objective: \quad & \underset{a \in a_{ad}}{minimize} \left[\underset{j=1,\ldots,J}{maximum} \left[U_j(a) \right] \right] , \\[2mm]
Subject\ to: \quad & \boldsymbol{K}(\boldsymbol{a})\boldsymbol{D}_j(\boldsymbol{a}) = \boldsymbol{R}_j , \quad j = 1,\ldots,J \\[2mm]
& V(\boldsymbol{a}) \leq \overline{V}
\end{aligned}
\tag{128}
$$

where J is the number of independent load cases for the structure, U_j the total elastic energy for the $j'th$ load case and other symbols are defined earlier. The non-differentiable min-max problem in Eq(128) can be readily cast in differentiable form by means of a so-called bound formulation as introduced by Bendsøe, Olhoff & Taylor[54]. The general idea behind such a formulation is to restate the min-max problem in an equivalent form where the non-differentiable objective function is substituted by a new differentiable objective function and a number of constraints. To this end, we introduce a new design variable β, which in a min-max formulation is used as an upper bound value for each of the functions comprised in the original min-max objective. The parameter β is at the same time taken to be the new objective function, and β is now minimized subject to the constraints of the original formulation as well as the set of constraints requiring that all functions included in the min-max objective must be less than or equal to β, i.e.,

$$
\begin{aligned}
Objective: \quad &\underset{\beta,\, a \in a_{ad}}{minimize} \quad [\ \beta\ ] \\[2ex]
Subject\ to: \quad &U_j(a) \leq \beta\ , \qquad\qquad j = 1,\ldots,J \\[1ex]
&K(a)D_j(a) = R_j\ , \qquad j = 1,\ldots,J \\[1ex]
&V(a) \leq \overline{V}
\end{aligned}
\tag{129}
$$

This formulation constitutes an equivalent, but differentiable statement of the initially non-differentiable problem in Eq(128), and the problem may be solved using ordinary mathematical programming methods and design sensitivity analysis. The linearized form of the optimization problem in Eq(130) implemented in ODESSY for solution by mathematical programming is as follows,

$$
\begin{aligned}
Objective: \quad &\underset{\beta,\, \Delta a}{minimize} \quad [\ \beta\ ] \\[2ex]
Subject\ to: \quad &U_j(a) + \nabla^T U_j(a)\Delta a \leq \beta\ , \qquad j = 1,\ldots,J \\[1ex]
&V(a) + \nabla^T V(a)\Delta a \leq \overline{V} \\[1ex]
&a + \Delta a \in a_{ad}
\end{aligned}
\tag{130}
$$

where all symbols have been defined earlier. Note that the constraint $a + \Delta a \in a_{ad}$ covers side constraints on all design variables as well as constraints on the moment variables $m_{1e}, \ldots, m_{4e}$ (see Sub-section 2.4.2) if the design parametrization used is based on the moment formulation.

If the present optimization problem is parametrized using second-rank layered microstructures, we will again encounter difficulties with local minima of the objective function with respect to the orientation of the orthotropic material, and usual gradient based optimization methods will in general fail in determining optimal orientations. In the previous two formulations this inherent difficulty was circumvented by adopting a two-level approach of redesign where the optimal orientations of the orthotropic second-rank microstructure were determined by means of an optimality criterion based approach. So far we have not been able to develop such an optimality criterion method for finding global optimal orientations of orthotropic materials in min-max formulated stiffness optimizations problems.

Hence, we only use the design parametrization where the material is modelled by a layered microstructure of arbitrary rank (moment formulation) for solution of min-max formulated multiple load case maximum stiffness layout design problems. Note again that this design parametrization yields a formulation which rules out the possibility of local minima with respect to the local anisotropy of the layered microstructure described by the set of moment variables $m_{1e}, \ldots, m_{4e}$. The optimization problem in Eq(128) may therefore also be solved by means of mathematical programming and design sensitivity analysis, and we should obtain global optimal solutions.

4.2 Eigenfrequency Maximization Problems

Optimization problems for disk, plate, and shell structures with the aim to determine the topology or layout of structures that maximizes natural frequencies of free vibrations represent another type of problem which has been treated quite recently by several authors. Diaz & Kikuchi[55] considered topology optimization problems for plane disks with the objective of maximizing a given frequency of vibration. In all the optimization problems addressed by Diaz & Kikuchi[55] only a single eigenfrequency was taken into account in the problem statement. Later, Ma, Kikuchi, Cheng & Hagiwara[56], Kikuchi, Cheng & Ma[57], and Soto & Diaz[18] presented different formulations for simultaneous maximization of several frequencies of free vibration. In all these papers, the multiobjective problem was considered in simplified form with a scalar objective defined as a weighted function of the eigenfrequencies. Hence, as was also the case for the multiple load case stiffness maximization problem, these authors did not attempt to solve the multiobjective optimization problem of direct practical significance, namely that of maximizing the smallest eigenfrequency from among a given set of eigenfrequencies. Moreover, problems associated with the non-differentiablity of possible multiple eigenvalues of free vibration were not taken into account in any of the works just cited. Only very recently, results have been presented for layout optimization where max-min formulations have been used, and where the possibility of multiple eigenvalues has been properly accounted for in the mathematical formulation of the optimization problem, see Krog & Olhoff[19],[58] (vibration problems) and Folgado, Rodrigues & Guedes[20] (stablility problems).

We shall now present a new general formulation for solution of eigenfrequency optimization problems which efficiently handles the difficulties associated with the non-differentiability of multiple eigenfrequencies. The formulation is particularly attractive as it allows us to solve eigenfrequency optimization problems with simple as well as multiple eigenfrequencies by means of ordinary procedures of mathematical programming and design sensitivity analysis.

4.2.1 Eigenfrequency Maximization, Max-Min Formulation

The maximum eigenfrequency layout design problems are implemented in ODESSY as problems of maximizing the minimum eigenfrequency from among a given set of frequencies subject to dynamic equilibrium, and subject to a constraint on the available volume $\overline{V}$ of structural material. In order to simplify the notation here, we shall only present the problem of maximizing the fundamental eigenfrequency of a structure. By denoting the vector of design variables for a layout design problem by $\boldsymbol{a} = \{a_1, \ldots, a_I\}^T$, and the admissible set of design variables by $\boldsymbol{a}_{ad}$, we have the following problem statement,

$$
\begin{aligned}
Objective: \quad &\underset{\boldsymbol{a} \in \boldsymbol{a}_{ad}}{maximize} \quad [\ \underset{j=1,\ldots,J}{minimum} \ [\ \lambda_j(\boldsymbol{a}) \ ] \ ] \\[2mm]
Subject\ to: \quad &(\boldsymbol{K}(\boldsymbol{a}) - \lambda_j(\boldsymbol{a})\boldsymbol{M}(\boldsymbol{a}))\,\boldsymbol{\Phi}_j(\boldsymbol{a}) = 0 \ , \qquad j = 1, \ldots, J \\[2mm]
&V(\boldsymbol{a}) \leq \overline{V}
\end{aligned}
\tag{131}
$$

Here $\lambda_j(\boldsymbol{a})$ denotes the eigenvalue of order j, $\boldsymbol{\Phi}_j(\boldsymbol{a})$ is the corresponding eigenvector, $\boldsymbol{K}(\boldsymbol{a})$ the finite element stiffness matrix, $\boldsymbol{M}(\boldsymbol{a})$ the mass matrix, and $V(\boldsymbol{a})$ is the total volume of structural material. The multiobjective optimization problem in Eq(131) is readily cast in equivalent scalar form by means of the bound formulation introduced by Bendsøe, Olhoff & Taylor[54] and described in Sub-section 4.1.3. Thus, we introduce a new design variable β which in an eigenfrequency maximization problem represents a variable lower bound on each of the eigenvalues in the set considered. This lower bound β, which we in turn adopt as the new objective function, is now maximized subject to the constraints that all the eigenvalues comprised in the original max-min objective be larger than or equal to β, i.e.,

$$
\begin{aligned}
Objective: \quad &\underset{\beta, \boldsymbol{a} \in \boldsymbol{a}_{ad}}{maximize} \quad [\ \beta \ ] \\[2mm]
Subject\ to: \quad &\lambda_j(\boldsymbol{a}) \geq \beta \ , \qquad\qquad\qquad\qquad j = 1, \ldots, J \\[2mm]
&(\boldsymbol{K}(\boldsymbol{a}) - \lambda_j(\boldsymbol{a})\boldsymbol{M}(\boldsymbol{a}))\,\boldsymbol{\Phi}_j(\boldsymbol{a}) = 0 \ , \qquad j = 1, \ldots, J \\[2mm]
&V(\boldsymbol{a}) \leq \overline{V}
\end{aligned}
\tag{132}
$$

If all the eigenvalues $\lambda_j, j = 1, \ldots, J$ were destinct the optimization problem in Eq(132) would now represent a differentiable problem which may be solved by an ordinary method of mathematical programming and design sensitivity analysis. However, possible multiple eigenvalues are not differentiable with respect to the design variables, and do not admit a usual linearization. As the max-min formulation of eigenfrequency problems considered here in fact prompts for the occurrence of multiple eigenvalues, the possibility of eigenvalues of this type must be properly accounted for in the mathematical formulation of the optimization problem. In order to do so, we recall from Sub-section 3.2.2 that the linear increments of a multiple eigenvalue $\tilde{\lambda}_m = \lambda_{r_m} = \ldots = \lambda_{R_m}$ associated with given design changes may be found as solutions to the following sub-eigenvalue problem for the eigenvalue increments $\Delta\lambda_j, \ j = r_m, \ldots, R_m$, where $\Delta\boldsymbol{a}$ is the given vector of design changes,

$$
det\left[\boldsymbol{f}_{sk}^T \Delta\boldsymbol{a} - \Delta\lambda\delta_{sk}\right] = 0 \ , \qquad s, k = r_m, \ldots, R_m
\tag{133}
$$

Here δ_{sk} denotes Kroneckers delta, and $\boldsymbol{f}_{sk}$ is a vector of gradients, see Eq(111). Furthermore, in Sub-section 3.2.2 it was realized that if the vector of design changes $\Delta\boldsymbol{a}$

is chosen such that all off-diagonal terms in Eq(133) vanish, i.e.,

$$f_{sk}^T \Delta a = 0 \ , \quad s \neq k \ , \quad s,k = r_m,\ldots,R_m \tag{134}$$

then the increments of the multiple eigenvalue will be simply given by the scalar product

$$\Delta \lambda_j = f_{jj}^T \Delta a \ , \quad j = r_m,\ldots,R_m \tag{135}$$

Finally, by comparison of the mathematical expression for the gradient f_{jj}, see Eq(116), with the usual expression in Eq(99) for the gradient $\nabla \lambda_j$ for a simple eigenvalue, we see that the two expressions are identical. This result has been implemented directly in a mathematical programming based procedure of redesign for solution of eigenvalue optimization problems with both simple and multiple eigenvalues. The general idea behind this algorithm is to enforce the search for an improving design change to directions where the conditions in Eq(135) is fulfilled for each set of multiple eigenvalues. Then the multiple eigenvalues may be treated as if they were simple, and the optimization problem may stated in a linear form which may be solved for improved values of the design variables by means of a usual mathematical programming method. In such an approach the linearized form of the optimization problem in Eq(132) takes the following form

$$
\begin{aligned}
Objective: \quad & \underset{\beta,\, \Delta a}{maximize} \ [\ \beta\ ] \\[2ex]
Subject\ to: \quad & \lambda_j(a) + f_{jj}^T(a)\Delta a \geq \beta \ , \quad j = r_m,\ldots,R_m \ , \quad m = 1,\ldots,M \\[1ex]
& f_{sk}^T(a)\Delta a = 0 \ , \quad\quad s \neq k \ , \ s,k = r_m,\ldots,R_m \ , \quad m = 1,\ldots,M \\[1ex]
& V(a) + \nabla^T V(a)\Delta a \leq \overline{V} \\[1ex]
& a + \Delta a \in a_{ad}
\end{aligned}
\tag{136}
$$

Here r_m and R_m denote indices of the first and the last member of a multiple eigenvalue (note that $r_m = R_m$ for a simple eigenvalue), M is the total number of simple and multiple eigenvalues considered, while $\nabla V = \{\partial V/\partial a_1,\ldots,\partial V/\partial a_I\}^T$ is the vector of gradients of the volume of structural material V, and again the constraint $a + \Delta a \in a_{ad}$ covers both the trivial upper and lower bound constraints on the design variables, and the non-trivial constraints which each set of moment variables $m_{1e},\ldots,m_{4e}$ must satisfy if a moment formulation is adopted, see Sub-section 2.4.2.

In relation to the optimization problem in Eq(136) it should be mentioned that the set of additional equality constraints Eq(134) that are imposed on the vector of design change in situations where multiple eigenvalues occur, may result in an over constrained optimization problem for which no feasible solutions exist. For each N-fold eigenvalue which is included in the optimization problem in Eq(136) the vector of design changes Δa is required to satisfy $N(N-1)/2$ additional equality constraints, and since each extra equality constraint reduces the number of free design variables by one, we must, if a feasible solution is to exist, require that the total number of design variables I satisfies the following inequality,

$$I > \sum_{m=1}^{M} N_m(N_m - 1)/2 \tag{137}$$

Here, the multiplicity N_m of each of the multiple eigenvalues are given by $N_m = R_m - r_m + 1$.

5 Examples of Layout Design Optimization

This section presentations a number of simple, but illustrative examples of the different types of topology and layout optimization problems which may be solved using the techniques developed in the foregoing. The section is divided into four sub-sections. In Sub-section 5.1 we study the effect of different formulations and design parametrizations for the solution of single and multiple load case maximum stiffness topology design problems for plane disk structures. In Sub-section 5.2 we then consider the effect of different configurations of the Mindlin plate microstructure on the performance and the design of stiffness optimized Mindlin plate structures subjected to a single set of static loads. In Sub-section 5.3 we then consider the solution of maximum eigenfrequency topology design problems for plane disk structures with multiple eigenfrequencies, and finally in Sub-section 5.4 we consider solution of maximum eigenfrequency layout design problems for Mindlin plates.

5.1 Effect of Formulation for Stiffness Maximization

The aim of the examples to be considered here is mainly to study the solution of single and multiple load case maximum stiffness layout design problems by means of the two design parametrizations described in Section 2, and using the different formulations and solution strategies described in Section 4.1. For this purpose, we solve a series of stiffness maximization problems for a plane disk structure applying different design parametrizations and applying different formulations of the optimization problem.

First Example Problem

In the first example we solve a simple single load case maximum stiffness layout design problem, using both the layered microstructure of second-rank and the layered microstructure of arbitrary rank. For this example we consider a rectangular design domain which is subjected to a single concentrated force and supported as shown in Figure 7. This design domain is discretized into a 60 x 20 mesh of eight-node isoparametric disk finite elements, and the material within each of the elements is modelled by either a second-rank layered microstructure as described in Section 2.4.1, or a layered microstructure of arbitrary rank as described in Section 2.4.2. In the optimization problem based on the second-rank layered microstructure we apply the material orientation

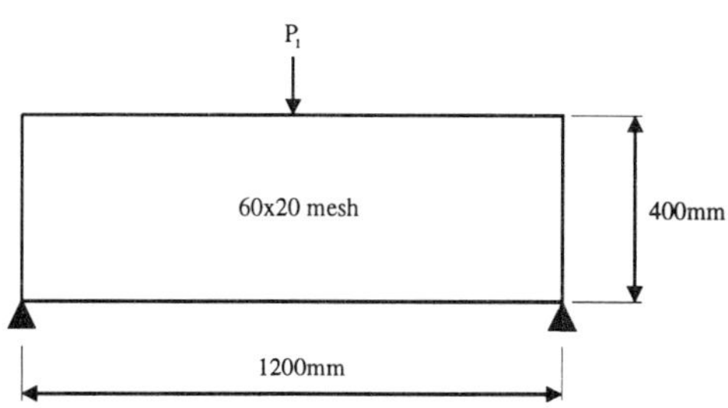

Figure 7: Design domain, load and support conditions for first example problem.

θ_e and the layer densities μ_{1e}, μ_{2e} for each of the elements as design variables, and in the optimization problem based on the more general layered microstructure of arbitrary rank we use the density variable ϱ_e and the set of moment variables $m_{1e}, \ldots, m_{4e}$ as design variables. The two stiffness maximization problems are formulated as minimization problems for the total elastic energy under a constraint on the available volume of material, and for the formulation based on the layered microstructure of arbitrary rank, we in addition have to consider two constraints for each set of moment variables. The optimization problem based on the second-rank microstructure has been solved

(a)

(b)

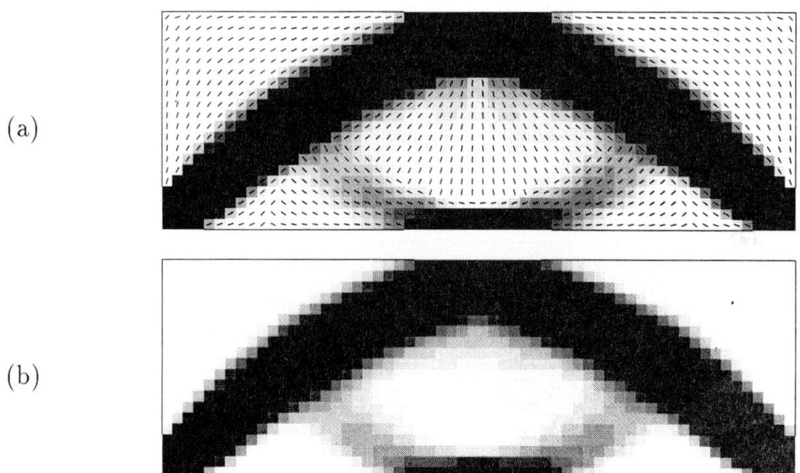

Figure 8: Single load case maximum stiffness designs. (a) Solution by second-rank layered microstructure. (b) Solution by layered microstructure of arbitrary rank. Black and white domains represent structure and void, respectively, while grey domains are composite. Figure (a) also shows principal material stiffness directions.

by an iterative two-level approach of redesign where the optimal values of the orientation variables are determined using the optimality criterion method mentioned in Section 4.1 and where the layer densities are updated by means of a mathematical programming approach. Contrary to this, the optimization problem based on the layered microstructure of arbitrary rank has been directly solved by means of a mathematical programming approach where all the design variables are updated simultaneously. Figure 8 shows the optimum designs obtained by the two approaches. In both cases the volume of material available for the design of the structure was taken to be 45% of the admissible design domain volume. The two optimum designs are found to have the same stiffness (total elastic energy) and are seen to be almost identical. This is not surprising, as the optimal microstructure for the solution of the present problem is a second-rank microstructure which is contained as a special case of the layered microstructure of arbitrary rank.

Second Example Problem

In the second example we shall use the traditional weighted sum formulation described in Sub-section 4.1.2, for the solution of a multiple load case stiffness maximization problem. For this purpose we consider the design of a plane disk structure subjected to five independent static load cases and supported as shown in Figure 9. As in the previous

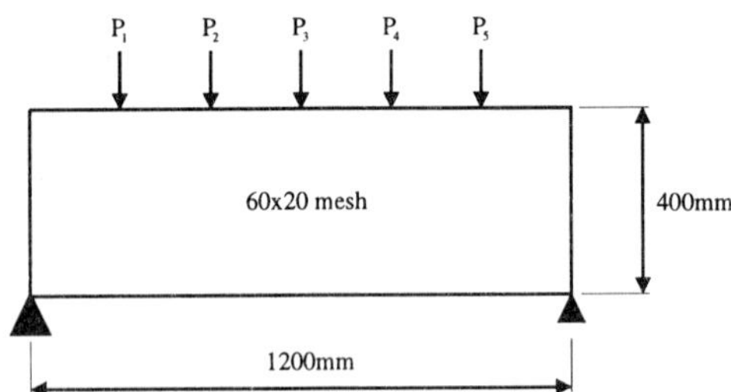

Figure 9: Plane disk subject to five independent inplane loading situations

example, we shall consider the solution of this problem by a design parametrization based on both the layered microstructure of second rank and the layered microstructure of arbitrary rank. The two multiobjective stiffness maximization problems are stated as minimization problems for a weighted sum of the total elastic energies associated with each of the five independent loads. A constraint is specified for the total available amount of material, and when using the formulation based on the layered microstructure of arbitrary rank we also need to consider two additional constraints for each set of moment variables. The two optimization problems have been solved by means of the same iterative redesign procedures as used in the preceding example. Figure 10 shows the optimal designs obtained by specifying equal weighting factors for the total elastic energy associated with each of the five load cases, and by specifying the total volume of material for the structural design to be equal to 45% of the design

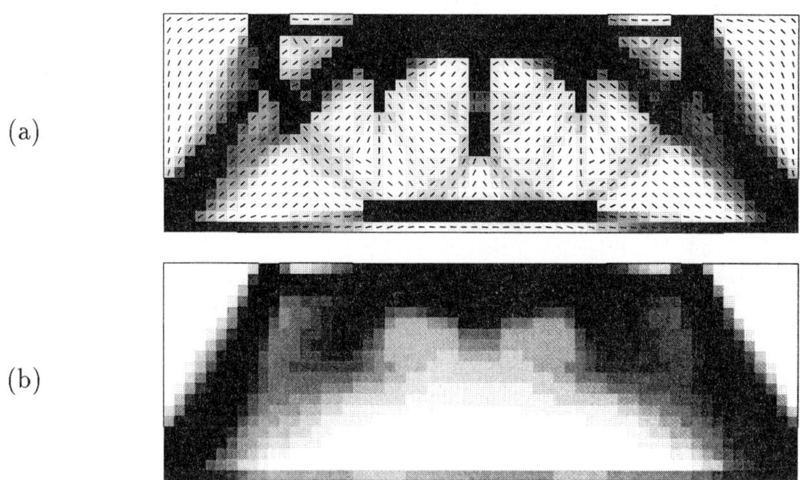

Figure 10: Multiple load case maximum stiffness designs obtained by minimizing a weighted sum of the total elastic energies associated with five given load cases. (a) Solution by second-rank layered microstructure, (b) Solution by layered microstructure of arbitrary rank. Black and white domains are structure and void, respectively, while grey domains are composite. Figure (a) also shows principal material stiffness directions.

domain volume. The solutions obtained by the two different design parametrizations are now remarkably different. We see that the solution obtained by means of the layered microstructure of arbitrary rank contains much more composite material than the solution obtained using the second-rank microstructure. Also, if we compare the values of the objective functions, i.e., the weighted sum of the total elastic energies for the five load cases, we find that the objective function for the design in Figure 10b is almost 5% lower than that of the design in Figure 10a. Again, this is no surprise, since the optimal microstructure for a multiple load case stiffness design problem is a third-rank layered microstructure with non-orthogonal layers. Figure 11 and Figure 12 show the iteration histories for the total elastic energies associated with the five load cases when using the second-rank layered microstructure and the layered microstructure of arbitrary rank, respectively. In both cases we observe a nice and stable convergence.

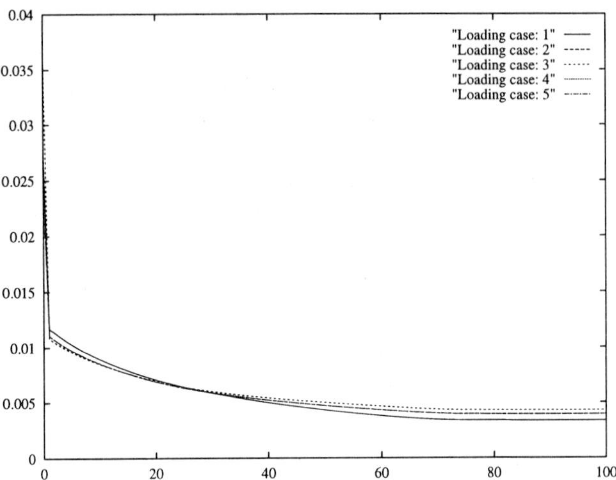

Figure 11: Iteration history for the total elastic energies associated with the five independent load cases obtained by minimization of a weighted sum of these total elastic energies. (Solution by second-rank layered microstructure)

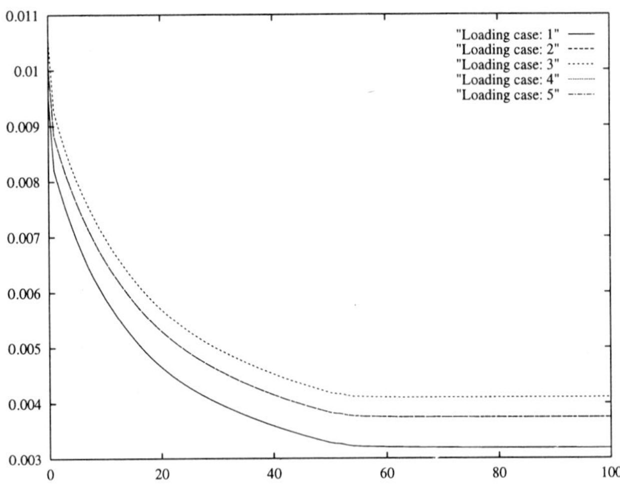

Figure 12: Iteration history for the total elastic energies associated with the five independent load cases obtained by minimization of a weighted sum of these total elastic energies. (Solution by layered microstructure of arbitrary rank)

Third Example Problem

Finally, as a last example of this sub-section we solve the same multiple load case problem as before, but now consider a min-max formulation of the problem. Recall that the min-max formulation for solution of multiple load case stiffness maximization problems covers minimization of the maximum total elastic strain energy from among each of the independent load cases. In Sub-section 4.1.3, this initially non-differentiable min-max problem was cast in differentiable form via a bound formulation, whereby the problem can be solved by means of standard mathematical programming methods. By applying such a formulation and adopting a design parametrization based on the layered microstructure of arbitrary rank, the multiple load case example problem considered earlier has here been solved by an iterative mathematical programming approach of redesign where all design variables are updated simultaneously.

Figure 13: Multiple load case stiffness optimum design obtained by minimizing the maximum total elastic energy associated with the five independent load cases for the structure. Black and white domains represent structure and void, respectively, while grey domains consist of composite. For the optimum design, the elastic strain energies associated with each of the five load cases are found to be equal to one another.

Figure 13 shows the optimum design obtained by solution of the min-max formulation of the problem. This solution is clearly seen to be different from the solution obtained for the traditional weighted sum formulation, and a comparison of the maximum total elastic energy for the design in Figure 13 relative to that of the design in Figure 10b shows nearly a 15% reduction. This result illustrates the superiority of the min-max problem formulation which may be conceived as "design for the worst case".

Figure 14 depicts the iteration history for the total elastic energies associated with the five independent load cases as obtained by solution of the min-max problem. Again, we observe a nice and stable convergence and it is worth noting that the min-max formulation results in assignment of equal amounts of energies to each of the five load cases.

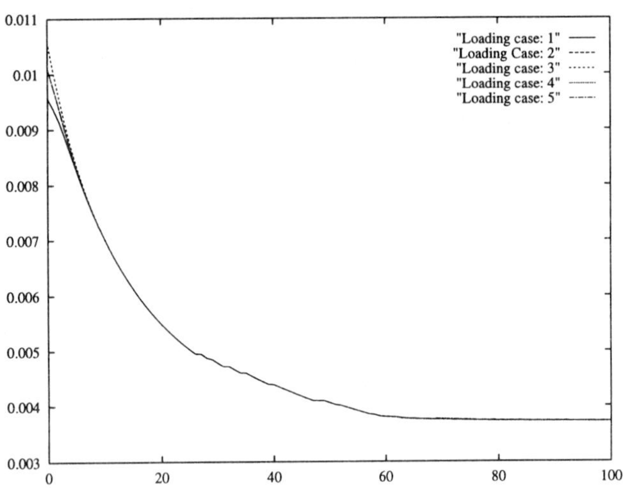

Figure 14: Iteration history for the total elastic energies associated with the five independent load case, obtained by minimizing the maximum of these energies.

5.2   Stiffness Layout Design Problems for Plates

The main purpose of the examples in this section is to study the effect of the type of plate section on the performance and on the optimal distribution of material/reinforcement in stiffness optimized Mindlin plates subjected to a single set of static loads. Hence, by application of the four different configurations of the Mindlin plate microstructure described in Sub-section 2.2, we shall consider in the following the solution of a series of single load case maximum stiffness layout design problems for perforated plates, rib stiffened plates and internally stiffened honeycomb and sandwich plates. The layout optimization problems are formulated and solved by adopting a design

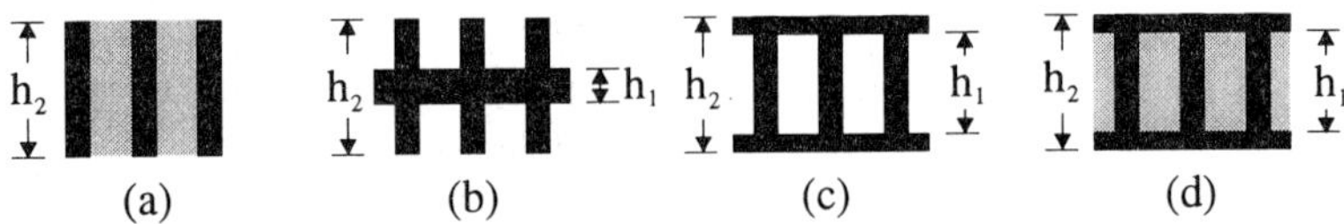

Figure 15: Mindlin plate microstructure configurations applied for layout optimization of: (a) Perforated plates: $h_2 = 10mm$. (b) Rib stiffened plates: $h_2 = 10mm$ and $h_1 = 2mm$. (c) Internally stiffened honeycomb plates: $h_2 = 10mm$ and $h_1 = 8mm$. (d) Internally stiffened sandwich plates: $h_2 = 10mm$ and $h_1 = 8mm$.

parametrization based on the second-rank Mindlin plate microstructure described in Sub-section 2.4.1, and Figure 15 defines the different sections for the different Mindlin plate microstructure configurations to be used in following. For all configurations in Figure 15, black domains represent a "stiff" isotropic material with a stiffness of 210MPa, while the grey domains in the perforated configuration represent an isotropic material with a very low stiffness (void) taken to be 2.1MPa, and the grey domains in the sandwich configuration represent an isotropic core material with a stiffness of 21MPa. All materials are assumed to have a Poisson's ratio of 0.3.

Two example problems are considered, a simply supported and a clamped quadratic plate under a single point load applied at the plate mid-point. Due to symmetry in terms of both the loading and the support conditions, only one quarter of the plate is considered, and Figure 16 shows the dimensions of the admissible design domain, and the load and support conditions for the simply supported and the clamped quadratic plate. The design domain is discretized into a 30 x 30 mesh of nine-node Mindlin plate

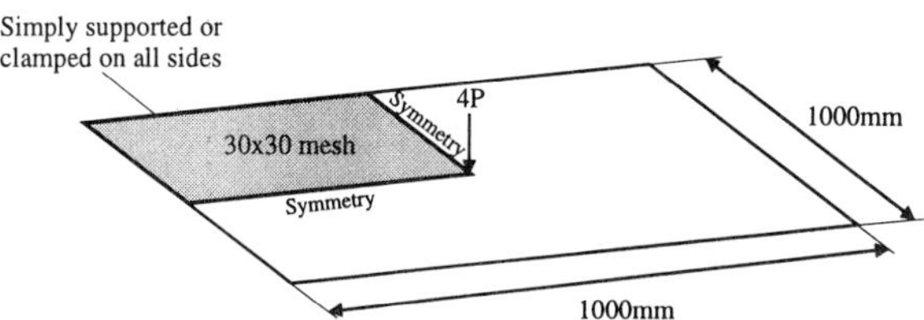

Figure 16: Admissible design domain, load and support conditions.

finite elements, the material within each finite element is modelled by a second-rank Mindlin plate microstructure with one of the plate configurations shown in Figure 15, and as design variables we apply the layer densities μ_{1e}, μ_{2e} and the orientation θ_e of the orthotropic microstructure in each of the finite elements. The layout optimization problems are formulated as problems of minimum total elastic strain energy, and in all examples we specify the total volume V of structural material to be equal to 50% of the admissible design domain volume, i.e., $V = 0.5$ ($10mm$ x $1000mm$ x $1000mm$). Also it should be mentioned that the optimization problems are solved by means of an iterative hierarchical procedure of redesign where the optimal orientations of the orthotropic material are determined using the optimality criterion approach mentioned in Sub-section 4.1 while the layer densities in the second rank microstructure are improved by means of mathematical programming and design sensitivity analysis.

Example with Simply Supported Plates

We start by considering the results for the simply supported plates. Figure 17 shows the optimal distributions of material/reinforcement obtained by using the different configurations of the Mindlin plate microstructure, while Figure 18 shows the corresponding iteration histories for the total elastic energy in the four simply supported plates. The optimal material/reinforcement distribution in the four different simply supported plates shown in Figure 17 are seen to be very similar. In all situations the

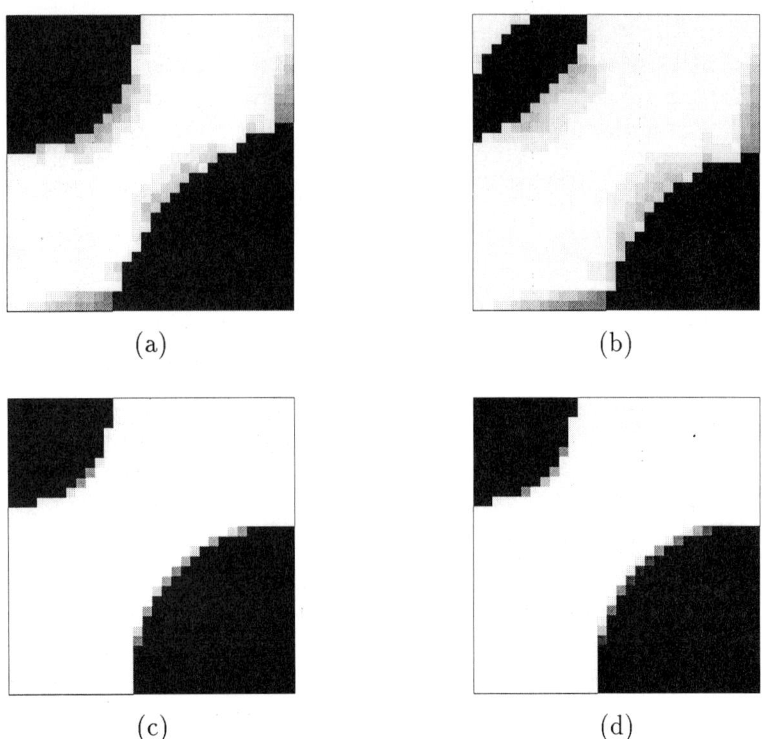

(a)   (b)

(c)   (d)

Figure 17: Stiffness optimal material/reinforcement distributions in simply supported quadratic plates with different sections. (a) Perforated plate. (b) Rib stiffened plate. (c) Internally stiffened honeycomb plate. (d) Internally stiffened sandwich plate. Only one quarter of the plates are shown, and in each figure the plate mid-point is located at the lower-right corner. In the pictures above, black areas are domains with full density of material/reinforcement, white areas are domains with zero density of material/reinforcement (void or base structure), while grey areas are domains with an intermediate density of material/reinforcement.

material/reinforcement is concentrated around the point force and around the corners of the simply supported plate. For the perforated plate shown in Figure 17(a) it should be emphasized that the white domains, which are associated with a very low stiffness, usually are taken to represent void. However in the present situation such an interpretation will not lead to a usable structural design, since the structure which is supposed to carry the load does not connect to the supports. This problem is not encountered in the solutions for the reinforcement problems shown in Figure 17(b)-(d), where the white domains have the stiffness of the base structure.

From the iteration histories in Figure 18 we see that the sandwich and honeycomb configurations yield the most rigid designs, then comes the perforated configuration and finally the rib stiffened configuration.   This is not surprising; in both the sandwich

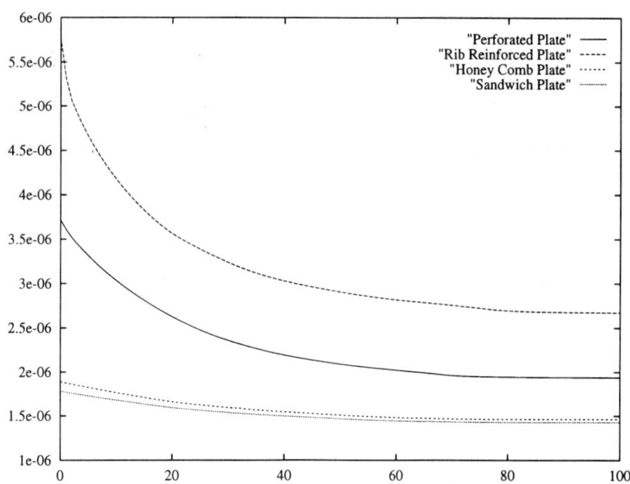

Figure 18: Iteration histories for the total elastic strain energies associated with the four different simply supported plates.

and the honeycomb configuration 40% of the material is used to form the two cover sheets which gives these plates a very high bending stiffness.   In the rib stiffened configuration the same 40% of material is used in a solid base plate with a rather low bending stiffness, and therefore the ribbed plate exhibits a rather poor performance. Finally, for the design of the perforated plate we can freely use all 100% of material to form the optimal design, and since no material is used to form a base structure with a quite low stiffness we obtain an intermediate performance. Also, it should be noted from Figure 18 that relative to the initial designs (with a uniform distribution of material/reinforcement), a large increase in stiffness is obtained for the perforated and the rib stiffened plates, while the gain in stiffness of the sandwich and the honeycomb plates is quite low relative to the a priori very efficient initial designs.

Example with Clamped Plates

Let us now consider the results obtained for the clamped plates. Figure 19 shows the optimum material/reinforcement distribution in the four clamped plates obtained by using the four different configurations of the Mindlin plate microstructure in Figure 15, and Figure 20 shows the corresponding iteration histories for the total elastic energies of the plates. The optimal material/reinforcement distributions for the four different configurations of the clamped plate are, as for the preceding simply supported plates, seen to be very similar.   However, in comparison with the simply supported plates

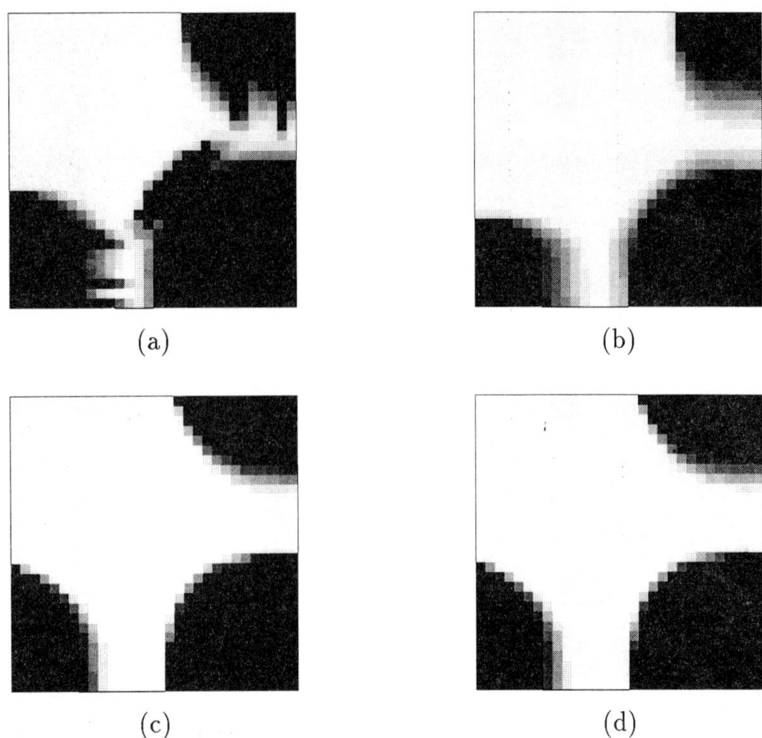

(a)   (b)

(c)   (d)

Figure 19: Stiffness optimal material/reinforcement distributions in clamped quadratic plates with different sections. (a) Perforated plate. (b) Rib stiffened plate. (c) Internally stiffened honeycomb plate. (d) Internally stiffened sandwich plate. Only one quarter of the plates are shown, and in each figure the plate mid-point is located at the lower-right corner. In the pictures above, black areas are domains with full density of material/reinforcement, white areas are domains with zero density of material/reinforcement (void or base structure), while grey areas are domains with an intermediate density of material/reinforcement.

the clamped edges may transfer a bending moment and this naturally causes change in design. For the perforated plate in Figure 19(a), it should again be noted that the usual interpretation of the very compliant material as void is not possible since such an interpretation yields a structure which does not connect to the supports.

The sandwich and honeycomb configurations are, like for the simply supported plates, seen to yield the stiffest designs, then comes the perforated configuration, and the rib stiffened configuration is the most compliant one. This is not surprising, for the reasons discussed earlier for simply supported plates. Again it is seen (Figure 20) that relative to the initial designs (with a uniform distribution of material/reinforcement),

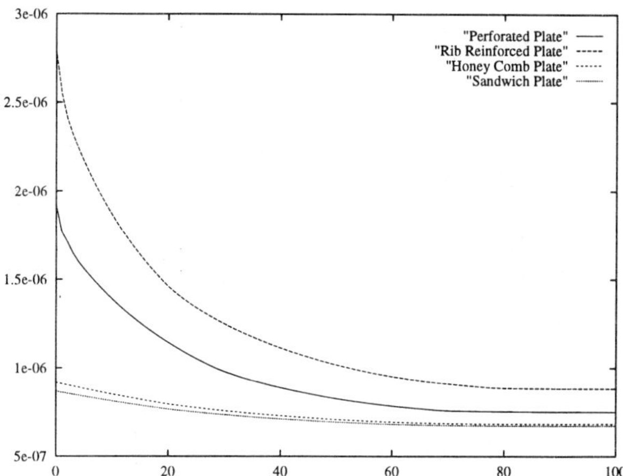

Figure 20: Iteration history for the total elastic strain energies of four different clamped plates.

a very small increase in stiffness is obtained for the sandwich and honeycomb plates, which get the major part of their bending stiffness from the two cover sheets, while a large increase in stiffness is obtained for both the perforated and rib stiffened plates. The examples considered here and in the previous section clearly illustrate that it is the correct choice of plate configuration rather than the optimal distribution of reinforcement that gives the primary contribution to the gain in stiffness.

5.3 Eigenfrequency Topology Design Problems for Plane Disks

We now consider the solution of maximum eigenfrequency topology design problems for planar disk structures, using a design parametrization where the material in each of the finite elements is modelled by a layered microstructure of arbitrary rank. In such a design parametrization for solution of topology design problems a very compliant material usually plays the role of "void". In the series of examples considered herein we shall study the effect of the stiffness of this compliant material on the solutions to maximum eigenfrequency layout design problems for plane disk structures. In all examples, the stiffness for the "stiff" material in the layered microstructure is taken to be $210MPa$, while the mass densities for the "stiff" and the "compliant" material is taken to be $7.8kg/m^3$ and zero, respectively. We now consider a rectangular admissible design domain which is clamped along two opposite boundaries and supported at the mid-point as illustrated in Figure 21, and in all examples the available amount of material for the structures to be designed is taken to be 40% of the admissible design domain volume. The design domain is discretized into a 20 x 100 mesh of isoparametric eight-node disk finite elements, and the design variables in the layout optimization

problems are the set of density and moment variables ϱ_e and $m_{1e}, \ldots, m_{4e}$ associated
with each of the finite elements of the discretized structure.

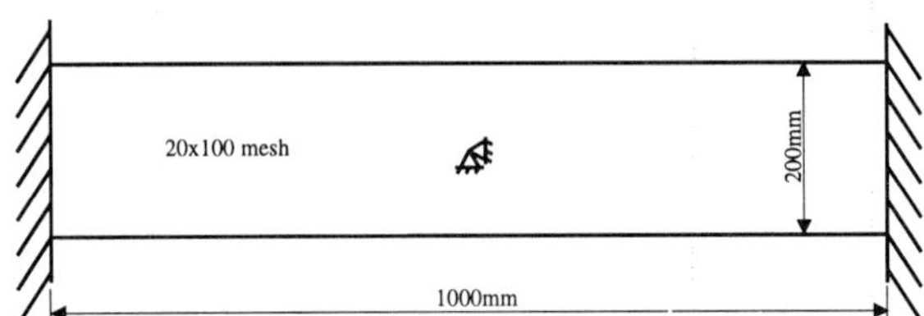

Figure 21: Admissible design domain and support conditions

The eigenfrequency optimization problems considered here are all stated as max-
min problems for the five lowest eigenfrequencies of the structure depicted in Figure
21, subject to an upper constraint on the available amount of stiff material, and in
the case of occurrence of multiple eigenfrequencies, also under a number of additional
equality constraints Eq(134) imposed in order to obtain a differentiable problem, see
Sub-section 4.2. Finally, since we apply arbitrarily high-rank microstructures, also the
four moment variables associated with each finite element in the discretized structure
must satisfy two simple inequality constraints. The max-min formulation for solution
of maximum eigenfrequency topology design problems parameterized by means of the
moment formulation is described in more detail in Sub-section 4.2.1, where the initially
non-differentiable max-min problem is cast in differentiable form via a bound formulation. The optimization problems considered here are all solved iteratively by means
of a two-level approach of redesign based on mathematical programming and analytical design sensitivity analysis. In each loop of redesign we first determine improving
updates of the moment variables and then solve for improving updates of the density
variables. This hierarchical redesign procedure appears to be necessary for the present
problem in order to obtain a numerically well-behaved problem which admits solution
by a sparse linear programming algorithm contained in ODESSY.

Figure 22 shows a series of eigenfrequency optimal topology designs found by the
use of different stiffnesses of the compliant material in the layered microstructure. The
five designs in Figure 22 are seen to be almost perfectly symmetric and very similar; in
all cases the major part of the material is concentrated in two solid cantilevers which
at their end points form a sort of hinge for a beam with a very low density of material
which connects to a solid lump of material concentrated around the simple support
at the center of the design domain. For the present series of problems, it must be
concluded that the finite stiffness of the compliant material does not appear to have a
significant influence on the optimal design.

It is a notable feature associated with the five solutions depicted in Figure 22
that the optimal designs associated with the three lowest stiffnesses of the compliant
material have a double fundamental eigenfrequency, while the optimal designs associ-

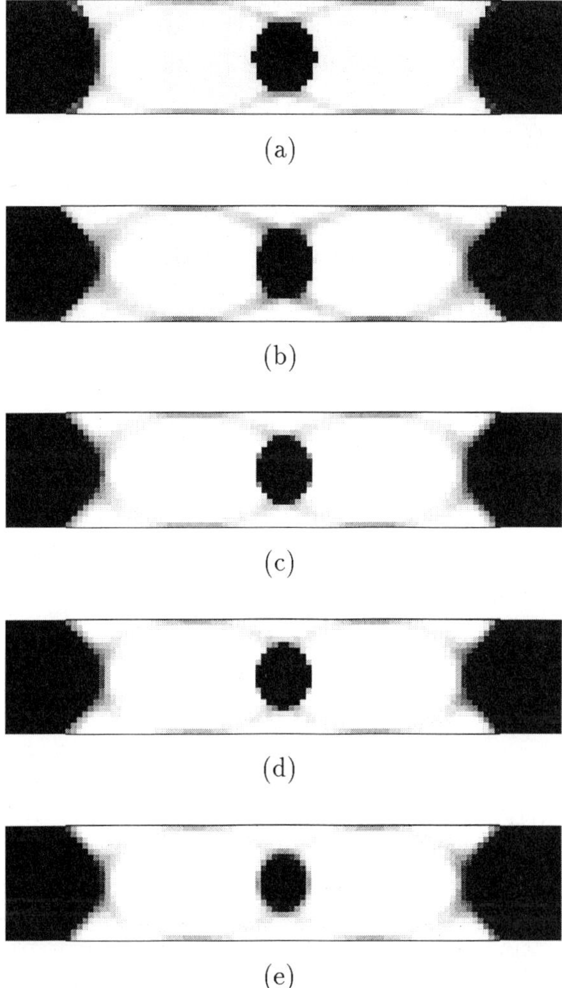

Figure 22: Eigenfrequency optimal topology designs obtained for different ratios between the stiffness of the "stiff" and the "compliant" materials in the layered microstructure. The stiffness ratios and the first three eigenfrequencies associated with each of the five designs are given below,

(a) Stiffness ratio= 25 $f_1 = 3473.605 Hz.$ $f_2 = 3667.614 Hz.$ $f_3 = 4268.937 Hz$
(b) Stiffness ratio= 50 $f_1 = 3331.995 Hz.$ $f_2 = 3338.801 Hz.$ $f_3 = 3468.466 Hz$
(c) Stiffness ratio= 75 $f_1 = 3305.411 Hz.$ $f_2 = 3306.412 Hz.$ $f_3 = 3357.239 Hz.$
(d) Stiffness ratio= 100 $f_1 = 3263.605 Hz.$ $f_2 = 3264.317 Hz.$ $f_3 = 3119.719 Hz.$
(e) Stiffness ratio= 200 $f_1 = 3179.707 Hz.$ $f_2 = 3179.061 Hz.$ $f_3 = 3211.421 Hz.$

ated with the highest stiffnesses of the compliant material have distinct fundamental eigenfrequencies. The iteration histories for the solutions associated with the highest and lowest stiffness of the compliant material are shown in Figure 23 and Figure 24, respectively.

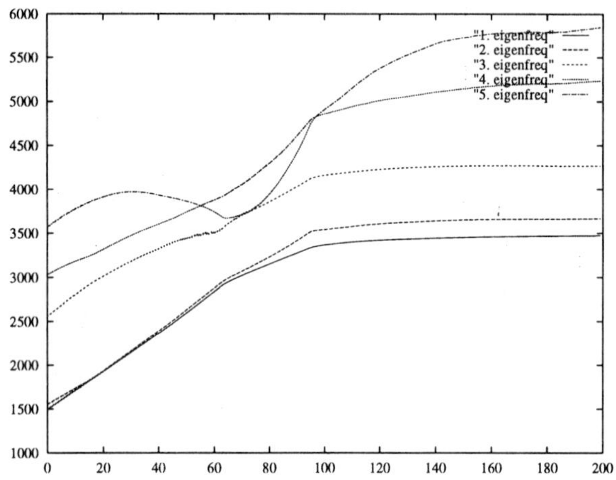

Figure 23: Iteration history obtained for ratio 25 between stiffnesses of "stiff" and "compliant" material.

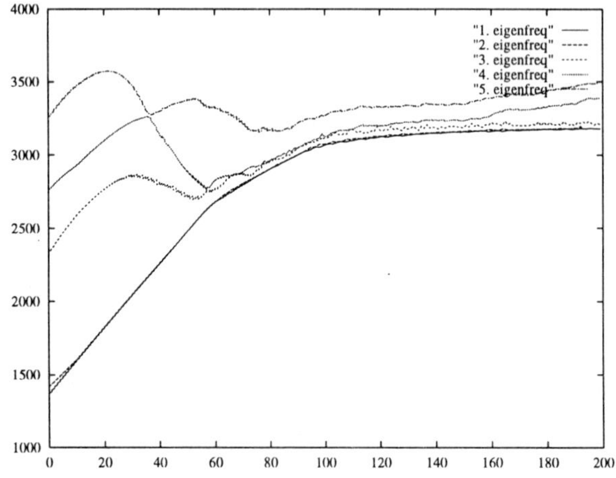

Figure 24: Iteration history obtained for ratio 200 between stiffnesses of "stiff" and "compliant" material.

From the two iteration histories we see a nice and stable convergence of the eigen-frequencies, and it is noted that final convergence for the two problems is very slow, which is not surprising since 10.000 design variables need to converge. Note also that the eigenfrequency spectrum is very dense for the solution with the lowest stiffness of the compliant material. The difference between the double fundamental eigenfrequency and the third eigenfrequency is only around $30Hz$. However, at this point it remains uncertain if a higher multiplicity than two may be obtained for the fundamental eigen-frequency of the present problem.

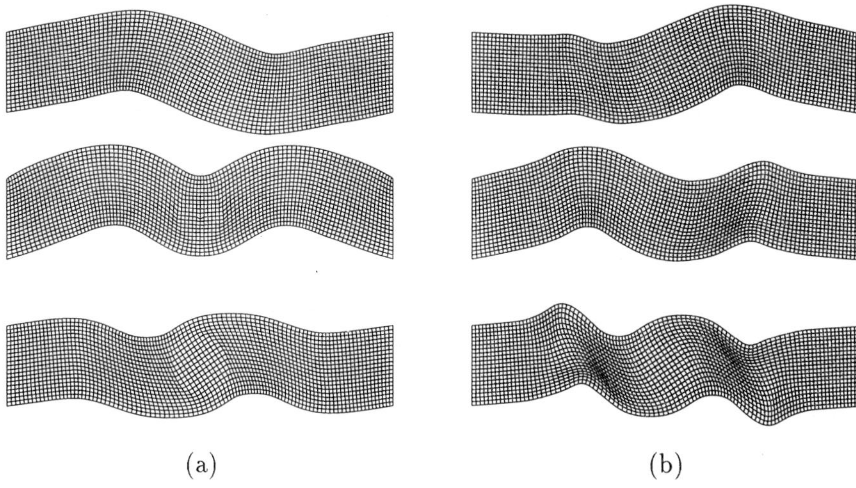

(a)   (b)

Figure 25: Eigenmodes associated with the three lowest eigenfrequencies for the opti-mum designs of a short clamped-clamped beam with a simple support at the mid-point. (a) Eigenmodes associated with the optimum design obtained for a stiffness ratio of 25. (b) Eigenmodes associated with the optimum design obtained for a stiffness ratio of 200.

Figure 25 shows the three first eigenmodes corresponding to the optimum designs in Figure 22(a) and Figure 22(e), respectively. The three eigenmodes in Figure 25(a) which all correspond to distinct eigenfrequencies, are seen to be either perfectly sym-metric or perfectly anti-symmetric modes, as expected due to the perfect symmetry of the corresponding design depicted in Figure 22(a). However, if we consider the first two eigenmodes in Figure 25(b) which correspond to the double fundamental eigenfre-quency of the optimum design in Figure 22(e), we see that these eigenmodes represent neither symmetric nor anti-symmetric modes. Furthermore, if we consider the third eigenmode in Figure 25(b) which corresponds to the distinct third eigenfrequency for the optimum design in Figure 22(e), we observe that this eigenmode is perfectly anti-symmetric. The explanation for the lack of symmetry/antisymmetry properties of the first two eigenmodes is that they are some linear combinations of a symmetric and an antisymetric eigenmode. In fact, a small numerical difference exists between the corresponding eigenfrequencies. How small this numerical difference between two eigenvalues should be before they should be considered a multiple, still remains uncertain. For all the examples in the present section a relative tolerance of 10^{-4} was used.

5.4　Eigenfrequency Layout Design Problems for Plates

In this section we consider examples of maximum eigenfrequency layout design problems for Mindlin plate structures. The aim of the examples is, on the one hand, to present a number of additional examples in which multiple eigenfrequencies occur, and on the other hand, like in Sub-section 5.2, to study the effect of the type of plate microstructure on the overall design and the performance of eigenfrequency optimized Mindlin plates. Hence, by use of the possibility to change the type of the layered Mindlin plate microstructure, see Sub-section 2.2, in the following we shall solve maximum eigenfrequency layout design problems for a perforated plate, a rib stiffened plate, and an internally stiffened honeycomb plate. The plate thicknesses h_1 and h_2 which define the different sections of the layered Mindlin plate microstructures used in the examples, are taken to be the same as those given in the caption of Figure 15. In addition to the data given there, it should be mentioned that in all exampes the mass density of the stiff material in the layered microstructure has been set to $7.8kg/m^3$, and that the mass density of the compliant material in the perforated configuration of the layered microstructure has been set equal to zero.

In all examples, a clamped rectangular plate of dimensions 1000mm x 1200mm is considered and the total amount of the material is taken to be 50% of the admissible design domain volume. The rectangular plate domain is discretized by 36 x 44 isoparametric nine-node Mindlin plate finite elements, and we adopt a design parametrization where the material within each of the finite elements is modelled by a layered microstructure of arbitrary rank. Hence, as design variables we use the density variable ϱ_e and the set of moment variables $m_{1e}, \ldots, m_{4e}$ associated with each of the elements in the discretized structure. The optimization problems are all stated as max-min problems for the second eigenfrequency of the rectangular plate under a constraint on the available amount of material, along with the constraints allowing for occurrence of multiple eigenfrequencies as discussed in Sub-section 4.2. Moreover, for each set of moment variables associated with each of the finite elements in the discretized structure, we must satisfy two simple constraint equations. The optimization problems considered here has, like the examples considered in the previous sub-section, been solved applying a two level approach of redesign where we in each loop initially determine improved values of the moment variables by means of mathematical programming and design sensitivity analysis, followed by update of the density variables by a similar approach. Finally, it should be mentioned that a relative difference of 10^{-3} between eigenfrequencies has been used as a practical threshold criterion for multiplicity.

Example with Perforated Plate.

Figure 26 shows the optimum design and the iteration history for the eigenfrequency optimized perforated plate. We see that the design is almost perfectly symmetric,

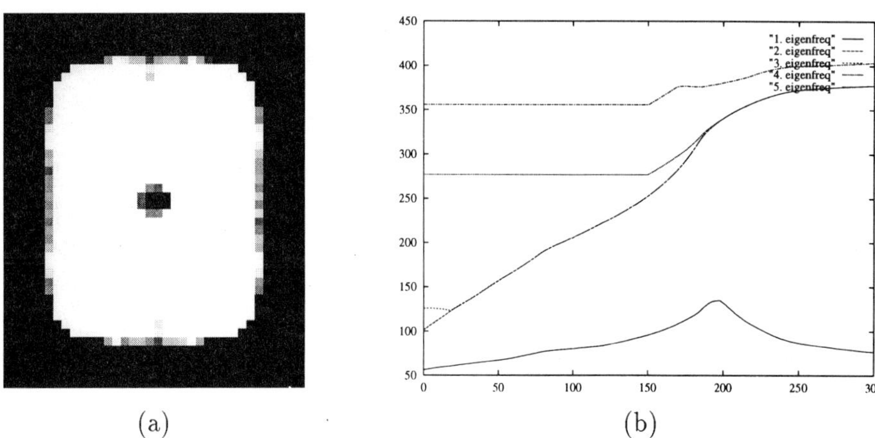
(a)   (b)

Figure 26: Maximization of the second eigenfrequency of a perforated plate. (a) Optimum distribution of material. Black areas are domains with full density of stiff material, white areas are domains with zero density of stiff material, and grey areas are domains with composites. (b) Iteration history for the eigenfrequencies.

and that there only exists very small subdomains with intermediate material densities. The optimum design takes full advantage of the rigid supports, and almost all the material has been moved to the plate boundaries. Regarding the iteration history, it should be noted that additional eigenfrequencies have been added to the optimization problem after 150 iterations, and we observe that the optimum design actually has a three-fold second eigenfrequency. The values of the three eigenfrequencies are given in the caption of Figure 27 that also shows the eigenmodes corresponding to the three-fold optimum eigenfrequency. From Figure 27, it is very difficult to see if the eigenmodes represent perfectly anti-symmetric modes, as expected due to the symmetric design of the structure, and it is also very difficult to see the difference between the third and fourth eigenmode.

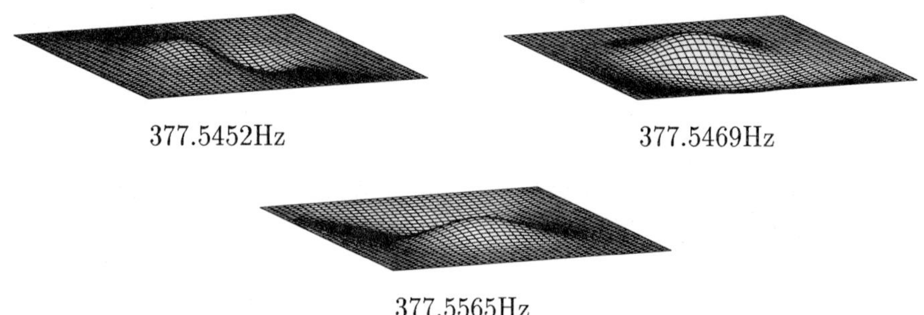

377.5452Hz   377.5469Hz

377.5565Hz

Figure 27: Second, third and fourth eigenmode for optimized perforated plate.

Example with Rib Stiffened Plate.

Figure 28 shows the optimum design and the iteration history for the eigenfrequency optimized rib stiffened plate. It should be noted that the topology of the optimum

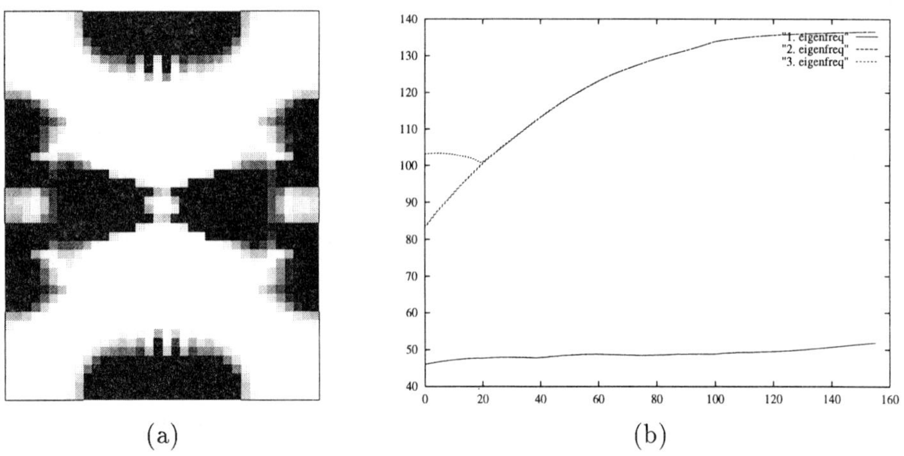

(a)   (b)

Figure 28: Maximization of second eigenfrequency of a ribbed plate. (a) Optimal distribution of exterior ribs. Black domains have full density of reinforcement, white domains no reinforcement, and grey domains an intermediate amount of reinforcement. (b) Iteration history for the first three eigenfrequencies

design has changed completely in comparison with the optimum design of the perforated plate. The design still takes advantage of the rigid supports, but now two large interior ribs have been formed. The iteration history shows the development of the first three eigenfrequencies, and we see that the optimum design has a double optimum second eigenfrequency. The two eigenmodes associated with the double second eigenfrequency for the optimum design are shown in Figure 29.

136.5583Hz   136.5594Hz

Figure 29: Second and third eigenmode for optimized rib stiffened plate.

Example with Internally Stiffened Honeycomb Plate.

Figure 30 shows the optimum design and the iteration history for the eigenfrequency optimized internally stiffened honeycomb plate.   The design again takes advantage

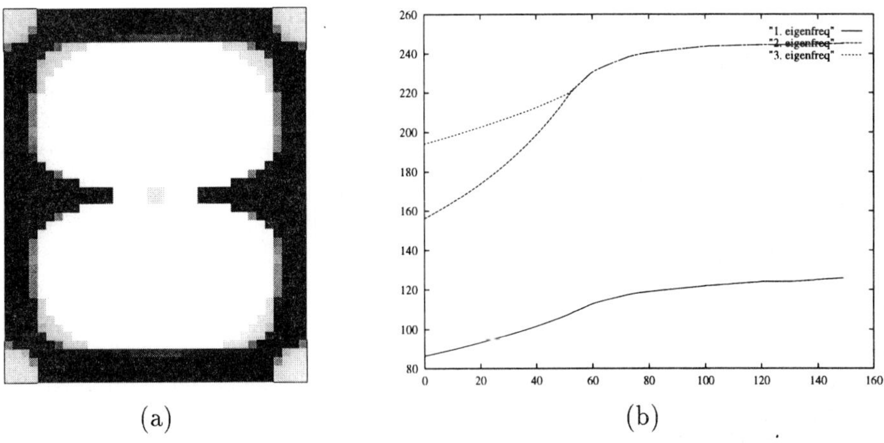

(a)   (b)

Figure 30: Maximization of second eigenfrequency of a honeycomb plate. (a) Optimal distribution of internal ribs. Black areas are domains with full density of reinforcing ribs, white areas are non-reinforced domains, and grey areas are domains with an intermediate integral reinforcement. (b) Iteration history for the first three eigenfrequencies.

of the rigid supports, and we here get a design which is sort of "in-between" the previous two designs. The iteration history shows the development of the first three eigenfrequencies, and as in the previous example the optimum design is associated with a double second eigenfrequency. The two eigenmodes associated with the double second eigenfrequency of the optimum design are shown in Figure 31.

245.1200Hz   245.1205Hz

Figure 31: Second and third eigenmode for optimized internally stiffened honeycomb plate.

Finally, if we study the performance of the three optimum structures, we see that the perforated plate has the highest eigenfrequency, then comes the honeycomb plate, and finally the rib stiffened plate. It is also noted that relative to the initial design with uniform distribution of material/reinforcement, the largest increase of the eigenfrequency is obtained for the perforated plate. In comparison with the results for stiffness optimized plates presented in Sub-section 5.2, it is interesting that we now also get a considerable gain in performance of the honeycomb plate.

6   Summary

This chapter has presented various formulations and methods of solution of different topology and layout optimization problems for disk, plate and shell structures. In particular, multiple load case stiffness maximization problems and eigenfrequency maximization problems have been discussed.

Two different design parametrizations based on the application of layered microstructures with a continously variable density of material/reinforcement were introduced in Section 2. In the first formulation a parametrization of the material/reinforcement distribution for a disk, plate or shell structure was obtained by modelling the material within each of the finite elements as a second-rank layered microstructure, and in the second formulation the parametrization was achieved by modelling the material in each element as a layered microstructure of arbitrary rank, using so-called moment variables. The effective stiffness properties of both the second-rank layered microstructure and the layered microstructure of arbitrary rank were derived applying a simple smear-out technique for layered materials in Mindlin plates. Furthermore, by allowing for four different configurations of the layered microstructures, a unified design parametrization was established for both topology optimization problems for planar disks and Mindlin plates and shells, and for reinforcement layout optimization of rib-stiffened Mindlin plates and shells and internally stiffened honeycomb sandwich Mindlin plates and shells.

In Section 3 different criteria for topology and layout optimization were considered. The total elastic energy was chosen as the objective function for the solution of maximum stiffness layout design problems, and different analytical expressions for the design sensitivities of the total elastic energy in a statically loaded linearly elastic structure were derived.

Next, criteria involving eigenvalues of vibration were discussed, and analytical design sensitivity expressions were derived for simple eigenvalues as well as for non-differentiable multiple eigenvalues. In particular, it was shown that the linear increment of both simple and multiple eigenvalues can be calculated applying the same expression, provided that some restrictions on the design pertubations are taken into account in the latter case. This result has been implemented in a very simple unified procedure for solution of optimization problems with simple as well as multiple eigenvalues where only gradients of simple eigenvalues and generalized gradients of multiple eigenvalues are needed. As is shown in Section 3, these gradient vectors can be calculated very efficiently from analytically derived design sensitivity expressions.

Section 4 presented several different formulations of maximum stiffness and maximum eigenfrequency topology and layout design problems, and described the solution strategies which have been implemented in ODESSY for the solution of the optimization problems. Here, the traditional single load case and "weighted sum formulated" multiple load case maximum stiffness layout design problems were discussed in detail together with corresponding topology optimization problems with eigenvalue objectives, and Section 4 subsequently presented significant new formulations and developments pertaining to multicriterion topology and layout optimization.

Thus, as a replacement of simple "weighted-sum" formulations of multiple load case stiffness maximization problems, the latter type of problem was directly formulated and solved as a minimization problem for the maximum total elastic energy from among several independent load cases, i.e., as a "worst case" type of problem. This was achieved by putting this initially non-differentiable multiobjective optimization problem in differentiable form via a bound formulation, whereby the problem can be solved directly by means of mathematical programming and design sensitivity analysis.

Similarly, it was shown that eigenfrequency layout design problems may be formulated and solved as a max-min problem of maximizing the smallest eigenvalue of vibration from among a given spectrum of eigenvalues. This multiobjective optimization problem was transformed to an equivalent single objective problem via a bound formulation, and in the case of occurrence of non-differentiable multiple eigenvalues, difficulties with the sensitivity analysis were avoided by considering some constraints on the vector of design changes. This way, the problem can be solved as a usual differentiable optimization problem by ordinary methods of mathematical programming and design sensitivity analysis.

Four different examples of topology and layout design optimization of disk, plate and shell structures were considered in Chapter 5, and we shall briefly summarize the main results.

In Section 5.1 we considered simple maximum stiffness layout design problems for plane disk structures under single and multiple load conditions, using design parametrizations based on the second-rank layered microstructure described in Section 2.4.1 and on the more general layered microstructure of arbitrary rank described in Section 2.4.2. The example clearly illustrated the importance of a proper design parametrization for solution of multiple load case maximum stiffness layout design problems, and also showed the importance of the formulation of the multi objective stiffness maximization problem. By applying a max-min formulation, we were able to increase the minimum stiffness of the structure by 15% in comparison with the solution obtained by using a traditional weighted sum formulation with equal weighting factors.

In the second example, the influence of the type of the applied plate microstructure on the optimum performance and on the optimum distribution of material/reinforcement of stiffness optimized simply suported and clamped quadratic plates was studied. Not surprisingly, we found that the sandwich and honeycomb plate models yield the most rigid plates while the rib stiffened plate is the more compliant one. Also, for the plates, load and boundary conditions considered in this example, the type of the plate model was seen to have very litle influence on how the material/reinforcement is being distributed.

In the third example, we considered the solution of maximum eigenfrequency layout layout design problems for plane disk structures. The optimization problems were formulated as max-min problems for the fundamental eigenfrequency, and the material was modelled by a layered microstructure of arbitrary rank. A simple parameter study was undertaken, in which we varied the stiffness of the compliant material in the layered microstructure. In the series of examples considered, no change occurred in the topology of the structure. However, a double fundamental eigenfrequency was obtained in the examples where the compliant material had very low stiffness, and the formulation for optimization with multiple eigenfrequencies was seen to apply very succesfully.

Finally, in the last example we considered the solution of maximum eigenfrequency layout design problems for rectangular clamped plates with different microstructures. Here, the optimization problems were formulated as max-min problems for the second eigenfrequency, and the material was modelled by a layered microstructure of arbitrary rank. For this example a three-fold second eigenfrequency was obtained for the perforated plate, while double second eigenfrequencies were obtained for the honeycomb and rib stiffened plates. Also in these examples the formulation for optimization with multiple eigenfrequencies was applied very succesfully. Moreover, in contrast to the stiffness layout design problems considered in Section 5.2, the type of the plate model used was seen to have significant influence on the optimum material/reinforcement distribution for the plates.

Acknowledgement

This work recieved support form the Danish Technical Research Council (Programme of Research on Computer Aided Design).

Appendix: Derivation of Effective Stiffness Tensors

A set of analytical expressions for the effective membrane, bending, and transverse shear stiffness tensors will be derived here for the four first-rank Mindlin plate microstructures illustrated in Figure 4. The derivations are performed by means of the smear-out technique for microstructurally layered Mindlin plates described in Section 2.3.1, and carried out simultaneously for all four microstructures using the concept of stiffness constants introduced in Eqs(4)-(9). It should be mentioned that many of the derivations performed herein can be also found in the work by Soto & Diaz[30] and Soto[31].

A.1 The Effective Membrane Stiffness Tensor

It is the aim of the smear-out technique in Section 2.3.1 to establish the membrane stiffness tensor in the constitutive relationship between direct averages of the membrane force and membrane strain tensors, i.e., to determine the effective membrane stiffness tensor $A_{\alpha\beta\kappa\gamma}^{R1}$ in the constitutive relationship

$$N_{\alpha\beta}^{avr} = A_{\alpha\beta\kappa\gamma}^{R1}\varepsilon_{\kappa\gamma}^{avr} \tag{138}$$

where $N_{\alpha\beta}^{avr}$ and $\varepsilon_{\alpha\beta}^{avr}$ represent direct averages of the membrane force and membrane strain tensors within some small rectangular plate domain Ω consisting of a finite number of thin parallel plate slices with membrane stiffness tensors $A_{\alpha\beta\kappa\gamma}^{+}$ and $A_{\alpha\beta\kappa\gamma}^{-}$ associated with subdomains Ω^{+} and Ω^{-}, respectively, see Figure 5. In the following we shall derive an analytical expression for this effective membrane stiffness tensor using the membrane stiffness constants A_{0}^{+} and A_{0}^{-}, given in Eqs(5)-(9), together with Eq(4) to characterize the membrane stiffness tensors $A_{\alpha\beta\kappa\gamma}^{+}$ and $A_{\alpha\beta\kappa\gamma}^{-}$ for the four first-rank Mindlin plate microstructures shown in Figure 4, and using the expressions presented in Eqs(11)-(16) for the smear-out process.

We start the derivation by introducing first the expressions for the average membrane force and membrane strain tensors given in Eq(13), and then the constitutive relationships for the subdomains Ω^{+} and Ω^{-}, given in Eq(12), into Eq(138), whereby the following relationship is obtained

$$\mu^{R1}N_{\alpha\beta}^{+} + (1 - \mu^{R1})N_{\alpha\beta}^{-} = A_{\alpha\beta\kappa\gamma}^{R1}\left(\mu^{R1}\varepsilon_{\kappa\gamma}^{+} + (1 - \mu^{R1})\varepsilon_{\kappa\gamma}^{-}\right)$$
$$\Updownarrow$$
$$\mu^{R1}A_{\alpha\beta\kappa\gamma}^{+}\varepsilon_{\kappa\gamma}^{+} + (1 - \mu^{R1})A_{\alpha\beta\kappa\gamma}^{-}\varepsilon_{\kappa\gamma}^{-} = A_{\alpha\beta\kappa\gamma}^{R1}\left(\mu^{R1}\varepsilon_{\kappa\gamma}^{+} + (1 - \mu^{R1})\varepsilon_{\kappa\gamma}^{-}\right) \tag{139}$$
$$\Updownarrow$$
$$\left(A_{\alpha\beta\kappa\gamma}^{+} - A_{\alpha\beta\kappa\gamma}^{R1}\right)\left(\mu^{R1}\varepsilon_{\kappa\gamma}^{+} + (1 - \mu^{R1})\varepsilon_{\kappa\gamma}^{-}\right) = (1 - \mu^{R1})\left(A_{\alpha\beta\kappa\gamma}^{+} - A_{\alpha\beta\kappa\gamma}^{-}\right)\varepsilon_{\kappa\gamma}^{-}$$

The next step is to establish a set of expressions which allows us to eliminate the strain tensors in the expression given above. Such a set of expressions is obtained by evaluating the discontinuity $\varepsilon_{\alpha\beta}^{+} - \varepsilon_{\alpha\beta}^{-}$ of the membrane strain tensor at the interfaces between adjacent subdomains Ω^{+} and Ω^{-}, i.e., by evaluating Eq(16) where β_1 and β_2 represent unknown scalars to be determined. For the derivation of the scalar β_1 we start by introducing the constitutive relationships for the subdomains Ω^+ and Ω^-, given in Eq(12), and then the discontinuity condition for the membrane strain tensor given in Eq(16), into the continuity condition for the normal component of the membrane force tensor, given in Eq(14), i.e.,

$$\left(N_{\alpha\beta}^+ - N_{\alpha\beta}^-\right) n_\alpha^{R1} n_\beta^{R1} = 0$$
$$\Updownarrow$$
$$\left(A_{\alpha\beta\kappa\gamma}^+ \varepsilon_{\kappa\gamma}^+ - A_{\alpha\beta\kappa\gamma}^- \varepsilon_{\kappa\gamma}^-\right) n_\alpha^{R1} n_\beta^{R1} = 0 \tag{140}$$
$$\Updownarrow$$
$$\left(A_{\alpha\beta\kappa\gamma}^+ \left(\varepsilon_{\kappa\gamma}^- + \beta_1 n_\kappa^{R1} n_\gamma^{R1} + \beta_2 \left(t_\kappa^{R1} n_\gamma^{R1} + n_\kappa^{R1} t_\gamma^{R1}\right)\right) - A_{\alpha\beta\kappa\gamma}^- \varepsilon_{\kappa\gamma}^-\right) n_\alpha^{R1} n_\beta^{R1} = 0$$

Since $A_{\alpha\beta\kappa\gamma}^+$ is an isotropic tensor with non-zero components given in Eqs(4)-(5), we have that $A_{\alpha\beta\kappa\gamma}^+ n_\alpha^{R1} n_\beta^{R1} n_\kappa^{R1} t_\gamma^{R1} = A_{\alpha\beta\kappa\gamma}^+ n_\alpha^{R1} n_\beta^{R1} t_\kappa^{R1} n_\gamma^{R1} = 0$, and that $A_{\alpha\beta\kappa\gamma}^+ n_\alpha^{R1} n_\beta^{R1} n_\kappa^{R1} n_\gamma^{R1} = A_0^+$, whereby the following expression is obtained for the scalar β_1

$$\left(A_{\alpha\beta\kappa\gamma}^+ \left(\varepsilon_{\kappa\gamma}^- + \beta_1 n_\kappa^{R1} n_\gamma^{R1}\right) - A_{\alpha\beta\kappa\gamma}^- \varepsilon_{\kappa\gamma}^-\right) n_\alpha^{R1} n_\beta^{R1} = 0$$
$$\Updownarrow$$
$$\beta_1 = -\frac{\left(A_{\alpha\beta\kappa\gamma}^+ - A_{\alpha\beta\kappa\gamma}^-\right) n_\alpha^{R1} n_\beta^{R1} \varepsilon_{\kappa\gamma}^-}{A_0^+} \tag{141}$$

The constant β_2 is determined in a very similar way. Here, we start by introducing the constitutive relationships for the subdomains Ω^+ and Ω^-, given in Eq(12), and then the discontinuity condition for the membrane strain tensor, given in Eq(16), into the continuity condition for the shear component of the membrane force tensor given in Eq(14), i.e.,

$$\left(N_{\alpha\beta}^+ - N_{\alpha\beta}^-\right) n_\alpha^{R1} t_\beta^{R1} = 0$$
$$\Updownarrow$$
$$\left(A_{\alpha\beta\kappa\gamma}^+ \varepsilon_{\kappa\gamma}^+ - A_{\alpha\beta\kappa\gamma}^- \varepsilon_{\kappa\gamma}^-\right) n_\alpha^{R1} t_\beta^{R1} = 0 \tag{142}$$
$$\Updownarrow$$
$$\left(A_{\alpha\beta\kappa\gamma}^+ \left(\varepsilon_{\kappa\gamma}^- + \beta_1 n_\kappa^{R1} n_\gamma^{R1} + \beta_2 \left(t_\kappa^{R1} n_\gamma^{R1} + n_\kappa^{R1} t_\gamma^{R1}\right)\right) - A_{\alpha\beta\kappa\gamma}^- \varepsilon_{\kappa\gamma}^-\right) n_\alpha^{R1} t_\beta^{R1} = 0$$

and since $A_{\alpha\beta\kappa\gamma}^+$ is an isotropic tensor with non-zero components given as shown in Eqs(4)-(5), we have that $A_{\alpha\beta\kappa\gamma}^+ n_\alpha^{R1} t_\beta^{R1} n_\kappa^{R1} n_\gamma^{R1} = 0$, and that $A_{\alpha\beta\kappa\gamma}^+ (n_\alpha^{R1} t_\beta^{R1} t_\kappa^{R1} n_\gamma^{R1} + n_\alpha^{R1} t_\beta^{R1} n_\kappa^{R1} t_\gamma^{R1}) = (1 - \nu)A_0^+$, whereby the following expression for the scalar β_2 is obtained

$$\left(A_{\alpha\beta\kappa\gamma}^+ \left(\varepsilon_{\kappa\gamma}^- + \beta_2 \left(t_\kappa^{R1} n_\gamma^{R1} + n_\kappa^{R1} t_\gamma^{R1}\right)\right) - A_{\alpha\beta\kappa\gamma}^- \varepsilon_{\kappa\gamma}^-\right) n_\alpha^{R1} t_\beta^{R1} = 0$$
$$\Updownarrow$$
$$\beta_2 = -\frac{\left(A_{\alpha\beta\kappa\gamma}^+ - A_{\alpha\beta\kappa\gamma}^-\right) n_\alpha^{R1} t_\beta^{R1} \varepsilon_{\kappa\gamma}^-}{(1 - \nu)A_0^+} \tag{143}$$

An expression for the discontinuity of the membrane strain tensor over the interfaces between adjacent subdomains Ω^+ and Ω^-, is now found by simply introducing Eq(141) and Eq(143) into the discontinuity condition for the membrane strain tensor, given in

Eq(16), i.e.,

$$\varepsilon^+_{\alpha\beta} - \varepsilon^-_{\alpha\beta} = \beta_1 n^{R1}_\alpha n^{R1}_\beta + \beta_2 \left(t^{R1}_\alpha n^{R1}_\beta + n^{R1}_\alpha t^{R1}_\beta \right)$$

$$= -\frac{(A^+_{\kappa\gamma\varepsilon\theta} - A^-_{\kappa\gamma\varepsilon\theta}) n^{R1}_\kappa n^{R1}_\gamma \varepsilon^-_{\varepsilon\theta}}{A^+_0} n^{R1}_\alpha n^{R1}_\beta - \frac{(A^+_{\kappa\gamma\varepsilon\theta} - A^-_{\kappa\gamma\varepsilon\theta}) n^{R1}_\kappa t^{R1}_\gamma \varepsilon^-_{\varepsilon\theta}}{(1-\nu) A^+_0} (t^{R1}_\alpha n^{R1}_\beta + n^{R1}_\alpha t^{R1}_\beta)$$

$$= -\frac{1}{A^+_0} \left(n^{R1}_\alpha n^{R1}_\beta n^{R1}_\kappa n^{R1}_\gamma + \frac{(t^{R1}_\alpha n^{R1}_\beta n^{R1}_\kappa t^{R1}_\gamma + n^{R1}_\alpha t^{R1}_\beta n^{R1}_\kappa t^{R1}_\gamma)}{1-\nu} \right) (A^+_{\kappa\gamma\varepsilon\theta} - A^-_{\kappa\gamma\varepsilon\theta}) \varepsilon^-_{\varepsilon\theta}$$

$$(144)$$

This expression is restated in the following more compact form

$$\varepsilon^+_{\alpha\beta} - \varepsilon^-_{\alpha\beta} = -\frac{1}{A^+_0} \Lambda^{A,R1}_{\alpha\beta\kappa\gamma} \eta^{A,R1}_{\kappa\gamma} \tag{145}$$

where

$$\Lambda^{A,R1}_{\alpha\beta\kappa\gamma} = n^{R1}_\alpha n^{R1}_\beta n^{R1}_\kappa n^{R1}_\gamma + \frac{(t^{R1}_\alpha n^{R1}_\beta n^{R1}_\kappa t^{R1}_\gamma + n^{R1}_\alpha t^{R1}_\beta n^{R1}_\kappa t^{R1}_\gamma + t^{R1}_\alpha n^{R1}_\beta t^{R1}_\kappa n^{R1}_\gamma + n^{R1}_\alpha t^{R1}_\beta t^{R1}_\kappa n^{R1}_\gamma)}{2(1-\nu)} \tag{146}$$

since indicies κ and γ can be interchanged due to obvious symmetry properties in Eq(144), and

$$\eta^{A,R1}_{\kappa\gamma} = (A^+_{\kappa\gamma\varepsilon\theta} - A^-_{\kappa\gamma\varepsilon\theta}) \varepsilon^-_{\varepsilon\theta} \tag{147}$$

We may now continue the derivation of our effective membrane stiffness tensor by combining Eq(139) with the set of expressions given in Eqs(145)-(147), whereby we obtain

$$\left(A^+_{\alpha\beta\kappa\gamma} - A^{R1}_{\alpha\beta\kappa\gamma} \right) \left(\mu^{R1} \varepsilon^+_{\kappa\gamma} + (1-\mu^{R1}) \varepsilon^-_{\kappa\gamma} \right) = (1-\mu^{R1}) \left(A^+_{\alpha\beta\kappa\gamma} - A^-_{\alpha\beta\kappa\gamma} \right) \varepsilon^-_{\kappa\gamma}$$

$$\Updownarrow$$

$$\left(A^+_{\alpha\beta\kappa\gamma} - A^{R1}_{\alpha\beta\kappa\gamma} \right) \left(\mu^{R1} \left(\varepsilon^-_{\kappa\gamma} - \frac{1}{A^+_0} \Lambda^{A,R1}_{\kappa\gamma\varepsilon\theta} \eta^{A,R1}_{\varepsilon\theta} \right) + (1-\mu^{R1}) \varepsilon^-_{\kappa\gamma} \right) = (1-\mu^{R1}) \left(A^+_{\alpha\beta\kappa\gamma} - A^-_{\alpha\beta\kappa\gamma} \right) \varepsilon^-_{\kappa\gamma}$$

$$\Updownarrow$$

$$\left(A^+_{\alpha\beta\kappa\gamma} - A^{R1}_{\alpha\beta\kappa\gamma} \right) \left(\varepsilon^-_{\kappa\gamma} - \frac{\mu^{R1}}{A^+_0} \Lambda^{A,R1}_{\kappa\gamma\varepsilon\theta} \eta^{A,R1}_{\varepsilon\theta} \right) = (1-\mu^{R1}) \left(A^+_{\alpha\beta\kappa\gamma} - A^-_{\alpha\beta\kappa\gamma} \right) \varepsilon^-_{\kappa\gamma}$$

$$\Updownarrow$$

$$\left(A^+_{\alpha\beta\kappa\gamma} - A^{R1}_{\alpha\beta\kappa\gamma} \right) \left(\left(A^+_{\varepsilon\theta\kappa\gamma} - A^-_{\varepsilon\theta\kappa\gamma} \right)^{-1} - \frac{\mu^{R1}}{A^+_0} \Lambda^{A,R1}_{\kappa\gamma\varepsilon\theta} \right) \eta^{A,R1}_{\varepsilon\theta} = (1-\mu^{R1}) \eta^{A,R1}_{\alpha\beta}$$

$$\Updownarrow$$

$$\left(\left(A^+_{\varepsilon\theta\kappa\gamma} - A^-_{\varepsilon\theta\kappa\gamma} \right)^{-1} - \frac{\mu^{R1}}{A^+_0} \Lambda^{A,R1}_{\kappa\gamma\varepsilon\theta} \right) \eta^{A,R1}_{\varepsilon\theta} = (1-\mu^{R1}) \left(A^+_{\varepsilon\theta\kappa\gamma} - A^{R1}_{\varepsilon\theta\kappa\gamma} \right)^{-1} \eta^{A,R1}_{\varepsilon\theta}$$

$$(148)$$

Finally, realizing that this equation must hold for arbitrary tensors $\eta^{A,R1}_{\varepsilon\theta}$, and applying the symmetry of the $\Lambda^{A,R1}_{\alpha\beta\kappa\gamma}$ tensor, we obtain

$$\left(\left(A^+_{\alpha\beta\kappa\gamma} - A^-_{\alpha\beta\kappa\gamma} \right)^{-1} - \frac{\mu^{R1}}{A^+_0} \Lambda^{A,R1}_{\kappa\gamma\alpha\beta} \right) = (1-\mu^{R1}) \left(A^+_{\alpha\beta\kappa\gamma} - A^{R1}_{\alpha\beta\kappa\gamma} \right)^{-1}$$

$$\Updownarrow$$

$$A^{R1}_{\alpha\beta\kappa\gamma} = A^+_{\alpha\beta\kappa\gamma} - (1-\mu^{R1}) \left(\left(A^+_{\alpha\beta\kappa\gamma} - A^-_{\alpha\beta\kappa\gamma} \right)^{-1} - \frac{\mu^{R1}}{A^+_0} \Lambda^{A,R1}_{\alpha\beta\kappa\gamma} \right)^{-1}$$

$$(149)$$

which is the final expression for the effective membrane stiffness tensor for our four first-rank Mindlin plate microstructures depicted in Figure 4.

A.2　The Effective Bending Stiffness Tensor

The effective bending stiffness tensor for a first-rank Mindlin plate microstructure can be derived by a procedure which is completely analogous to that just applied for the derivation of the effective membrane stiffness tensor. The effective bending stiffness tensor $D^{R1}_{\alpha\beta\kappa\gamma}$ is defined through the constitutive relationship

$$M^{avr}_{\alpha\beta} = D^{R1}_{\alpha\beta\kappa\gamma}\kappa^{avr}_{\kappa\gamma} \tag{150}$$

where $M^{avr}_{\alpha\beta}$ and $\kappa^{avr}_{\alpha\beta}$ represent direct averages of the bending moment and curvature tensors within some small rectangular plate domain Ω consisting of a finite number of thin parallel plate slices with bending stiffness tensors $D^{+}_{\alpha\beta\kappa\gamma}$ and $D^{-}_{\alpha\beta\kappa\gamma}$ associated with subdomains Ω^+ and Ω^-, respectively, as shown in Figure 5.

Proceeding in analogy with the derivation in the foregoing sub-section, we obtain the following analytical expression for the effective bending stiffness tensor for the set of first-rank Mindlin plate microstructures depicted in Figure 4,

$$D^{R1}_{\alpha\beta\kappa\gamma} = D^{+}_{\alpha\beta\kappa\gamma} - (1 - \mu^{R1})\left(\left(D^{+}_{\alpha\beta\kappa\gamma} - D^{-}_{\alpha\beta\kappa\gamma}\right)^{-1} - \frac{\mu^{R1}}{D_0}\Lambda^{D,R1}_{\alpha\beta\kappa\gamma}\right)^{-1} \tag{151}$$

A.3　The Effective Transverse Shear Stiffness Tensor

The derivation of the effective transverse shear stiffness tensor for the first-rank Mindlin plate microstructure can also be performed in analogy with the derivations in Sub-section A.1. However, since we are here dealing with tensors of lower order than those of the two preceding sub-sections, the derivation will be presented in some detail.

Thus, we apply the smear-out technique described in Section 2.3.1 to obtain the effective transverse shear stiffness tensor $S^{R1}_{\alpha\beta}$ in the following constitutive relationship

$$Q^{avr}_{\alpha} = S^{R1}_{\alpha\beta}\gamma^{avr}_{\beta} \tag{152}$$

where Q^{avr}_{α} and $\gamma^{avr}_{\alpha\beta}$ represent direct averages of the transverse shear force and transverse shear strain tensors within some small rectangular plate domain Ω consisting of a finite number of thin parallel plate slices with transverse shear stiffness tensors $S^{+}_{\alpha\beta}$ and $S^{-}_{\alpha\beta}$ associated with subdomains Ω^+ and Ω^-, respectively, see Figure 5. This effective transverse shear stiffness tensor shall now be derived using the transverse shear stiffness constants S^{+}_0 and S^{-}_0, given in Eqs(5)-(9), together with Eq(4) to characterize the transverse shear stiffness tensors $S^{+}_{\alpha\beta}$ and $S^{-}_{\alpha\beta}$ for the four first-rank Mindlin plate microstructures shown in Figure 4, and using the expressions presented in Eqs(11)-Eq(16) for the smear-out process.

We first introduce the expressions for the average transverse shear force and transverse shear strain tensors given in Eq(13), and then the constitutive relationships for the subdomains Ω^+ and Ω^-, given in Eq(12), into Eq(150), and get the following relationship

$$\mu^{R1} Q_\alpha^+ + (1 - \mu^{R1}) Q_\alpha^- = S_{\alpha\beta}^{R1} \left(\mu^{R1} \gamma_\beta^+ + (1 - \mu^{R1}) \gamma_\beta^- \right)$$

$\Updownarrow$

$$\mu^{R1} S_{\alpha\beta}^+ \gamma_\beta^+ + (1 - \mu^{R1}) S_{\alpha\beta}^- \gamma_\beta^- = S_{\alpha\beta}^{R1} \left(\mu^{R1} \gamma_\beta^+ + (1 - \mu^{R1}) \gamma_\beta^- \right) \qquad (153)$$

$\Updownarrow$

$$\left(S_{\alpha\beta}^+ - S_{\alpha\beta}^{R1} \right) \left(\mu^{R1} \gamma_\beta^+ + (1 - \mu^{R1}) \gamma_\beta^- \right) = (1 - \mu^{R1}) \left(S_{\alpha\beta}^+ - S_{\alpha\beta}^- \right) \gamma_\beta^-$$

As a next step we determine the discontinuity in the transverse shear strain tensor over the interfaces of the microstructure as given by Eq(16), where β_5 is an unknown scalar to be determined. The scalar β_5 is found by introducing first the constitutive relationships for the subdomains Ω^+ and Ω^- given in Eq(12), and then the discontinuity condition for the transverse shear strain tensor given in Eq(16), into the continuity condition for the normal component of the transverse shear force tensor given in Eq(14), whereby

$$\left(S_{\alpha\beta}^+ \gamma_\beta^+ - S_{\alpha\beta}^- \gamma_\beta^- \right) n_\alpha^{R1} = 0$$

$\Updownarrow$

$$\left(S_{\alpha\beta}^+ \left(\gamma_\beta^- + \beta_5 n_\beta \right) - S_{\alpha\beta}^- \gamma_\beta^- \right) n_\alpha^{R1} = 0 \qquad (154)$$

$\Updownarrow$

$$\beta_5 = - \frac{\left(S_{\alpha\beta}^+ - S_{\alpha\beta}^- \right) n_\alpha^{R1} \gamma_\beta^-}{S_0^+}$$

Note that the derivation of the scalar β_5 was performed using the expressions for the non-zero components of the transverse shear stiffness tensor $S_{\alpha\beta}^+$ given by Eqs(4)-(5). The expression for the scalar β_5 may now be substituted back into the discontinuity condition for the transverse shear strain tensor, given in Eq(16), to yield the following set of expressions

$$\gamma_\alpha^+ - \gamma_\alpha = -\frac{1}{S_0^+} \Lambda_{\alpha\beta}^{S,R1} \eta_\beta^{S,R1} \qquad (155)$$

where

$$\Lambda_{\alpha\beta}^{S,R1} = n_\alpha^{R1} n_\beta^{R1} \qquad (156)$$

and

$$\eta_\beta^{S,R1} = (S_{\beta\theta}^+ - S_{\beta\theta}^-) \gamma_\theta^- \qquad (157)$$

Finally, by combining Eq(153) and Eqs(155)-(157) we obtain the following analytical expression for the effective transverse shear stiffness tensor for the set of first-rank Mindlin plate microstructures depicted in Figure 4.

$$S_{\alpha\beta}^{R1} = S_{\alpha\beta}^+ - (1 - \mu^{R1}) \left(\left(S_{\alpha\beta}^+ - S_{\alpha\beta}^- \right)^{-1} - \frac{\mu^{R1}}{S_0} \Lambda_{\alpha\beta}^{S,R1} \right)^{-1} \qquad (158)$$

References

[1] Rozvany, GIN; Bendsøe, MP; Kirsch, U. Layout Optimization of Structures. *Applied Mechanics Reviews*, 48:41–119, 1995.

[2] Olhoff, N; Bendsøe, MP; Rasmussen, J. On CAD-integrated Structural Topology and Design Optimization. *Comp. Meths. Appl. Mech. Engrg.*, 89:259–279, 1991.

[3] Bendsøe, MP; Kikuchi, N. Generating Optimal Topologies in Structural Design using a Homogenizaiton Method. *Comp. Meth. Appl. Mechs. Engrg.*, 71:197–224, 1988.

[4] Bendsøe, MP. Optimal Shape Design as a Material Distibution Problem. *Struct. Optim.*, 1:193–202, 1989.

[5] Olhoff, N; Lurie, KA; Cherkaev, AV; Fedorov, A. Sliding Regimes of Anisotropy in Optimal Design of Vibrating Axisymetric Plates. *Int. J. Solids Struct.*, 17(10):931–948, 1981.

[6] Cheng, G; Olhoff, N. Regularized Formulation for Optimal Design of Axisymetric Plates. *Int. J. Solids Struct.*, 18(2):153–169, 1982.

[7] Cheng, G; Olhoff, N. An Investigation Concerning Optimal Design of Solid Elastic Plates. *Int. J. Solids Struct.*, 17:305–323, 1981.

[8] Kohn, RV; Strang, G. Optimal Design and Relaxation of Variational Problems. *Comm. Pure Appl. Math.*, 39:1–25(part I), 139–182(partII) and 353–357(partIII), 1986.

[9] Kohn, RV; Strang, G. Optimal Design in Elasticity and Plasticity. *Num. Meths. Engrg.*, 22:183–188, 1986.

[10] Avellaneda, M. Optimal Bounds and Microgeometries for Elastic Two-Phase Composites. *SIAM J. Appl. Math.*, 47(6):1216–1228, 1987.

[11] Lipton, R. On Optimal Reinforcement of Plates and Choice of Design Parameters. *Control and Cybernetics*, 23(3):481–493, 1994.

[12] Diaz, A; Lipton, R; Soto, CA. A New Formulation of the Problem of Optimum Reinforcement of Reissner-Mindlin Plates. *Comp. Meth. Appl. Mechs. Engrg*, 1994.

[13] Diaz, A; Bendsøe, MP. Shape Optimization of Structures for Multiple Loading Situations using a Homogenization Method. *Struct. Optim.*, 4:17–22, 1992.

[14] Olhoff, N; Thomsen, J; Rasmussen, J. Topology Optimization of Bi-Material Structures. In Pedersen, P, editor, *Optimal Design with Advanced Materials*, pages 191–206, Lyngby, Denmark, 1993. Elsevier, Amsterdam, The Netherlands, 1993.

[15] Suzuki, K; Kikuchi, N. Layout Optimization using the Homogenization Method: Generalized Layout Design of Three-Dimensional Shells for Car Bodies. In Rozvany, GIN, editor, *Optimization of Large Structural Systems*, pages 110–126, Berchtesgaden, Germany, 1991. Nato-ASI, Lecture Notes, 1991.

[16] Soto, CA; Diaz, AR. Optimum Layout of Plate Structures using Homogenization. In Bendsøe, MP; Mota Soares, CA, editors, *Topology Design of Structures*, pages 407–420, Sesimbra, Portugal, 1993. Nato ASI, Kluwer Academic Publishers, Dordrecht, The Netherlands, 1993.

[17] Soto, CA; Diaz, AR. A Model for Layout Optimization of Plate Structures. In Pedersen, P, editor, *Optimal Design with Advanced Materials*, pages 337–350, Lyngby, Denmark, 1993. Elsevier, Amsterdam, The Netherlands, 1993.

[18] Soto, CA; Diaz, AR. Layout of Plate Structures for Improved Dynamic Responce using a Homogenization Method. In Gilmore et. al., , editor, *Advances in Design Automatization*, pages 667–674, Albuquerqe, New-Mexico, USA, 1993. ASME, 1993.

[19] Krog, LA; Olhoff, N. Topology Optimization of Integral Rib Reinforcement of Plate and Shell Structures with Weighted-Sum and Max-Min Objectives. In Bestle, D; Schiehlen, W, editors, *Optimization of Mechanical Systems.*, pages 171–179, Stuttgart, Germany, 1995. Kluwer Academic Publishers, Dordrecht, The Netherlands, 1996.

[20] Folgado, J; Rodriques, H; Guedes, JM. Layout Design of Plate Reinforcements with a Buckling Load Criterion. In Olhoff, N; Rozvany, GIN, editors, *Structural and Multidisciplinary Optimization*, pages 659–666, Goslar, Germany, 1995. ISSMO, Pergamon, Oxford, UK, 1995.

[21] Bendsøe, MP. *Optimization of Structural Topology, Shape, and Material.* Springer-Verlag, Heidelberg, Germany, 1995.

[22] Bendsøe, M.P.; Mota Soares, C.A., editors. *Topology Design of Structures.* NATO ASI, Kluwer Academic Publishers, Dordrecht, The Netherlands, 1993.

[23] Pedersen, P, editor. *Optimal Design with Advanced Materials.* Elsevier, Amsterdam, The Netherlands, 1993.

[24] Olhoff, N; Rozvany, GIN, editors. *Structural and Multidisciplinary Optimization*, Goslar, Germany, 1995. Pergamon, Oxford, UK, 1995.

[25] Krog, LA. Layout Optimization of Disk, Plate, and Shell Structures. Ph.D thesis, Special Report No. 27, Institute of Mechanical Engineering, Aalborg University, Denmark, 1996.

[26] Lurie, KA. *Applied Optimal Control Theory of Distributed Systems.* Plenum Press, New York ,USA, 1993.

[27] Rozvany, GIN; Olhoff, N; Cheng, KT; Taylor, JE. On the Solid Plate Paradox in Structural Optimization. *J. Struct. Mech.*, 10(1):1–32, 1982.

[28] Wang, C-M; Rozvany, GIN; Olhoff, N. Optimal Plastic Design of Axisymmetric Solid Plates with a Maximum Thickness Constraint. *Computers & Structures*, 18(4):653–665, 1984.

[29] Wang, C-M; Thevendran, V; Rozvany, GIN; Olhoff, N. Optimal Plastic Design of Circular Plates: Numerical Solutions and Built-In Edges. *Computers & Structures*, 22(4):519–528, 1986.

[30] Soto, CA; Diaz, A. On Modelling of Ribbed Plates for Shape Optimization. *Stuct. Optim.*, 6:175–188, 1993.

[31] Soto, CA. Shape Optimization of Plate Structures using Plate Homogenization with Applications in Mechanical Design. Ph.D thesis, Department of Mechanical Engineering, Michigan State University, USA, 1993.

[32] Thomsen, J. Optimization of Properties of Anisotropic Materials and Topologies of Structures (in Danish). Ph.D thesis, Institute of Mechanical Engineering, Aalborg University, Denmark, 1992.

[33] Avellaneda, M; Milton, GW. Bounds on the Effective Elastic Tensor of Composites based on Two-Point Correlations. In Hui, D; Kozik, TJ, editors, *Composite Material Technology*, pages 89–93. ASME, 1989.

[34] Vinson, JR; Sierakowski, RL. *The Behaviour of Structures Composed of Composite Materials.* Martinus Nijhoff Publishers, Dordrecht, The Netherlands, 1987.

[35] Lipton, R. Optimal Design and Relaxation for Reinforced Plates Subject to Random Transverse Loads. *J. Probabilistic Engineering Mechanics*, 9:167–177, 1994.

[36] Pedersen, P. On Optimal Orientation of Orthotropic Materials. *Struct. Optim.*, 1:101–106, 1989.

[37] Pedersen, P. Bounds on Elastic Energy in Solids of Orthotropic Materials. *Struct. Optim.*, 2:55–63, 1990.

[38] Pedersen, P. On Thickness and Orientational Design with Orthotropic Materials. *Struct. Optim.*, 3:69–78, 1991.

[39] Francfort, GA; Murat, F. Homogenization and Optimal Bounds in Linear Elasticity. *Arch. Rat. Mech. Anal.*, 94:307–334, 1986.

[40] Lipton, R; Diaz, A. Moment Formulations for Optimum Layout in 3D Elasticity. In Olhoff, N; Rozvany, GIN, editors, *Structural and Multidisciplinary Optimization*, pages 161–168, Goslar, Germany, 1995. Pergamon, Oxford, UK, 1995.

[41] Krein, MG; Nudelman, AA. *The Markov Moment Problem and Extremal Problems.* Translation of Mathematical Monographs, 50. American Mathematical Society, 1977.

[42] Masur, EF. Optimum Stiffness and Strength of Elastic Structures. *J. Engng. Mech. Div, ASCE*, 96:621–640, 1970.

[43] Courant, R; Hilbert, D. *Methods of Mathematical Physics*, volume 1. Interscience Publishers, New York, USA, 1953.

[44] Wittrick, WH. Rates of Change of Eigenvalues, With Reference to Buckling and Vibration Problems. *Journal of Royal Aeronautical Society*, 66:590–591, 1962.

[45] Lancaster, P. On Eigenvalues of Matrices Dependent on a Parameter. *Numerische Mathematik*, 6:377–387, 1962.

[46] Seyranian, AP; Lund, E; Olhoff, N. Multiple Eigenvalues in Structural Optimization Problems. *Struct. Optim.*, 8(4):207–227, 1994.

[47] Lund, E. Finite Element Based Design Sensitivity Analysis and Optimization. Ph.D thesis, Institute of Mechanical Engineering, Aalborg University, Denmark, 1994.

[48] Bratus, AS; Seyranian, AP. Bimodal Solutions in Eigenvalue Optimization Problems. *Appl. Math. Mech.*, 47:451–457, 1983.

[49] Seyranian, AP. Multiple Eigenvalues in Optimization Problems. *Applied Mathematics in Mechanics*, 51:272–275, 1987.

[50] Haug, EJ; Rousselet, B. Design Sensitivity Analysis in Structural Mechanics II. *Journal of Structural Mechanics.*, 8(2):161–186, 1980.

[51] Masur, EF. On Stuctural Design Under Multiple Eigenvalue Constraints. *Int. J. Solids Struct.*, 20:211–231, 1984.

[52] Masur, EF. Some Additional Comments On Stuctural Design Under Multiple Eigenvalue Constraints. *Int. J. Solids Struct.*, 21:117–120, 1985.

[53] Haug, EJ; Choi, KK; Komkov, V. *Design Sensitivity Analysis of Structural Systems.* Academic Press, New York, 1986.

[54] Bendsøe, MP; Olhoff, N; Taylor, JE. A Variational Formulation for Multicriteria Structural Optimization. *J. Struct. Mech.*, 11:523–544, 1983.

[55] Diaz, A; Kikuchi, N. Solution to Shape and Topology Eigenvalue Optimization Problems using a Homogenization Method. *Int. J. Num. Meth. in Engng.*, 35:1487–1502, 1992.

[56] Ma, ZD; Kikuchi, N; Cheng, HC; Hagiwara, I. Topology and Shape Optimization Technique for Free Vibration Problems. In *ASME Winter Annular Meeting*, pages 127–138, Anaheim, California, USA, 1992. ASME, 1992.

[57] Kikuchi, N; Cheng, HC; Ma, ZD. Optimal Shape and Topology Design of Vibrating Structures. In *Advances in Structural Optimization*, pages 189–222. Kluwer, The Netherlands, 1995.

[58] Krog, LA; Olhoff, N. Topology Optimization of Plate and Shell Structures with Multiple Eigenfrequencies. In Olhoff, N; Rozvany, GIN, editors, *Structural and Multidisciplinary Optimization*, pages 675–682, Goslar, Germany, 1995. Pergamon, Oxford, UK, 1995.